Antonia Lavis

Bibliography of the geology and eruptive phenomena of the south Italian volcanoes that were visited in 1889 as well as of the submarine volcano of A.D. 1831

Antonia Lavis

Bibliography of the geology and eruptive phenomena of the south Italian volcanoes that were visited in 1889 as well as of the submarine volcano of A.D. 1831

ISBN/EAN: 9783337229689

Printed in Europe, USA, Canada, Australia, Japan

Cover: Foto ©berggeist007 / pixelio.de

More available books at **www.hansebooks.com**

BIBLIOGRAPHY

OF

THE GEOLOGY AND ERUPTIVE PHENOMENA

OF THE SOUTH ITALIAN VOLCANOES

THAT WERE VISITED IN 1889

AS WELL AS OF THE SUBMARINE VOLCANO OF A. D. 1831

Compiled by

MADAME ANTONIA LAVIS

&

D.ͬ JOHNSTON-LAVIS

FROM

THE SOUTH ITALIAN VOLCANOES etc.

NAPLES

PRINTED BY FERRANTE — Vico Tiratojo 25

1891

CHAPTER VII.

BIBLIOGRAPHY

of

THE GEOLOGY AND ERUPTIVE PHENOMENA

OF THE SOUTH ITALIAN VOLCANOES

THAT WERE VISITED IN 1889

AS WELL AS OF THE SUBMARINE VOLCANO OF A. D. 1831.

Compiled by

MADAME ANTONIA F. LAVIS AND D.ʀ H. J. JOHNSTON-LAVIS

MADAME ANTONIA F. LAVIS AND D.ʀ H. J. JOHNSTON-LAVIS

The constant progress of geological and vulcanological research in this interesting region is marked by the appearance of numerous memoirs and notes , many of which are published in proceedings of learned societies and often are overlooked or are unknown to other workers at the same subject. In 1881 the committee of organization of the International Geological Congress undertook the production of a Geological and Paleontological Bibliographical list of Italy. For many unavoidable reasons that list was an imperfect one, besides which during the last ten years a great number of new studies have been published. It was therefore thought advisable to place before the geologists who

g

visited this region a fairly complete bibliography which may be
a requisite should they or others be tempted to study any of the
districts. In undertaking this compilation we had little compre-
hended the difficulties and the very long and tedious work
necessary to bring it to a fairly successful termination.

The subject has been divided according to the different
volcanic groups, and where a book or memoir describes more
than one such, the title is found repeated in the separate divisions
concerned. The last district is that of the volcanic group of the
Alban Hills, to which we do not pretend to give a complete
bibliography for the following reasons. We possess no intimate
acquaintance with its literature, nor is it possible to easily divide
this one from the volcanic district immediately to the N. of
Rome, and finally the ground has been already covered by the
important publications of a similar nature by Signor B. Contarini,
Prof. R. Meli, Signor P. Zezi and others.

Of Roccamonfina and its neighbourhood we believe this is
the first separate list of works referring to that volcano that
has appeared.

The region, generally known as the Campi Phlegræi, is one
that is as yet but little understood so that each year numerous
additions are made to the literature of a district as classic from
a vulcanological as from a historical point of view. Many older
publications bearing thereon have been added, so that the list has
been much lengthened as well as a considerable number of cor-
rections of errors have been made.

In regard to Vesuvius we cannot do better than quote the
words of Comm. L. Riccio as follows. — " The eruptive period
that commenced on Dec. 16th 1631 had not terminated when
already V. Bove at pp. 47 and 48 of the pamphlet published by
Mormile, gave the first list of no less than 56 published accounts
on that occasion. A few years after Ferrante Bucca recorded a
much more important and numerous catalogue. During the last
century various authors considered it of use to publish lists,
especially of the 1631 eruption of Vesuvius , as Majone (1703),
Lasor a Varca (*Savanarola)* (1713) , Morhof (1714), P. G. M.
della Torre (1755), Ab. Galiani (1772), Vetrani (1780), Soria (1781),
Giustiniani (1793) and Duca della Torre (*Senior*) (1796) ; and in
the present century Scacchi (1847), Palmieri (1859), De Blasiis
(1875); and again Scacchi in 1883. " To these we may add L.
M. Greco, J. Roth the price lists of books by G. Dura, Napoli,
1866, F. Furchheim (1879), and Hoepli, Milan, 1879.

It is however to Comm. L. Riccio that so much is due. In the

first place, his untiring exertions during many years, has re-
sulted in the bringing together of the most complete collection
of works generally referring to Vulcanology and Seismology
and especially to Vesuvius and the other Neapolitan volcanoes
which is now deposited in the rooms of the Neapolitan Section
of the Italian Alpine Club. The foundation of this splendid library cosisted of about 2000 books and pamphlets belonging to
the celebrated French seismologist Alexis Perrey. By Signor
Riccio's exertions the number is at present raised to about 7000,
amongst which are many valuable manuscripts. Unfortunately
this unique library is practically lost, for the difficulties of get-
ting at the looks for study are so ·great that the most diligent is
prevented from succeeding. We also owe much to Cav. L. Riccio
for searching out manuscripts relating to Vesuvius that are stored
in other libraries, and many of which he has had copied or has
published with suitable comments. The Vesuvian bibliography
published in 1881 is the work of Signor Zezi and contains 650
titles. The list we now present is much more extensive, a considerable portion of which has been obtained from the catalogue
of the Alpine Club library.

For bibliographical lists of Etna, we are endebted to P, G.
A. Massa, Sartorius Von Waltershausen, Von Lassaulx, and O.
Silvestri the latter bringing the lists up to 850 entries That catalogue however, includes besides the province of Catania, which
we here have eliminated, the Lipari Islands, of which we have
made a separate list, adding much new material, as also with
that of Graham's Island.

No doubt there still remain a considerable number of omis-
sions, and not a few errors and unperfections which we shall he
very grateful to have communicated to us, and which will be
either utilized by us in some future edition or transmitted to
some other competent persons, as future circumstances shall
decide.

Abbreviations

(B. N.) Biblioteca Nazionale di Napoli.
(O. V.) Osservatorio Vesuviano.
(C. A.) Club Alpino, Sezione di Napoli.

Where no distinct name of an author or responsible editor
is to be found in a publication, it is catalogued under the title of
ANONYMOUS where it is arranged according to the actual or

presumed date of its publication. Those works without dates are placed in alphabetical order at the end. The same system is adopted with the different publications of any one author, but when a memoir is written by two or more persons, the title is put at the end of the list of the first author's works.

The lists include also earthquakes where these appear to be of a volcanic nature or limited to the immediate vicinity of a volcanic district The number of our additions may be judged of, by comparing the catalogue prepared by the International Geological Congress of 1881 with the present.

Int. Geol. Cong.		*Present list*
730	{ Lipari Island 110 { Grahams Island 28 (Etna 880	1027
667	Vesuvius	1552
200	. { Campi Phlegræi 539 { Roccamonfina 33	572
125	Alban Hills.	210
1821	Total	3361

Finally we have to thank Dr. L. Sambon for several additions and help in many ways as also to Prof. F. Borsari. Our thanks are likewise due to Mr. F. Furchheim who has kindly corrected the German and to Conte de la Ville for aid in our bibliographical search.

TOPOGRAPHICAL AND GEOLOGICAL

MAPS, CHARTS, PLANS, AN MODELS

OF THE

SOUTH ITALIAN VOLCANOES.

1. CARTA (COROGRAFICA) D' ITALIA — 1:1 000 000, in 7 sheets. 3 editions. 1st in three colours with the mountains shaded iu brown and water in blue, 2nd mountains in grey, 3rd without mountain shading. See sheets 3 (part of Alban Hills) 4 (part of Alban Hills, Roccamonfina, Campi Phlegræa, and Vesuvius). 6 (Etna, and Lipari Islands). — *Istituto Geografico Militare Ital. 1889.*

2. CARTA (COROGRAFICA) D' ITALIA — 1:800 000, in 6 sheets. 2 Editions. 1st in three colours as in precedent 2nd Without the mountains shaded. See sheets as in precedent. — *Ibid, 1889.*

3. CARTA COROGRAFICA DEL REGNO D'ITALIA E DELLE REGIONI ADIA-CENTI — 1:500 000, in 35 sheets. 3 editions. 1st in three colours ; 2nd in two colours; 3rd in black , without mountain shadeing. See sheets 18 (Rome), 24 Naples and Vesuvius, 29 Lipari Islands, 34 Etna. — *Ibid. 1889.*

4. CARTA DELLA SICILIA — 1:500 000, in 1 sheet. In black ; mountains shaded. Includes Etna and Lipari Islands.— *Ibid, 1885.*

5. CARTA TOGRAFICA DEL REGNO D'ITALIA — 1:100 000, in 277 photo-engraved sheets, in course of publication (1889). The orography is shown by contour lines of 50 m. as well as by zenith-light shading. See sheets 150 Alban Hills ; 160 , 161 , 171, 172 Roccamonfina ; 183 , 184 Ischia and Campi Phlegræi; 184 185 Vesuvius; 261 , 202 , 269 , 270. Etna; 244 Lipari Islands. — *Ibid.*

6. CARTA TOPOGRAFICA DEL REGNO D'ITALIA — 1:100 000. Chromo-lithographic edition in three colours without line shading of mountains. Same divisions as last. — *Ibid. 1889.*

7. CARTA TOPOGRAFICA DEL REGNO D' ITALIA — 1:75 000. Economic edition similar to N.° 5.

8. CARTA TOPOGRAFICA DELLA LOMBARDIA, DEL VENETO E DELL'ITA-
LIA CENTRALE — 1:75 000, in 159 half sheets. Mountains shown
G 15, II 15, G 16, II 16 (Alban Hills); — *Ibid. 1829-1889.*

9. TAVOLETTE RILEVATE PER LA COSTRUZIONE DELLA CARTA DEL
REGNO D'ITALIA — Part to the scale of 1:50 000 and part
1:25 000. See sheets 150 I-IV (1:25 000) (Alban Hills). 160,
161, 171, 172 I-IV (1:50 000) (Roccamonfina); 183 II, 184 I-IV
Ischia and Campi Phlegræi. 184 I-II. 185 III-IV (Vesuvius);
244 I-IV (Lipari Islands); 261, 262, 269, 270 (Etna); All
1:50 000 — *Ibid. 1873-1879.*

10. CARTA TOPOGRAFICA DI ROMA E DINTORNI — 1:100 000, in 1
sheet Similar to N.° 5. Alban Hills. — *Ibid. 1883.*

11. CARTA TOPOGRAFICA DI NAPOLI E DINTORNI — 1:100 000, in 1
sheet Similar to N.° 5. Campi Phlegræi and Vesuvius — *Ibid.
1885.*

12. CARTA TOPOGRAFICA DEL MONTE VESUVIO — 1:10 000, in 6 sheets.
Contour map with contours of 5 m. — *Ibid. 1876.*

13. CARTA DELLA PROVINCIA DI NAPOLI E PARTE DELLE CONTIGUE
DI CASERTA, SALERNO E BENEVENTO — 1:10 000, in 6 sheets
in copper engraving. See sheets 2 (Roccamonfina); 3 (Campi
Phlegræi and Vesuvius); — *Napoli, 1861-1875.*

14. CARTA TOPOGRAFICA E IDROGRAFIGA DEI CONTORNI DI NAPOLI —
1:25 000, in 15 sheets, in copper engraving. See sheets 8
(Campi Phlegræa); 9 (Vesuvius); 10 (Ischia and Procida). —
Napoli, 1818-1870.

15. RAISED MODEL OF ETNA AND NEIGHBOURHOOD — 1:50 000 horizontal and 1:25 000 vertical scale, cast in zinc and plated
with copper. — *Istituto Geografico Militare Ital. 1876.*

16. RAISED MODEL OF VESUVIUS AND NEIGHBOURHOOD — 1:50 000
horizontal, 1:20 000 vertical scale, cast in zinc and plated
with copper. — *Ibid. 1878.*

17. L'ITALIA NEL SUO ASPETTO FISICO, BY CAV. CES. POMBA —
1:100 000 horizont. and vert. Raised map on section of globe. —
*Pub. G. B. Paravia e C., Torino, Roma, Milano, Firenze,
Napoli, 1890.*

18. CARTA FISICA DELL'ITALIA, BY CAV. C. CHERUBINI — 1:750 000
horizontal, 1:150 000 vertical Raised map. — *Ibid. 1876.*

19. ITALIA; CARTA FISICA, BY ING. D. LOCCHI — 1:200 000. Raised.
map. — *Ibid.*

20. ROMA E DINTORNI, BY ING. D. LOCCHI — 1:100 000. Raised
map coloured in two editions — 1st physical and political. 2nd
geological (Alban Hills). — *Ibid.*

21. NAPOLI E DINTORNI, BY ING. D. LOCCHI — 1:100 000. Raised

map coloured in two editions — 1st physical and political.
2nd geological. (Campi Phlegraei and Vesuvius). — *Ibid.*

22. ISOLA D'ISCHIA — 1:15 000. Raised map. — *Ibid.*

23. CARTA IN RILIEVO DELL'ITALIA, BY CAV. G. ROGGERO—1:2800 000
horizontal, 1:320 000 vertical. Raised map. — *Ibid.*

24. CARTA GEOLOGICA DELL'ISOLA D'ISCHIA, BY C. W. C. FUCHS.—
1:25 000. — *Ufficio Geologico Ital. Firenze, 1873.*

25. CARTA DELLE ISOLE PONZA, PALMAROLA E ZANNONE, BY C.
DOELTER — 1:20 000· — *Ibid. Roma, 1876.*

26. CARTA GEOLOGICA DELLA SICILIA — 1:100 000. See sheets 244
(Lipari Islands); 261, 262, 269, 270 and section II (Etna). —
Ibid. Roma, 1890.

27. CARTA GEOLOGICA DELLA SICILIA — 1:500 000 — Etna and Lipari Islands. — *Ibid,*

28. DESCRIZIONE GEOLOGICA DELL'ISOLA DI SICILIA CON UNA CARTA
GEOLOGICA, TAVOLE IN ZINCOTIPIA ED INCISIONI, DELL'ING.
BALDACCI. — *Ibid.*

29. CARTA DELLA CAMPAGNA ROMANA E REGIONI LIMITROFE —
1:100 000, in 6 sheets with 1 section. See sheet " Rome ,,
-   for Alban Hills — *Ibid.*

30. SAGGIO DI CARTA GEOLOGICA DELLA TERRA DI LAVORO, BY
PROF. G. TENORE — 1:20 000. Includes Roccamonfina. — *Napoli, 1872.*

31. GEOLOGICAL MAP OF MONTE SOMMA AND VESUVIUS ENTIRELY
CONSTRUCTED BY H. J. JOHNSTON-LAVIS, DURING THE YEARS
1880-88 — 1:10 000. with short explanation. 2 Editions one in
English the other in Italian. — *Pub. by G. Philip & son. 32
Fleet Street. London, 1891.*

32. PIANO D'ISCHIA E PROCIDA — 1:25 000 — *Ufficio Idrografico
della R. Marina Italiana, 1889.*

33. CARTA DEL GOLFO DI POZZUOLI — 1:20 000 — *Ibid. 1887.*

34. CARTA, DALLA GAJOLA A TORRE DEL GRECO (PIANI PORTE GRA-
NATELLO E TORRE DEL GRECO — 1:20 000 — *Ibid. 1885.*

35. PIANO DELLA RADA DI CASTELLAMMARE — 1:20 000 (*pubblica-
zione provvisoria — Ibid. 1889.*

36. CARTA DEL MARE JONIO E MAR TIRRENO — 1:1000 000 — *Ibid.
1878.*

37. CARTA DELLE ISOLE EOLIE — 1:150 000 — *Ibid. 1881.*

38. PIANO DEGLI ANCORAGGI DI VULCANO, LIPARI E PANARIA —
1:25 000. — *Ibid. 1882.*

LIPARI OR EOLIAN ISLANDS

ABICH H. — Eine Excursion am Crater des Stromboli im Juli 1836. — *Zeitsch. d. Deut. geol. Ges. B, IX, Seit. 392-406. Berlin, 1836. map. 1.*

ABICH H. — Besuch des Kraterbodens von Stromboli am 25 Jul. 1856. — *Zeits. d. Deutch. Geol. Gesell. 1856-57.*

AGATIO DI SOMMA. — Historico racconto dei terremoti della Calabria dell'anno 1638, etc. — *Napoli, 1641, pp. 189.*

ALEXANDER C. — Practical Remarks on the lavas of Vesuvius, Etna and the Lipari Islands. — *Proceed. Scient. Soc. London, Vol. I. London, 1839.*

AMICO ET STATELLA. — Lexicon topographicum Siculum, etc. — *Catanae, 1759-60, Vols. III, in 4°. See Vol. III, pl. 1, pp. 45-52.*

ANDERSON T. — The Volcanoes of the two Sicilies.—*Geol. Mag., Dec. III, Vol. V, p. 473.*

ANDERSON T. — 1888. — See *Johnston-Lavis.*

ARAGO F. — Liste des Volcans actuellement enflammés. — *Annu. d. Bur. d. Longit. année 1824, pp. 167-189.* (C. A.).

ANONYMOUS. — Extrait du « Journal d'Angleterre » contenant une description curieuse de la montagne d'Eole en Italie. — *Jour-nal des Sçavans, 1685, in 12°, pp. 419-420.* (C. A.).

1*

98 LIPARI

ANONYMOUS. — Breve descrizione geografica del Regno di Sicilia.—*Palermo*, 1787, *in 4°, p. 293. See pp. 272 and follow. Isole Eolie, etc.*).

ANONYMOUS. — Cenni sull'Etna e sulle attuali sue eruzioni, con breve sunto dell'opera del Bar. W. Sartorius von Waltershausen. — ?

ANONYMOUS.—Vulcani di Europa. — *Il Propagatore delle Scienze Nat. Anno I, Pt. II, pp. 328.*

BALTZER A. — Eruption von tridimitischen Aschen am Insel Vulcano den 7 sept. 1873.—*Neue Züricher Zeitung, N. 5, 1875.— Referate ans der Naturw. Ges in Zürich den 4 Januar 1875. — Boll. d. R. Com. geol. d'Italia, Vol. VI, pag. 197, Roma, 1875.*

BALTZER A. — Geognostisch chemische Mittheilungen ueber die neuesten Eruption auf Vulcano und die-producte derselben.— *Zeitsch. d. Deutsch. Geol. Gesell. Berlin, 1875.*

BLAKE J. F. — A Visit to the Volcanoes of Italy. — *Proceed. Geol. Assoc., London, 1889, Vol. IX, pp. 145-176.*

BORNEMANN J. G. — Ansichten von Stromboli.—*Zeits. d. d. Geol. Gesells. I, 1842, pp. 696-701, Pl. 4. (C. A.).*

BORNEMANN J. G. — Sur l'état des volcans d'Italie pendant l'été de 1856. — *Translation of De Perrey from Tageblett. der 32 Versam. Deutch. Naturf. und Aertze in Wien, 1856, pp. 114-141, Original M.'S. pp. 4. (C. A.).*

BUCCA L. — Le andesiti dell'isola di Lipari. Studio micrografico. — *Boll. Com. Geol. 1885. N. 9 and 10, pp. 16.*

CAMPI P. — Istoria di Lipari. — *(Cit. par Mongilore and Mercalli.) M. S. is in the Ex-convento dei Capucini al Lipari.*

CHAIX C. — The Past History of Vulcano.—*Bull. Amer. Geogr. Soc.. Vol. XX, pp. 464-469.*

CHIRONE V. — Le terme di S. Calogero nell'isola di Lipari. — *Napoli, 1880.*

CIANCIO A. — Ragionamento sulla privativa del Marchese Nunziante nella fabbricazione dell'Allume Vulcanico. — *Napoli ? in 4°, p. 60.*

CLUVERUS PH. — Sicilia antiqua et Insulae adiacentes. — *Lugduni Balavorum, 1723. Also: Thes. Siculae, Vol. I.*

CORDIER — Rapport sur le voyage de M. Constant Prevost à l'île de Julia, à Malta, en Sicile, aux îles Lipari, et dans les environs de Naples. — *Nouv. Ann. des Voy. 2me Sér. T. X, pp. 43-80, Avril, 1836. (C. A.).*

CORTESE E. — L'Eruzione dell'Isola Vulcano veduta nel Settembre 1888. — *Boll. Com. Geol. Vol. IX, pp. 214-223. Roma, 1888.*

COSSA A. — Su minerali e roccie dell'Isola di Vulcano. — *R. Acc. d. Lincei. Ser. 3ª, An. CCLXXV. Vol. II, 1877-78.*

COZZA P. — Un incendio sconosciuto del Vesuvio. — *Archivio Stor. Province Napoletane, An. XV. fasc. III, pp. 642-646.*

CROTTI C. — Viaggio. per la Sicilia eseguito nell'anno 1830. Poemetto. — *Napoli, 1830, in fol. pp. 20. At p. 6: Stromboli, Vulcano, etc.*

DESNOYERS — Notice sur l'ile Julia, le Stromboli, les colonnes du temple de Pozzuoli. — *Bull. d. l. Soc. de Géol. 1831, pp. 220.*

DEVILLE CH. (S. CLAIRE). — Sur quelques produits d'émanations de la Sicile. — *Compt. Rend. Vol. XLIII. Paris, 1856.*

DEVILLE CH. (S. CLAIRE) — Sur la nature des éruptions actuelles du Volcan de Stromboli. — *Bull. d. l. Soc. géol. d. France. 2.ᵉ Sér. Tom. XV, pp. 345-362, Paris, 1858.*

DOLOMIEU (DE) D. — Voyage au iles de Lipari fait en 1781. — *Paris, 1783, in 8°, pp. VIII, + 208.*

DONATI. — Notice sur l'ile de Stromboli. — *Bull. d. l. Soc. de Géol. d. France, T. I, pp. 242-245. 1831. (C. A.).*

FERRARA F. — Campi flegrei della Sicilia e delle isole che le sono attorno. Messina, 1810. — *Atti d. Acc. Gioenia, Ser. 1ª, Vol. II, Catania, in 4°, fol. 2, pp. XIX + 424, maps.*

FOUQUÉ F. — Voyage aux îles Eoliennes. — *Compt. rend. d. l'Ac. d. Sc. Tom. LX, pp. 1185; Tom. LXII, pp. 616 et 1366. Ann. d. Mission scient. et litt. 2ª Ser. Tom. III, pp. 165. Paris, 1865-1867.*

FUCHS K. — Vulkane u. Erdbeben. Leipzig 1877. — *In French: Les Volcans et les tremblements de Terre, Vol. I, Paris, 1866. (Iles Lipari, pag. 215)*

FULCHER L. W. — A visit to the Lipari Islands and Mount Etna. — *Journ. of the City of London College Science Soc., N.° 16, April, 1890, pp. 9.*

FULCHER L. W. — Vulcano and Stromboli. — *Geol. Mag., Dec. III, Vol. VII, 1890, pp. 347-353.*

GALVANI D. — Memoria geologica e mineralogica su le Isole Eolie classificazione de' prodotti volcanici delle medesime. — *Nuovi Ann. d. Sc. Nat. d. Bologna, T. VI, 1841, pp. 18, in 8°. (C. A.).*

GENOVESI. — Sull'acque termo-minerali e sulla Grotta di Lipari, 1879. — *Idrol. Med. del dott. Chiminelli. Fasc. XIV, 15. ottobre nov. 1880, pp. 94-95.*

HAMILTON W. Observations on M. Vesuvius, M. Etna and other Volcanoes of the two Sicilies. — *London, 1772.*

HAMILTON W. — Campi Phelegrei; observations sur les vulcans des Deux Siciles (French and English) — *Naples, 1776-79, in fol., col. pl. 54.*

HOFFMAN F. — Ueber die geognostische Beschaffenheit der Liparen. — *Poggd. Ann. Bd. XXVI, Seit. 31 u. folg. 1832, in 8°, pp. 88, pl. 4.*

HOFFMAN F. — Mémoire sur les terrains volcaniques de Naples, de la Sicile, et des îles de Lipari. — *Bull. d. l. Soc. géol. de France, Vol. III, 1833, pp. 170-180*

HOUEL J. — Voyage Pittoresque des Isles de Sicile, de Malte et de Lipari. — *Paris, 1782, Vol. IV in fol. Numerous rich engravings.*

HOUEL J. — Reisen durch Sicilien u. Malta etc.. übersetz t. v. J. L. Heerl, mit Kpft. — *Vol. I-IV, Gotha, 1797-1809, in 8°.*

IDDINGS P. AND PENFIELD S. L. — Fayalite in the obsidian of Lipari. — *Amer. Jour. Sc. Vol. XL, 1890, pp. 75-78.*

JERVIS G. — Tesori sotterranei dell'Italia. — *4 Vols. in 8°, Torino, 1874-1888, numerous plates.*

JOHNSTON-LAVIS H. J. — The island of Vulcano and Stromboli. — *Nature, Vol. XXXVIII, pp. 13-14. London, 1888.*

JOHNSTON-LAVIS H. J. — The Recent Eruption at Vulcano.—*Ibid. p. 173.*

JOHNSTON-LAVIS H. J. — The Eruption of Vulcano Island. — *" Nature," Vol. XLII, 1890, pp. 78-79.*

JOHNSTON-LAVIS H. J. — The Conservation of Heat in Volcanic Chimneys. — *Brit. Assoc. Reports, 1888, pp. 2.*

JOHNSTON-LAVIS H. J. — The State of the Active Sicilian Volcanoes in September 1880. — *Scottish Geograph. Mag., Vol. VI, N.° 3, March 1890, pp. 145-150.*

JOHNSTON-LAVIS H. J. — Further Notes on the Late Eruption of Vulcano Island, — *" Nature," Vol. XXXIX, 1889, pp. 109-111*

JOHNSTON-LAVIS H. J. AND ANDERSON T. — Notes on the Recent Volcanic Eruption in the Island of Vulcano. — *Brit. Assoc. Reports, pp. 3, 1888.*

JUDD W. J. — Contributions to the Study of Volcanoes. — *Geol. Mag. Vol. II, London, 1876. pp. 1, 56, 145, 206, 245, 308, 348, 388.*

LYELL C. — Principles of Geology. — *London, Numerous editions. Principes de Géologie; Traduction Française sur la sixième édition anglaise. Lyon, 1846. (Stromboli. pt. III, pp. 357).*

MALLET R. — The mechanism of the active Volcano of Stromboli.—*Proceed. of the R. Soc. London. N. 155. 1874. Translated in Ital. by Prof. O. Silvestri, Bull. d. Vulc. ital. Fasc. VII, VIII. IX, X. Roma, 1876.*

MARGALLE. — 1866. — *See Zurcher.*

MASSA G. A.—La Sicilia in prospettiva.—*Palermo, 1709. 2 Vols. in 4°. Parte I, pp: 12-359, Parte II, pp. 503.*

MERCALLI G. — Contribuzioni alla geologia delle isole Lipari. — *Atti d. Soc. It. d. Sc. Nat. Vol. XXII, Milano, 1879, pp. 14.* (C. A.).

MERCALLI G. — Le ultime eruzioni dell'Isola di Vulcano. — *Bull. d. Vulcan, Ital. Ann. IV, p. 28. Roma, 1879.*

MERCALLI G. — Contribuzioni alla Geologia delle Isole Lipari. — *Atti d. Soc. Ital. d. Sc. Nat. Vol. XXII, Milano, 1879.— Estratto dal Bull. d. R. Comit. Geol. d'Italia, Vol. I, Ser. 2ª, p. 315. Roma, 1880.*

MERCALLI G. — Natura delle eruzioni dello Stromboli ed in generale dell'attivita sismo-vulcanica nelle Eolie. — *Atti d. Soc. It. di Sc. Nat. Vol. XXIV, Milano, 1881, in 4°, pp. 30.* (C. A.).

MERCALLI G. — La fossa di Vulcano e lo Stromboli dal 1884 al 1886. — *Atti d. Soc. It. d. Sc. Nat. Vol. XIX, Milano, 1886, in 4°, pp. 9.* (C. A.).

MERCALLI G. — L'Isola Vulcano e lo Stromboli dal 1886 al 1888.— *Atti Soc. Ital. Sc. Nat., Vol. XXXI, Milano, 1888.*

MERCALLI G. — L' eruzione dell' isola Vulcano. — *Rassegna Nazionale, Ann. X. Firenze, 1889, in 8°, pp. 18.*

MERCALLI G. — Sopra alcune lave antiche e moderne dello Stromboli. — *Rend. R. Ist. Lombard, Ser. II, Vol. XXIII, fasc. XX. 1891, pp. 11.*

MINÀ PALUMBO F. — Cenno topografico delle isole adiacenti alla Sicilia. — *L' Empedocle, anno I, fasc. 7 e 8 (Nov. Dec.) 1851, pp. 419-436, 465-492.*

PALMIERI L. — Intorno ad una recente eruzione nell'Isola di Vulcano ed alla continuazione del terremoto di Corleone. — *Rend. R. Accad. Sc. Fis. Mat. An. XV, Napoli, 1876, p. 123.*

PAPARCURI S. (1743). — Discorso fisico-matematico sopra la variazione de'venti pronosticata ventiquattr'ore prima dalle varie e diverse qualità, ed effetti dei Fumi di Vulcano. — *Opusc. di Autor. Sic. V, pag. 76-120.*

PAREIRA A. — Sieben Tage auf den Eolischen Inseln. 1880. — *1881?*

PAYAN D. — Notice sur quelques volcans de l'Italie méridionale.— *Bull. Soc. Stat. Arts Utiles, Sc. Nat. du Départ. d. l. Drôme, t. III, in 8°, 1842, pp. 145-163.* (C. A.).

PILLA L. — Parallelo tra i tre Volcani ardenti delle Sicilie. — *Atti d. Acc. Gioenia 1837. — Cit. Jahrb. f. Min. p. 347. Stuttgart, 1836.*

PILLA L. — Sur les Iles Eoliennes et sur les localités volcaniques de la Sicile. — *Ann. des mines, 3ª Sér. Vol. XVIII, p. 127. Paris, 1840 (?).*

PILLA L, — Osservazioni fisiche sopra il Vulcano di Stromboli.— *Il Lucifero, anno 1°, Napoli, pp. 30, 54, 89. 106.*

PLATANIA GAET. — Sui projettili squarciati di Vulcano (Isole Eolie) nell' eruzione del 1888-1890. — *Ann. dell' Ufficio Cent. Meteor. Geodin., Pt. IV, Vol. X, 1891, pp. 7.*

PLATANIA GIOV. — Éruptions volcaniques aux îles Lipari, du 3 au 6 août 1888. — *La Nature, 16ᵉ ann. Paris, 1888, N. 795, pp. 198-199.*

PLATANIA GIOV. — Eruption volcanique à l'île Vulcano.—*La Nature, 16ᵉ Ann. 1888. N. 805, pp. 359-363, fig.*

PLATANIA GIOV. — I fenomeni sottomarini durante l'eruzione di Vulcano (Eolie) nel 1888-1889. — *Atti, Rend. Acc. Sc. Let. Art. Acireale, N. Ser. Vol. I, 1889, pp. 16, tables 3.*

PLATANIA GIOV. —Stromboli e Vulcano nel Settembre del 1889.— *Boll. d. Osserv. Meteor, R. Ist. Nautico di Riposto, An. XV, fasc. 9-12, pp. 14, Riposto, 1889.*

PLATANIA GIOV. — La récente éruption volcanique à l'île Vulcano, 1888-90) — *La Nature, 19ᵉ ann. Paris, 1891, N.° 927. pp. 211-214, fig.*

PRESTANDREA E CALCARA P. — Breve cenno sulla geognosi ed agricoltura delle isole di Lipari e Vulcano. — *Giorn. di Comiss. d'Agric. e Pastor. per l. Sicilia, 1858.*

QUATREFAGES DE. — Sur l'état du Cratère supérieur de Stromboli en Juin 1844. — *Compt. Rend. Ac. Sc. Paris, Vol. XLIII, 1845, p. 610.*

QUATREFAGES DE. — Souvenirs d'un Naturliste, Milazzo, Stromboli. — *Revue des Deux Mondes, Vol. XVII, Chap. IV, Paris, 1847, pp. 120.*

RAMMELSBERG C. —Ueber die Natur der gegenwärtigen Eruptionen des Vulkans von Stromboli. — *Zeits d. d. Geol. Gesells XI, 2, 1859, pp. 103-107.* (C. A.).

RICCIARDI L. — Sull'allineamento dei vulcani italiani, etc. — *Reggio-Emilia, 1887. in 8°, pp. 10, col. map. 1.*

SALINO F. — Le isole Lipari. — *Boll. d. Club. Alp. Ital. Torino, 1874, pp. 135.*

SALINO F. — Le eruzioni di Vulcano. — *Cosmos, di G. Cera, 1890, pp. 45-56.*

SAMBON L. — Eolie — *Napoli, 1891, pp. 60, See also, Pro Patria for 1890, Napoli.*

SCACCHI A. — Sabbia eruttata da Vulcano dal dì 11 al 26 gennaio 1886. — *Boll. Oss. Coll. Carlo Alberto, S. II, Vol VI. n.° 8 — Moncalieri, 1886.*

SCHMIDT G. F. J. — Vulkanstudien. — *Leipzig, 1874, in 8°.*

SCROPE G. P. — Volcanoes. The character of their phenomena, etc. — *London, 1862. Paris, 1864, Berlin, 1872.*

SCROPE G. P. — The mechanism of Stromboli. — *Geol. Mag. N. 126. London, 1874.*

SECCHI P. A. — Lezioni di fisica terrestre. — *Torino è Roma, 1867.*

SEGUENZA G. — Dell'arsenico nei prodotti vulcanici delle isole Eolie. — *Eco Peloritano. An. III. Fasc. VII, Messina, 1856, in 8°, pp. 8.*

SEGUENZA G. — Di certe rocce vulcaniche interstratificate fra rocce di sedimento. — *Rend. d. R. Acc. d. Sc. Fis. e Mat. d. Napoli, 1876.*

SILVESTRI O. — Fenomeni eruttivi dell'isola di Vulcano e Stromboli nel 1874. — *Boll. d. Vulcan. Ital. Fasc. IX e X, pp. 117. Roma, 1874.*

SILVESTRI O. — Il meccanismo nel Vulcano attivo di Stromboli. Translation into Italian of R. Mallet. — *Boll. d. Vulcan. Ital. etc. Fasc. VII, VIII, IX, X. Roma, 1876.*

SILVESTRI O. — Sull'attuale eruzione di Vulcano nelle isole Eolie incominciata il 3 agosto 1888. — *Annali dell'Ufficio Centrale di Meteorologia e Geodinamica, Parte IV. Vol. IX, 1887, Roma, in 4°, pp. 13.*

SILVESTRI O. — Etna, Sicilia ed isole vulcaniche adiacenti, sotto il punto di vista dei fenomeni eruttivi e geodinamici presentati durante l'anno 1888. — *Atti Accad. Gioenia. S. IV. Vol. I. N.° 2. Catania. Annuario Meteor. Anno IV.*

SILVESTRI O. — L'Eruzione dell'isola di Vulcano. — *Boll. Oss. R. Coll. Carlo Alberto, S. II, Vol, VIII, n.° 10, Torino, 1888.*

SILVESTRI O. — L'isola di Vulcano ed il suo risveglio eruttivo. — *Nuova Antologia, Vol. XXI, fasc. 11. Roma, 1889.*

SILVESTRI O. — Sur l'éruption recente de l'île de Vulcano. —

Comptes Rend. Acad. Sc. T. CIX. 6, Paris, Août, 1889, pp. 3.

SILVESTRI O., CONSIGLIO PONTE S., SILVESTRI A. — Sulla attuale eruzione scoppiata il dì 3 Agosto 1888 all'Isola Vulcano nell'arcipelago Eolie (Sunto). — *Bull. mens. Acc. Gioenia d. Sc. nat., Nuova serie, fasc. VIII, Catania, 1889.*

SILVESTRI O. E ARCIDIACONO S. — Etna, Sicilia, ed Isole Vulcaniche adiacenti, sotto il punto di vista dei fenomeni eruttivi e geodinamici durante l'anno 1889. — *Boll. Mens. del Osserv. Cent. di Moncalieri, Ser. II, Vol. X, N.° 2, Febbraio 1890.*

SOMMA DI AGATIO. — V. Agatio di Somma.

SPALLANZANI L. — Viaggi alle due Sicilie e in alcune parti dell'Appennino. — *Pavia, Vol. I-IV, 1792. Also in German. Bd. I-VIII. Leipzig, 1794-96.*

SPALLANZANI L. — Travels in the Two Sicilies and some parts of the Apennines. — *Translated from the Original Italian.— 4 vols. with 11 plates. London, 1798.*

STAGNO S. F. — Ragionamento sopra il nascimento dell'isola di Vulcano. — *Opusc. di Autori. Sic. Vol. II, in 4°, pp. 93-121, Palermo, 1759, in 12°.*

STOPPANI A. — Corso di Geologia. — *Vol. I-III Milano, 1873 (Lipari Isles etc. Vol. I, pp. 680).*

STRUEVER. — Ematite di Stromboli. — *Atti Accad. Lincei, S. IV, Vol. IV, Fasc. 9°, p. 626.*

TACCHINI P. — Sulle attuali eruzioni di Vulcano e Stromboli. — *Rendicont. R. Accad. Lincei, S. IV, Vol. V, 2° sem. Roma.*

THOMAS T. H. — A visit to the Lipari Isles and Etna. — *Trans. Cardiff Naturalist's Soc. Vol. XXII, pt. 1, 1890, pp. 16.*

TROVATINI G. M. — Dissertazione chimico-fisica sull'analisi dell'acqua minerale dell'isola di Vulcano nel Porto di Levante detta volgarmente l'acqua del Bagno. — *Napoli, 1786, in 4°.* (B. N.).

VARENIUS B. — Geografia generalis, in qua affectiones generales telluris explicantur. (Lib. I. Cap. X: on the island of Vulcano and Etna). — *Amstelodami, 1664.*

ZURCHER AND MARGALLE. — Volcans et tremblements de terre (Etna et Stromboli). — *Hachette et Cie. Paris, 1866.*

GRAHAM'S ISLAND

ISOLA FERDINANDEA

OR

ISOLA GIULIA

ANONYMOUS. — Breve ragguaglio del novello vulcano. — *Effemer. Scient. e Letter. per Sicilia* , *T. I. Palermo, 1832, in 8e, pp. 31, pl. 1. (C. A.).*

ANONYMOUS. — Réapparition de l'île Ferdinandea (ou Julia) dans la Méditerranée. — *Bull. Soc. Géog. France,* 2.me *sér.,* T: *I, N. 1, 1834, fol. 100. (C. A.).*

ARAGO F. — Considération sur la manière dont se forma dans la Méditerranée , en Juillet 1831 une île qui a été tour à tour appelée Ferdinandea, Hotham, Graham, Nerita, et Julia. — *Compt. Rend. Acad. Sc. Paris, T. IV., pp. 753-757. (C. A.)*

BERGHAUS. — Annalen (Insel Julia). — *Bd. IV, Seit. 365 u. Bd. V, Seit. 124, N. 198. — Same in French: Ann. d. Sc. Nat. Vol. XXIV, pag. 103, Paris.*

CAPOCCI E. — Un nuovo vulcano in Sicilia. Dialogo. — *Il Propagatore delle Sc. Nat. Napoli, 1847-48, Pt. 1, pp. 185-186,*

CORDIER. — Rapport sur le voyage de M. Constant Prevost à l'île de Julia, à Malta, en Sicile, aux îles Lipari et dans les environs, de Naples. — *Nouv. Ann. des Voy.* 2nd *série, T. X, p. 43-80, Avril 1836. (C. A.).*

DAUBENY C. — Note on a paper by J. Davy entitled : Notice of

2.

the remains of the recent volcano in the Mediterranean. —
Phil. Trans. 1833. pp. 545-548. (C. A.).

DAVY AND DAUBENY. — Account of a new Volcano in the Mediterranean. — *London, Paris, 1831, in 4.°.*

DAVY H. — Isle of Julia. — *Philos. Trans. Part. I, pag. 143,
and Part. II, pag. 237. London 1833.*

DAVY J. — Further Notice of the New Volcano in the Mediterranean. — *Phil. Trans. 1832, pp. 251-254.* (C. A.).

DAVY J. — Notice of the Remains of the Recent Volcano in the
Mediterranean. — *Phil. Trans. 1833, pp. 143-146.* (C. A.).

DAVY J. — Some Remarks in Reply to Dr. Daubeny's Note on the
Air Disengaged from the Sea over the Site of the Recent
Volcano in the Mediterranean. — *Phil. Trans. 1834 , pp.
551-554.* (C. A.).

DESNOYERS. — Notice sur l'île Julia , le Stromboli , les colonnes
du temple de Pozzuoli. — *Bull. d. l. Soc. de Géol. 1831,
T, II, pp. 238-242.*

FUCHS K. — Vulkane u Erdbeben. — *Leipzig, 1875. French edition: Les Volcans et les Tremblements de Terre. Vol. I. Paris, 1866. (Isola Ferdinandea, pp. 220).*

GELCICH E. — Die Insel Ferdinandea. — *Deutsche. Rundsch. f.
Geogr. u. Statistik, Vol. VIII, pp. 225-228.*

GEMMELLARO C. — Relazione de' fenomeni del Nuovo Vulcano sorto
dal mare fra la costa di Sicilia, e l'isola Pantelleria nel mese
di luglio 1831. — *Atti dell'Accad. Gioenia Sc. Nat. Catania
Vol. VIII, pp. 271-298. Catania, 1831, in 8.°, pp. 48, Appendix, pp. XXIV, pl. 11.*

GRAVINA C. (PRINCIPE DI VALSAVOYE). — Poesie. La eruzione del
Vulcano sotto-marino tra Sciacca e Pantelleria nel luglio del
1831. Sonetto. — *Catania, 1834, in 12°, pp. 72. See p. 11.*
(B. N.).

HOFFMANN F. — Intorno al nuovo vulcano presso la città di Sciacca. Lettera al Duca di Serradifalco. — *Giorn. d. Sc. Lett.
ed Arti. N. 101. Palermo, 1831.*

MAZZOLLA B. — Descrizione dell'isola Ferdinandea al mezzogiorno
della Sicilia. — *Napoli, 1831, in 12°, oblong, pp. 4, pl. 8.
Other Editions.* (C. A.).

PILLA L. — Phénomènes volcaniques récents dans un des points
de la mer qui baigne les côtes de la Sicile. Lettre à Mons.
Arago. — *Compt. Rend. Acad. Sc., T. XXIII, Paris, 1846,
pp. 988-990.* (C. A.).

PREVOST C. — Observations sur le nouvel ilot volcanique qui s'est

-formè en juillet 1831 dans la mer de Sicile. — *Bull. Soc. Geol. France. t. II, 1831, pp. 32-38.* (C. A.).

PREVOST C, — Description de l'ile volcanique sortie récemment du sein de la Méditerranée. — *Nouv. Ann. des Voyages, série 2, T. XXII, 1831, pp. 288-303.* (C. A.).

PREVOST C. — Introduction au rapport fait à l'Académie Royale, des Sciences sur le voyage à l'ile Julia en 1831 et 1832. — *Paris, in 8°, pp. 47.* (C. A.).

PREVOST C. — Notes sur l'ile Julia pour servir à l'histoire de la formation des montagnes volcaniques. — *Mém. Géol. Soc. France, T. II, pp. 91-124, pl. 3.* (C. A.).

RICCO A. — L'ile Ferdinandea, le soleil bleu et les crépuscules rouges de 1831. — *Compt. Rend. Ac. Sc. Paris, Vol. CII, pp. 1060-1063.*

RUSSO FERRUGGIA S. — Storia dell'isola Ferdinandea sorta nella costa meridionale della Sicilia. — *Trapani, 1831, in 4°, pp. 58.* (C. A,).

SAINT-LAURENT DE. — Détails sur l'ile volcanique nouvellement apparue dans la Méditerranée. — *Bull. Soc. Géol. t. XVI, N.° 102, 1831, pp. 185-188.* (C. A.).

SMYTH W. H. — Some Remarks on an Error respecting the Site and origin of Graham's Island. — *Phil. Trans. 1832, pp. 255-258, pl. 1.* (C. A.).

ETNA

ABICH II. — Vues illustr. des phénomènes géologiques du Vésuve et de l'Etna, etc. — *Paris, 1837.*

ABICH H. — Vulkanische Phaenomene am Aetna. — *Briefl. Mitth. Jahrbuch f. Miner. Geogn. u. Geol. Seit. 551, Stuttgart 1839.*

ABICH H, — Erlaüternde Abbildungen der geol. Erscheinungen beobachtet am Vesuv u. am Aetna in den Jahren 1833 und 34.—*Berlin, 1837. Braunschweig, 1841 (text in French and German) in fol. pl. 10.*

ABICH A.—Geol. Beobachtungen über die Vulkan-Erscheinungen— *V. Bd. Lief. I. Ueber die Natur und den Zusammenhnang der vulkanischen Bildungen. Braunschweig, 1841.*

ABICH II. — On some Points in the History and Formation of Etna. — *Quart. Journ. Geol. Soc. Vol. XIV, 1858.* (C. A.).

ABU-HAMID DA GRANADA. — In the « Biblioteca Arabo-Sicula » by Amari. — *(Pag. 74. Erupt. of XII century).*

ACARIUS DE SÉRIONNE. — Dissertation sur le mont Etna. — *Paris MDCCXXXVI, in 12°, pp. 179-223.* (C. A.).

ACARIUS DE SÉRIONNE. — L'Etna dep. Cornelius Severus et les sentences de Publius Syrus. Traduit en francais avec des remarques des dissertations critiques, historiques, géographiques. — *A Paris chez Chaubert et Clousier, MDCCXXXVI, in 12°, 2 maps, pp. 358.* (C. A.).

AELIAN. — (συμμιχτος ιστορια).—in Stobaeus Flor. 78-39. (Erupt. of 693. B. C.).

AESCHYLUS. — Prometheus. — (Erupt. of 364. B. C.).

ALBERTI L. — Descrizione di Sicilia. — (Cit. by Massa).

ALESSI G. — Elogio del cav. Gius. Giocni. — Palermo, 1824, in 4°. (B. N.).·

ALESSI G.—Storia critica delle Eruzioni dell'Etna. Otto discorsi.— Atti d. Acc. Gioenia. Ser. 1ª, Vol. III, pp. 17-75 ; IV, pp. 23-73; V, pp. 43-72; VI, pp. 85-114; VII, pp, 21-66; VIII, pp. 99-149; IX, pp. 121-216. Catania, 1824-32.

ALESSI G. — Sopra gli ossidi di Silicio ed i silicati appartenenti alla Sicilia.—Atti d. Acc. Gioenia, Ser. 1ª, Tom. V. Catania, 1827.

ALEXANDER C. — Practical Remarks on the lavas of Vesuvius, Etna, and the Lipari Islands. — Proceed. Scient. Soc. London, Vol. I, London, 1839.

AMARI M. — Storia dei Musulmani in Sicilia (Erupts. of VII-IX cent. — I. Biblioteca Arabo-Sicula, pp. 85-218.

AMICO C. — Cronologia Universale (Cit. by Ferrara and by Amico in: — Catana Illustr. T. IV, p. 252, (Unpublished).

AMICO V. M. —Catana Illustrata sive sacra et civilis urbis Catanae Historia. — Catania, 1740, in fol.

AMICO V. M. — 2nd edit of Fazellus. — Catanae, 1749, See Fazellus.

AMICO V. M. ET STATELLA. — Lexicon topographicum Siculum, etc. — Panormi, et Catanae, 1759-60, Vols III, in 4°. See Vol. III, pt. 1, pp. 45-52.

ANDERSON T. — The Volcanoes of the two Sicilies. — Geol. Mag., Dec. III, Vol. V, pp, 473.

ANDREAE J. L. — De montibus ignivomis dissertatio inauguralis.— Alldorf, 1710.

ANNA A. D'. — Eruption of mount Etna. — London, June, 1800. A col. transparency on a steel engraving in R. fol. (In the collection of M.r L. Sambon, Naples).

ANONYMOUS DI SCIACCA. — (Cit. by Recupero). — (Erupt. de cendres, 1408).

ANONYMOUS. — Scriptores Rerum Sicularum in unum corpus nunc primum congesti.—Francoforti A.-M. A. Wechel, 1579. (See respective authors.)

ANONYMOUS. — A Chronological account of several incendiums or fires of Mt. Aetna. — Philos. Transactions. Vol. IV, p. 967. London, 1669.

ANONYMOUS. — An answer to some inquiries concerning the eruptions of Mt. Aetna, 1669, communicated by some inquisitive English merchants now residing in Sicily. — *Phil. Trans. Vol. IV, p. 1028. London, Sept. 1669.—Coll. Accad. T. I, pl. 2 pp. 201-205.— Gibelin, T. I, pp. 4-13.*

ANONYMOUS. — Plan du Mont Etna, communément dit Mont Gibel en l'Isle de Sicile, et l'incendie arrivé par un tremblement de terre le 8 Mars 1669. — *Bibl. Nationale d. Paris.*

ANONYMOUS. — Relazione dell'incendio di Mongibello dell'anno 1669. — *(Cit. by Massa: Etna).*

ANONYMOUS. — Relatione (vera) del nuovo incendio della Montagna di Mongibello cavata da una lettera scritta da Tauramina ad un Signore dimorante in Roma. — *Roma and Napoli, 1669, in 12°, fol, 4.* (C. A.).

NONYMOUS. — Archivio dei Benedettini in Catania. — *Arca I. Lit. B., pag. 100. (Erupt. 1536, 1682 and others.)*

ANONYMOUS. — Extrait du Journal d'Angleterre contenant une relation chronologique des embrasemens du Mont Etna. — *Journal des Sçavans, 1683, in 12°, pp. 103-105.* (C. A.).

ANONYMOUS. — Lagrimoso Spectacolo della misera città di Catania nell'Isola di Sicilia, la quale fu distrutta li 15 Gennaio del corrente anno 1693 da un spaventoso terremoto, etc. — *Viterbo, 1693, in 16°, fol. VI.* (C. A.).

ANONYMOUS. — Della Sicilia, grand'isola del Mediterraneo, in prospettiva il Mont'Etna, o Mongibello, esposto in veduta da un religioso della compagnia di Gesù. — *Palermo, 1708, in 8°, pp. VIII, 126.* (C. A.).

ANONYMOUS. — Manuscriptum ex libro in Ecclesia majori Nicholosorum asservato, etc. — *(With notices on the erupt. of 1766).*

ANONYMOUS. — Relation (an exact) of the famous Earthquake and Eruption of Mt. Etna. — *London, 1775.*

ANONYMOUS. — Untergang der Stadt Messina. Ingleichen eine kurze Beschreibung von den beiden Feuerspeyenden Bergen Vesuv und Aetna. — *? 1783.*

ANONYMOUS. — Breve descrizione geografica del Regno di Sicilia.—*Palermo, 1787, in 4°, p. 293. (pp. 209-211 l'eruzioni del Mongibello, etc.).*

ANONYMOUS. — Compendio delle Transazioni filosofiche, ecc. — *Giornale Letterario di Napoli, Napoli, 1793. Vols 112, in 8°, Vol. V, pp. 78-89.*

ANONYMOUS.—Descrizione della eruzione dell'Etna di quest'anno.— *Gazz. Britannica, N. 73, mercoledì 13 Nov. 1811.*

ANONYMOUS. — Monte San Simone und die Eruption von 1811.—

.*Morgenblall, N. 138. Seil. 551. (Cit. by Hoff. Geschichte der naturl. Veränderungen, etc. II, pag. 241).*

ANONYMOUS. — Eruption de l'Etna. Poussière rougeâtre transportée d'au-delà des mers en Italie par le vent. — *Bull. Soc. Géog. d. F. T. XIII, pp. 307-308. June 1830.* (C. A.).

ANONYMOUS. — Kurze Beschreibung des Actnaausbruches im November 1832. Fröb. Notiz. XXXVI. 1833. — *Neu, Jahrb. f. Min. Geogn. u. Geol. Seit. 583. Stuttgart, 1833.*

ANONYMOUS. — Descrizione di Catania e delle cose notevoli dei dintorni di essa. — *Catania, 1841, in 8°, pp. 277.*

ANONYMOUS. — Giornale della presente eruzione dell'Etna. — *Rend. R. Accad, Sc. Fis. Mat. Napoli, T. I, pp. 466-468. 1842-1845*

ANONYMOUS. — L'Etna et ses éruptions. — *L'Univers Illustré, 15 février 1865, pp. 99-100, with 3 figs.* (C. A.).

ANONYMOUS. — Eruption of Etna 1865. — *Am. Journ. Sc. 2 nd. Ser. Vol. XL. N.° 118, July 1865, p. 122.* (C. A.).

ANONYMOUS. — Eruzione dell' Etna dell'anno 1879. — *Gazetta di Catania, May and June 1879, 16 numbers.* (C. A.).

ANONYMOUS. — Eruzione dell'Etna del 1879. — *Gazetta di Messina 17 numbers.* (C. A.).

ANONYMOUS. — La fine della eruzione dell' Etna 1879. Disegno e testo sopra comunicazioni di Nicola Lazzaro.— *Illustrazione Italiana. Sem. 2°, pag. 5, Milano, 1879.*

ANONYMOUS. — La salita degli Alpinisti Romani sull'Etna. — Disegni di D. Paolucci, sopra schizzi originali di Fornari e testo di Martinori — *Illustrazione Italiana. Sem. 2°, pag. 123.— Milano 1879.*

ANONYMOUS. — Sulla eruzione dell'Etna del 1879. Tre disegni sopra schizzi autentici del Prof. Orazio Silvestri.—*Illustrazione Italiana. Sem. 2°, p. 5 Milano, 1879.*

ANONYMOUS. — Sulla eruzione di fango a Paternò nelle adiacenze dell'Etna. Disegno e testo sopra comunicazione autentica del Prof Orazio Silvestri. — *Illustrazione Italiana. Sem. 1°, pag. 113, Milano, 1879.*

ANONYMOUS. — Relazione degli Ingegneri del R. Corpo delle Miniere addetti al rilevamento geologico della zona solfifera di Sicilia, sulla eruzione dell'Etna avvenuta nei mesi di maggio e giugno 1879. — *Roma, 1879, in 4°, pp. 7, map. 1.* (C. A.).

ANONYMOUS. — Relazione della Commissione governativa. Eruzione dell' Etna del 26 Maggio 1879. — *Gazz. Uff. d. Regno Vol, I, 152, luglio 1879.* — *Boll. d. Soc. Geogr. Ital. pp.*

550-60, N. 8, con carta, 1879. — *Boll. d. R. Com. Geol. d'Italia, Vol. X. Roma, 1879.*

ANONYMOUS. — Chronicon Siciliae complectens accuratam regni Siciliae historiam. — *Thes. Sic.*

ANONYMOUS. — Vulcani di Europa. — *Il Propagatore delle Scienze Nat. Anno. I, Pl. II, pp. 328.*

ANONYMOUS. — De gestis Gallorum et Aragonensium. (Cit. by Carrera. —*Mem. Stor. d. Città d. Catania. I, 2, c. 2.*

APPIANUS A. — Bellor. civil. (Erupt. 34 B. C.). — *Edit. Amstelodami, 1660.*

APRILE Z. — Cronaca di Sicilia. — *Vol. I, Palermo, 1725.*

ARABICUS CH. — (Cit. by Carrera).

ARACRI G. — Relazione della pioggia di cenere avvenuta in Calabria ulteriore il dì 27 marzo 1809. — *Atti d. Acc. Pontan. Napoli, 1810, Vol. I, pp. 167-170.*

ARADAS A. — Brevissimo sunto della conchiliologia etnea. — *Atti d. Soc. It. d. Sc. Nat. Vol. XII, Fasc. III, Milano, 1869, in 8°, pp. 9.* (C. A.).

ARADAS A. — Un'abbozzo del panorama Etneo. Discorso dell' A. come Presidente del Congresso dei Naturalisti italiani tenuto in Catania nel 1805. — *Atti d. Soc. It. d. Sc. Nat. Vol. XII, Fasc. III, Milano, 1869, in 8°, pp. 36.*

ARADAS A. — Sulle variazioni delle acque del golfo di Catania rimpetto al littorale. — *Atti d. Acc. Gioenia, Catania, 1881.*

ARAGO F. — Liste des Volcans actuellement enflammés) — *Annu. d. Bur. d. Longit. année 1824, pp. 167-189.* (C. A.).

ARDINI L. — Carta agronomica dell'Etna. This map is not published but can be obtained from the author. — *Catania, 1878.*

ARETII M. — De situ Siciliae. — *See Bibl. Historica Regni Siciliae Joan. Bapt. Carusii, Tom. I.* — *Also: Thesaurus Antiquitatum et historiae Siciliae Gravii et Burmanii Lugduni Batavorum, 1723. (Aetna eaux minérales, etc.)*

ARISTOTELES. — περὶ κοσμον. — *page 365. (general treatise).*

ARISTOTELES. — πεμασίσνν. ἀκουσματάθι Ϩαν — *C. 38, page 832. Edit. Boeckh. (Erupt. of IV century. (B. C.).*

AULUS GELLIUS. — Noct. Att. — *lib. XVII, cap. 10.*

AURIA DON V. — Storia Cronologica dei Vicerè di Sicilia. 1400-1597. — *Palermo, Coppola, 1797. Pp. 143. Erupt. of 1669, earthquake of 1663.*

AURIA DON V. — Diario delle cose occorse nella città di Palermo e nel regno di Sicilia 1631-74. — *Bibl. de Marzo, Vol. III and V). Erupt. 1669.*

3*

AUTORI SICILIANI (Opuscoli di...). — Opuscoli I-XX. — *Palermo 1785-88.* (*For some, see the respective authors*).

AZOUR A. — Sulla materia dei fuochi etnei. — *Giorn. d. Litt. in Roma, p. 183, 1676.* (*Cit. by Massa, Etna*).

BACCHI A. — Elpidiani, Civis Romani. De Thermis, etc. — *1588. Venetiis, in fol. fig. (pp. 48-492-1).*

BACCI A. — De thermis libri septem, etc. — *Venetiis, 1571, in fol.* (B. N.). *Also in 1588.*

BACCIO A. — De Thermis, veterum. — *See Graveti Thesaurus antiquit. et Hist. R. Sic. Vol. XII. Ludguni Batavorum,1783.— Venetiis, 1737, with the indication of the mineral sources of Paternò near Catania).*

BALDACCI L. MAZZETTI L. E TRAVAGLIA R. — Relazione sull' eruzione dell'Etna. — *Boll. d. R. Com. Geol. d'Italia, Vol. X, pp. 195. Roma, 1879.*

BALTZER A. — Wanderungen am Aetna. Sep. Abdr. des Jahrbuchs des Schweiz. Alpen-Clubs. IX Jahrgang. — *Zürich, 1874.* (*Mit Ansicht auf den Aetna von Nicolosi und topographischer Karte der Val del Bove).*

BARBAGALLO J. — Descriptio montis Aetnei ignam vomentis 1766 die Aprilis 27 (In hexameter verses). — *Catanae, 1766.*

BARDI G. — Sommario cronologico. — (*Cit. by Mongitore*).

BARONIUS (Cardinalis). — Historiae annales. — *Romae 1675, pp. 667-682.* (*Erupt. 1169, etc.*).

BARTOLOMEO (DON) A PATERNIONE. — Chronica in monasterio Sanctae Mariae de Licordia Auctore don Bartolomeo a Paternione. — (*Erup. of XV century.*).

BASILE G. — Note di fenomeni vulcanici presentati dall'Etna dal settembre 1884 a tutto l'anno 1875. — *Atti d. Acc. Gioenia. Ser. 3ª, Vol. X. 1875.*

BASILE G. — L'elefante fossile nel terreno vulcanico dell'Etna.— *Atti d. Acc. Gioenia, Ser. 3ª, Vol. XI. Catania, 1876.*

BASILE G. — Le bombe vulcaniche dell'Etna. — (*Att. Acc. Gioenia Sc. Nat. S. III, Tomo XX). Catania, 1888, pp. 82.*

BAUDRAND. — Article Etna — *Geographia. Tom. I,* (*Cit. by Massa*),

BEAUMONT (DE) E. — Sur la structure et sur l'origine du mont Etna. — *Ann. d. Mines, Vol. IX, Paris, 1836.*

BEAUMONT (DE) E. — L'Eruption de 1865. — *Compt. rend. de l'Acad. d. Sc. 20 Mars. Paris, 1865.*

BEAUMONT (DE) E. — Origine et structure de l'Etna. — *In: Mémoires pour servir à une description géol. de la France, Vol. IV, pag. 25-26).*

BEAUMONT (DE) E. — Carte du relief de l'Etna. — *Ecole d. Mines. Paris.*

BELLA PRIMA (PRINCIPE) P. — Sopra i Basalti globulari pel Murgo. — *Atti d. Acc. Gioenia, Ser. 1*, Vol. XIV, 1837.*

BELLA PRIMA (PRINCIPE) P. — Sopra il terreno di Lognina, Aci-Trezza e Castello. — *Atti d. Acc. Gioenia, Ser. 1*, Vol. XV, Catania, 1839.*

BELLA PRIMA (PRINCIPE P.) — Osservazioni geognostico-geologiche sul poggio di S. Filippo e sui dintorni in Militello. — *Atti d. Acc. Gioenia, Ser. 2.*, Vol. I. Catania, 1844.*

BELLEVUE (FLEURIAN DE). — Mémoire sur l'action du feu dans les volcans. — *Journ. d. Phys. Vol. LX, pag. 446. Paris.*

BEMBUS P. — Omnia opera. — *Tom. I-III, Basiliae, 1567.*

BEMBUS P. — De Actna, Dialogus. — *Venetiis, 1495 and 1530 (erupt. 1494 and preceding ones. — 2.* Edit. Amstelodami, 1703. (B. N.). Also 1818.*

BENOIT L. — 1870. — *See Aradas A.*

BERGH TH. — Die Eruptionen des Aetna. — *Philologus, 1873, pag. 138 (Proofs of the erupt. 693. B. C.).*

BIANCONI G. — Storia naturale dei terreni ardenti, dei vulcani fangosi, etc. — *Bologna 1840, in 8.°*

BISCARI E. G. — Memoria sul suolo di Catania. — *Catania, 1771.*

BISCHOFF G. — Lehrbuch der chemischen u. physikalischen Geologie. — *1° Auflage, Bonn. 1874, Bd. I-III, — 2° Auflage Bonn. 1863-66. Bd. I-III. Supplementband herausgegeben von F. Zirkel, darin besonders Cap. XIII, XIV, XV. Ueber Vulkanerscheinungen, Aetnalaven, Schlaken u. s. w.*

BLAEU CAESIUS (WILLEM JANSZOON). — 1571-1638. Atlas major.— *Tom. I-II. 1662. (Cit. by Massa, Etna).*

BLAKE J. F. — A Visit to the Volcanoes of Italy. — *Procced. Geol. Assoc. London, 1889, Vol. IX, pp. 145-176.*

BLASI (DI) G. E. — Storia cronologica (Erupt. of Etna of 1787 to 1842). — *Vol. I, with append.*

BLESENSIS P. — Cronica manoscritta dei re di Sicilia; epistola 46 ad Richard. (Erupt. 1100. Cit. by Massa and Amico). — *Syracus. Episcop. Parisiis 1667.*

BLUNDUS P. — De Siciliae mirandis. — *I. 3, cap. 20. (Cit. by Massa).*

BOCCARDO G. — Le terre e le acque dell'Italia. — *Milano, 1865 (Erupt. 1865, p. 49 and follow).*

BOCCARDO G. — Sismopirologia, terremoti, vulcani e lente oscillazioni del suolo. — *Genova, 1869, (Etna, p. 215.).*

BOCCONI P. S. — L'embrasement du Mont Etna. — *Paris, 1672.*

110   E T N A

BOCCONI P. S. — Museo di fisica, di esperienze. (Eruption and earthquakes in 1693), — *4° Tav. I-XVIII, Venetia, 1697.*

BURIGNY, — Histoire générale de Sicile. — *Vol. II, pag, 4, La Haye, 1745.*

BURIGNY. — Storia della Sicilia continuata da SCASSO e BOREL-I.O. — *Vol. I-II. Palermo, 1786-94, in 4.° — Also printed separately the: " Descrizione Geografica della Sicilia. ,, Palermo, in 4.°*

BURMANNUS. — Thesaurus Scriptorum Siciliae, Sardiniae, etc. 1723. — *See Graevius.*

BYLANDT PLASTERCAMP (DE) A. — Théorie des volcans. — *Vols. III, in 8.° With Atlas of 17 pl. in fol. Paris, 1835. (Etna, Vol. II, p. 187 and pl.).*

CABAEUS. — Meteorologia. — *Vol. I, Tent. LXIII, Qu. II, Romae 1664. (Cit. by Massa, Etna).*

CAESARIUS HEISTERBACHENSIS. — Illustria Miracula. Erupt. of 1200. — *L. 12, p. 857. 1599 , Colon. Agrip. Offic. Birck-mannica.*

CAESII B. — Mutimensis è Soc. Jesu. Mineralogia, sive naturalis philosophiae Thesauri, etc. — *Lugduni, in fol. pp. 16-626-69, (pp. 118-122. Etna, etc.).*

CAFIERO F. — Eruzione dell'Etna. — *Boll. Soc. Meteor. It.*

CAFICI I. - Stazione della età della pietra a S. Cono in Provincia di Catania. — *Boll. d. Paleont. Ital. N.° 3-4. 1879.*

CAJETANI SYRACUSANI P. OCTAVII. — Isagoge in Historiam Siculam. — *Thes. Sicul. Vol. II.*

CALLEJO Y ANGULO, (DEL) P. — Description de l'isle de Sicile. — *1734.*

CAMILIANO C. — Descrizione della Sicilia. — *(Published in " De Marzo, Opere storiche." Vol. XXVI).*

CAMPAILLA T. — L'Apocalissa, Opuscoli filosofici, etc. — *Siracusa, 1784, in fol. pp. 26-466, pl. 4,*

CAMPANELLA TH. — (1568-1630). Discorso sull'incendio dell' Etna e del come si accende. — *Palermo, 1738, Milano, 1750.*

CAMPANELLA TH. — Mutinensis dessertatio. — *Cit. by Amico.*

CAPECELATRO F. — Historia di Napoli. — *Vol. I-II, Napoli, 1724.*

CAPOCCI E. — Three Memoirs. Catalogo dei tremoti avvenuti nella parte continentale del regno delle due Sicilie posti in raffronto con le eruzione vulcaniche ed altri fenomeni cosmici e meteorici. — *1st M., R. Ist. d. Incoraggiamento d. Napoli. T. IX, pp. 335-378, Sept. 22nd 1859, 2nd M. , T. IX, pp. 379-422, 3rd M., T. X, pp. 393-327, 1863. (C. A.).*

CARAFFA P. — Notucae Descriptio. — *Thes. Sic.*

CARCACI. — Descrizione di Catania. — 2ª ediz. Catania, 1847.

CAREGA DI MURICCE F. — Etna, Conferenza tenuta presso il Club Alpino di Bologna. — Bologna, 1877.

CAREGA DI MURICCE F. — Il nuovo Monte Etneo (Umberto-Margherita) studiato e descritto dal prof. Orazio Silvestri. — Rass. d. Alp. Ann. II, N. 1, pag. 7, Rocca San Casciano (Firenze) Marzo, 1880.

CARNEVALE G. — Historia e descrizione del regno di Sicilia. — Napoli, Salviani, 1591.

CARRERA P. — Il Mongibello descritto in tre libri nel quale oltre diverse notitie si spiega l'historia degl'incendii e la cagione di questi. — Catania, 1636, in 4°.

CARRERA P. — Poesie pertinente alle materie di Mongibello , e del sacro vito della gloriosa S. Agatha. — Catania, 1636, in 8°, pp. 177-203. (C. A.).

CARRERA P. — Memorie storiche della città di Catania. — Vols. II, in fol., Catania, 1639. Thes. Sic. Vol. X.

CARRERA P. — Descriptio urbis Catanae. — Lugd. Bat. Thes. Sicul. Vol. IX, 1723.

CARRERA P. — Aetnae descriptio. — Thes. Sic. Vol. IX.

CARREY E. — L'Etna. — Extrait du "Moniteur". 1863, fol. XXVIII (C. A.).

CARUSO G. B. — Bibliotheca historica regni Siciliae, etc. — Panormi, 1723.

CARUSO G. B. — Memorie storiche (Till 1654). — Palermo, 1742-45.

CASTIGLIONE C. — Panormitani terrae motus descriptio (carmen). — Panormi, Aiccardo, 1726.

BOECKII A. — Corpus Inscriptionum Graecarum (Erupt. 475. B. C.). — Berolini, 1843. II. pp. 302.

BOECKII A. — Edit. Pindar. Pyth. — Odes. I. p. 224.

BOLLANDUS. — Acta Sanctorum. (Erupt. of 252).·

BOLANUS. — De igne aetneo. — (Dissertation lost; Cit. by Amico).

BOLTSHAUSER A. — Nouveau Guide de Catane. — Catane, 1874.

BOMARE (DE). — Sopra il Vesuvio ed altri vulcani. Dei Vulc. o monti ignivomi. — See Anonymous, Livorno, 1775.

BONFILII J. ET CONSTANTII. — Messanae descriptio. Aetnae iucendium, 1537. — Pp. 35.

BONITO M. — Terra tremante, ovvero continuazione dei terremoti dalla creazione del mondo fino al tempo presente. — Napoli, 1691, in 4.°

BORCH (DE). — Lithographie Sicilienne. — Rome, 1777.

BORCH (DE). — Lithologie et minéralogie Sicilienne. — Rome, 1778, in 4.°

BORCH (DE). — Lettres sur la Sicile 1777. — *Turin, 1782.* — *German by Werther, Bern, 1796.*

BORMANS (DE). — Collation des 167 premiers vers de l'Aetna de Lucilius junior, avec un fragment manuscrit du XIᵉ siècle. — *Bull. Acad. R. de Belgique T. XXI, N.° 8, pp. 258-379.* (C. A.).

BORNEMANN J. G. — Aetnakrater. — *Zeitschr. d. Deutsch. geol. Ges. Bd. VIII, s. 535. Berlin, 1855.*

BORNEMANN J. G. — Sur l'état des volcans d'Italie pendant l'été de 1856. — *Translation of De Perrey from Tageblatt der 32. Versam. Deutsch. Naturf. und Aertzt in Wien 1856, pp. 114-141, original M. S. pp. 4. (C. A.).*

BORELLI J. A. — Historia et Meteorologia incendii Actnaei anno 1669 ac responsio ad censuras Honoratii Fabri contra librum de vi percussionis. — *Regio Julio, 1670, pp. 162 + VI, 1 pl. 1 map.*

BOSIO G. — Historia Gerosolimitana. — *(Cit. by Mongitore).*

BOTTONE D. M. — De immani Trinacriae terraemotu idea historico-physica. — *Messanae, 1718, Ital. transl. by Marcello Malpighi.*

BOTTONE D. M. — Pyrologia topografica, idest de igne dissertatio juxta loca cum eorum descriptionibus. — *Neapolis, 1691, in fol. pp. 40-217, pl. 3. Messanae, 1721.*

BOURDELOT. — Risposta alle lettere del Boccone.—*(Cit. by Massa).*

BOURQUELOT J. — Viaggio in Sicilia. — *Bibl. d. Viaggi. Vol. X, Milano, Frat. Treves, 1873.*

BREISLACK S. — Institutions Géologiques. — *Vol. I-III. Milan, 1818.*

BRIETII PII. — Annales mundi. — *Venetiis, 1692. (Fiovanti. Description of the earthquake of 1169).*

BROCCHI G. B. — Sulle diverse formazioni di rocce della Sicilia.— *Bibl. Ital. Vol. XII. — L' Iride, Giorn. d. Sc. Lett. e Art. An. I, Tom. II, N. 7, p. 21. Palermo, 1822.*

BROCCHI G. B. — Osservazioni naturali fatte alle isole de' Ciclopi e nella contigua spiaggia di Catania. — *Bibl. Ital. Vol. X, p. 217.*

BROCCHI G. B. — Antichità dell'Etna. — *Bibl. Ital. Vol. XXVII, pag. 53.*

BRYDONE P. — Tour through Sicily and Malta. — *Vol. I, and II. London, 1773, French transl. by M. Demeunier. Amsterdam, 1775, 2 Vols. in 8.° Vol. I, pp. XVI+419, Vol. II, pp. 404. Another edition. Neuchatel, 1776, 2 in Vol. I, in 8.°, pp. 272, pl. 1.*

Buch (von) L. — Zusammenstellung der noch thätigen Vulkane. — *Pogg. Ann. Bd. V. Leipzig, 1827.*

Buch (von) L. — Ueber die Zusammensetzung der Basalt. Inseln u. über Erhebungskratere. — *Abh. Akad. d. Wiss. z. Berlin, 1820.* — Gesammelte Schriften, Bd. III, Seit. 15 Aetna. Berlin, 1877.

Buch (von) L. — Physikalische Beschreibung der Canarischen Inseln. — *Berlin, 1825. Description physique des Iles Canaries, traduite par C. Boulanger. Paris, 1836.*

Buch (von) L. — Ueber Erhebungskratere u. Vulkane. — *Pogg. Ann. Bd. XXXVII. Leipzig, 1836.*

Buch (von) L. — Gesammelte Schriften. — *III. Bd. Seit. 229, u. folg. (Darin der Aetna Pag. 513-16). Berlin, 1877.*

Buda J. (Lombardo-Catanensis). — Vulcania Litholosylloge Aetna in classes digesta. — *Nuova Raccolta. Vol. III, pp. 145-170. Catania, 1789.*

Buonfiglio G. — Historia Siciliana. — *Vol. I-II. Messina, 1738-39.*

Burgis. — Lettre du 10 avril 1536 sur la dernière éruption. — *(Cit. by Palgrave).*

Burgos P. A. — Distinta relazione dello spaventoso eccidio cagionato dai terremoti ultimamente con replicate scosse, accaduto a 9 e 11 Gennaro 1693 nel Regno di Sicilia. — *Napoli, 1693, in 4.° (C. A.).*

Burgos P. A. — Sicilia piàngente su le rovine delle sue più belle città atterrate da' tremuoti agli undici di Gennaio dell' anno 1693, etc. — *Palermo, 1693, in 4.°, pp. 19. (C. A.).*

Burgos A. — Descriptio Terrae Motus Siculi anni 1193. — *Thes. Sic. Vol. IX, p. 88 (from Carrera).*

Castone C. — Viaggio della Sicilia. — *Palermo, 1828, in 12°, pp. 3+240, I pl. P. 163 and follow. dell'Etna, etc. (B. N.).*

Cattaneo L. — L'eruzione dell'Etna del 1879. — *Illustr. Ital. d. Treves. pag. 371. 1° Sem. Milano, 1879.*

Chiavetta B. Abb. Basiliano. — Memoria dell' ultima eruzione dell'Etna accaduta il 27 marzo 1809. — *Messina, 1809.*

Chircherio. — Mundus. Subterraneus. — *Vol. I-II. with plan. 1678.*

Chisari V. — Breve notizie sulle acque termali di Paternò da lui scoperte. — *Catania, 1736, in 8°, (Cit. by Scina) Prospetto della storia letteraria di Sicilia, V. I, p. 137, Palermo, 1824).*

Christ W. v. — Der Aetna in der griechischen Poesie. — *?, 1888, pp. 50.*

Cicero M. T. — De natura deorum lib. II, cap. 38, 96. — *(after;*

Cajetani Siracusani Octavii Isagoge ad hist. Sic. cap. XII, *pag. 55.*

CIMARELLI A. — Risoluzioni filosofiche (at Cap. 12, pag. 104, description of the mineral springs of Paternò.—*Cit. by De Gregorio*).

CLAUDIANIUS C. — Opus de Raptu Proserpinac , etc. — *Nicolae Bizzi Nobili Bergomensi 2° Editio. Lucae, 1751, pp. 656.*

CLUB ALPIN FRANÇAIS (Bulletin Trimestriel).—Congrès des Clubs Alpins à Catane en 1880 et ascension de l'Etna. — *3e Trim. Paris, 1880.*

CLUVERUS PH. — Geologia. De creatione et formatione Globi terrestris. — *Lugd. Bat. 1619, in 4.°*

CLUVERUS PH. — Sicilia antiqua et Insulae minores adiacentes.— *Lugduni Batavorum, 1723, (Also: Thes. Sic. Vol. I).*

COCO G. — L'Etna. Saggi Poetici. — *Acireale, 1859, in 8°, pp. 90 93 (C. A.).*

CODEX Diplomaticus Siciliae. (Erupt. 760.A.D.)—*Dipl. CCLXXIV, Panormi, 1743.*

COGNATUS NOZERENUS (Gilbertus). — De incendio Actnae Anni MDXXXVI. — (*De Sylva narrationem, pp. 35-39) Basle, in 4°, (small). (C. A.).*

COLLINI M. — Considérations sur les Montagnes Volcaniques , etc. — *Mannheim, 1781, in 4°, pp. VIII+ 61, pl. 1.*

CONSTANTIUS. — See Bonfilii.

CONTEJEAN CH. — Une ascension de l' Etna. — *Turin, 1884 , in 8°, pp. 14. (C. A.).*

CONTI C. — Sull'eruzione dell'Etna incominciata il giorno 19 maggio 1886. — (*Boll. Com. Geol. 5-6) Roma, 1886, Vol. VII. pp. 149-156, map. 1.*

CORNELII S. P. — Actna et quae supersunt fragmenta. — *Amesterdam, 1703, in 8°, fol. 3, pp. 186, fol. 12 (index), pl. I. see Bembus P., Aetna, Trans. Ital. by Gargiulli, Venice, 1701. (C. A.). Trans. in French. Paris, 1736, in 12°, pp. 8 + 358, pl. 2.*

CORONELLI P. — Isolarium Atlantis Veneti. — *Erupts. of Etna 1535-1683, Pars, I, Venetii, 1696.*

COSSA A. — Osservazioni chimico-microscopiche sulla cenere dell'Etnà e sulla lava raccolta a Giarre il 2 giugno 1879. — *R. Acc. d. Lincei. Trans. 3, III, Roma, 1879. Boll. d. R. Com. geol. d' Italia, Vol. X, pag. 329, Roma, 1879.*

COSSA A. — Sur la cendre et la lave de la récente éruption (1879) de l'Etna. — *Compt. rend. d. l'Acad. d. Sc. d. Paris, 1879.*

COSSENTINI F. — Colpo d'occhio sulle produzioni vegetali dell'Etna. — *Atti d. Acc. Gioenia, Vol. IV, Catania, 1879 ?*

COVELLI N. — 1873. — *See Monticelli.*

CREMONENSIS L. (Ludovico Cremonensis). Annales 1808. (Erupt. 1570. — *See: Thes. Sic. Cit. also by Alessi*).

DAL VERME. — Una escursione al nuovo cratere sull'Etna. — *Boll. Soc. Geogr. Ital., Vol. XI, ser. 2, p. 679.*

DAUBENY CH. — Sketch of the Geology of Sicily. — *Jameson's Philos. Journ. Vol. XIII, pag. 197 and 254: With geological map. London, 1825.*

DAUBENY CH. — Description of active and extinct Volcanoes, Earthquakes and Thermal springs. — *With 4 plates, 18 maps and woodcuts. 1st Edit. London, 1826. 2nd Edit. London, 1848.*

DAUBENY CH. — 1831. — *See Davy.*

DAUBENY CH. — Die noch thätigen und erloschenen Vulkane; Bearb. von G. Leonard. Cap. XIV. Aetna. — *Stuttgart, 1851.*

DAUBRÉE. — La Chaleur intérieure du Globe. Conférences populaires faites à l'Asile Impérial de Vincennes. — *Paris, Hachette et C.ie, 1866.*

DAUBRÉE. — Note accompagnant le Rapport de M. Silvestri, sur l'éruption de l'Etna, des 18 et 19 mai 1886. — *Compt. Rend. Ac. Sc., Vol. CII, pp. 604-607.*

DAVY H. — On the phenomena of Volcanoes. — *Phil. Trans. pag. 241-50, London, 1828. — Ann. d. Chim. et d. Phys. Vol. XXXVIII, pag. 133, Paris, 1828.*

D. C. G. G. — Relazione dell'eruzione dell'Etna nel mese di Luglio 1787, scritta da D. C. G. G. — *Catania, 1787, in 4°, fol. 1, pp. 40. (C. A.).*

DE LUCA P. — Eruzione dell'Etna in Novembre del 1843 e suoi effetti nell'industria de'Brontesi. —?, *18-14, pl. 1.*

DELUC G. A. — Nouvelles Observations sur les Volcans et sur leurs laves. — *Journ. d. Mines. Vol. XVI, 1804. — Vol. XX. Paris, 1806.*

DELUTHO ABBÉ. — L'Etna, poème de Cornelius Severus. Traduction Nouvelle. — *Paris, 1842, in 8°, pp. XXVII+105.* (C. A.).

DEL VISCIO G. — Il Gargano in mezzo ai moti sismici d'Europa ed alle eruzioni dell'Etna. — *Bull. Soc. Meteor. It. 1888.*

DENON. — Voyage en Sicile. — *Paris, 1788.*

DERVEIL. — Voyage en Sicile et Malte. — *Vol. I-II, 3° Mil. Kupfern. Neufchâtel, 1776.*

DE SAINT-NON R. — Voyage pittoresque à Naples et en Sicile. —

4*

*Paris, 1781-86. — V Vols. Also, 1829, in 8°, Vol. IV, pp.
LXXIX+445, 572, 544, 570 and atlas.* (C. A.).

DEVILLE CH. (SAINTE-CLAIRE). — Relation sur l' Etna en Archiac
II. D. Histoires des progrès de la géologie. — *Vol. I , pp.
579. Paris, 1847.*

DEVILLE CH. (SAINTE-CLAIRE). — Sur les produits des Volcans de
l'Italie méridionale. — *Compt. Rend. Vol. XLII. Paris, 1856.*

DEVILLE CH. (SAINTE-CLAIRE). — Sur quelques produits d'émanations de la Sicile. Deux lettres à M. Dumas. — *Compt. Rend.
Acad. Sc. T. XLI, pp. 889-894. Nov. 19, 1855 et T. XLIII,
pp. 359-370, Aoûl 18, 1856.* (C. A.).

DEVILLE CH. (SAINTE-CLAIRE). — Sur les émanations volcaniques.
2 Pts. Paris, 1850-62. — *Compt. Rend. Acad. Sc. Vol. LXIV.
1857, Vol. LXIX, 1862. — See also le Bull. d. l. Soc. Géol.
d. France, Vol. XIV, Paris, 1857...*

DEVILLE CH. (SAINTE-CLAIRE). — Gaz de la Salinelle de Paternò.—
*Ann. d. Chim. et Phys. 3e Série. Vol. LII, page 51. Paris
1858.*

DEVILLE CH. (SAINTE-CLAIRE) ET LEBLANC. — Sur la composition
chimique des Gaz rejetés par les émanations volcaniques de
l'Italie méridionale. — *Paris, 1858.*

DEVILLE CH. (SAINTE-CLAIRE). — Extrait de deux lettres de M. O.
Silvestri sur l'éruption de 1865. — *Compt. Rend. Ac. Sc.
juillet 31, Paris, 1865.*

DEVILLE H. (SAINTE-CLAIRE) ET GRANDJEAN. — Analyse de la
lave de l'Etna. — *Compt. Rend. Vol. XLVIII, page 21, Paris, 1859.*

DIACONO P. (PAOLO DIACONO). — Historiae Miscellanea. — *Basi-
leae, 1505.*

DICKERT TH. — Relief à couleurs de l'Etna. — *Les couleurs Géologiques d'après la carte de M. Sartorius de Waltershausen.*

DOGLIONI N. — Anfiteatro d' Europa, in cui si ha la descrittione
del mondo celeste, etc., etc., — *Venetia, 1623, in 4°, with
portrait, p. 72-1377. (P. 993: Dell' Ethna detto Mongibello
e sua historia; P. 694 Del Monte di Somma, e sua historia.*

DIODORUS SICULUS. — *Eruptions of 475, 425, 394. B. C., Vol.
III, pag. 6, Vol. XIV, pag. 59.*

DOLOMIEU (DE) D. — Mémoire sur les Iles Ponces et catologue
des produits de l'Etna; descript. de l'éruption de l'Etna du
mois de juillet 1787. — *Paris, 1788, pl. 2.*

DOLOMIEU (DE) D. — Sur une éruption de l' Etna. — *Journ. d.
phys. Vol. XL, Paris, 1792.*

DOLOMIEU (DE) D. — Distribution méthodique de toutes les ma-

tières dont l'accumulation forme les montagnes volcaniques.—
Journ. d. phys. Vol. XLIV, 1794, Vol. XLV, Paris, 1795.

DOMNANDO. —· Constitution et soulèvement de l'Etna au 10 Octobre
1834. — *Bull. d. l. Soc. géol. d. France, Vol. VI, page 124,
Paris, 1835.*

DRYDEN J. — Voyage to Sicily and Malta, etc. — *London, 1776,
in 8.°*

DURIER C. — L'Etna — *Compt. rend. d. l. réun. d. Clubs Alpins
à Genève dans l'Août 1879, Genève, 1880.*

EICHWALD (VON) E. C. — Beitrag zur vergleichenden Geognosie
auf einer Reise durch die Eifel, Tyrol, Italien, Sicilien u.
Algier. — *mit 4 Kupfl, Moskau, 1851, in 4.°*

ELIDRIS SHERIFF. — Descrizione della Sicilia, tradotta da Franc.
Tardia. — *Opuscoli di autori Siciliani, Vol. VIII, pag. 233,
1788.*

ERRICO S. — Ode di Mongibello. — *(Cit. by Massa).*

FALB. R. — Gedanken und Studien über den Vulkanismus, etc.—
*Gratz. 1875, Cap. III, Seit , 46, der Ausbruch des Aetna
am 29 August, 1874.*

FALCANDUS H. — Historia de rebus gestis in Sicilia regno, — *Parisiis , 1550 , in 4.° (copied several times since 1129-1166
under the title: De Calamitate Siciliae) (Earthquake of 1169).*

FAUJAS DE ST. FOND. — Minéralogie des Volcans.—*Paris, 1784.*

FAUJAS DE ST. FOND. — Discours sur les Volcans brûlants. —
Vol. I, Fol. Paris, 1778.

FAUJAS DE ST. FOND.—Essai de Géologie.— *Tom. I, Paris, 1809.*

FAYE. — Sur les orages volcaniques. — *Compt. Rend. d. l'Acad.
d. Sc. Tom. XCI, pp. 708, Paris, 1880.*

FAZELLI T. — De Actna monte et ejus ignibus. — *De Rebus Siculis, Cap. 4th , lib. II, 1558, in 4°, pp. 616 (C. A.).*

FAZELLUS TH. — De rebus Siculis.—*Decades II, Panormi, 1557,
Ibidem 1560. (Also Fazellus in Thes. Sic. Vol. VI, the best
edition of the work of V. Amico, Catania, 1749-50, with notes
and appendix till 1700) Aetna, and mineral springs, etc.).*

FAZELLUS TH. — Dell' Istoria di Sicilia tradotta da M. Remigio ,
data in luce per Martino Lafarina. — *Palermo, 1628.*

DU PÉROU — Notice sur l'Etna, formation et composition de son
massif. Eruption de Février 1865, précédée d'une histoire des
anciennes éruptions, etc. — *Catane, 1865.*

FERRARA F. — Storia generale dell' Etna, etc. — *Catania, 1793,
in 8°, pp. XLIV+359, pl. V.*

FERRARA A. — Memoria sopra le acque della Sicilia, loro natura, analisi e usi. — *Londra, 1811.*

FERRARA F. — In the almanach « Fa per tutti » (eruption 1811).— *Catania, 1812*.

FERRARA F. — Storia naturale della Sicilia, compr. la Mineralogia. — *4°, Catania, 1813*.

FERRARA F. — Descrizione dell'Etna con la Storia delle eruzioni e il Catalogo dei prodotti. — *Palermo, 1818, in 8°, pp. XVI+256, pl. V*.

FERRARA F. — Guida dei viaggiatori agli oggetti più interessanti a vedersi in Sicilia. — *Palermo, 1822, in 8°, pp. 304, pl. VIII, (B. N.)*.

FERRARA F. — Campi Flegrei della Sicilia e delle isole che le sono attorno. — *Messina , 1810 , in 4°, fol. II, pp. XIX+424, maps. Atti d. Acc. Gioenia, Ser. 1ª, Vol. II. Catania, 1825*.

FERRARA F. — Della influenza dell'aria alla sommità dell'Etna sopra l'economia animale. — *Giornale di Sc. Lett. ed Arti per la Sicilia, N.° XXVI. Palermo, 1825, in 8.° pp. 17. (C. A.)*.

FERRARA F. — Memoria sopra i tremuoti della Sicilia in marzo 1823. — *Palermo, 1823, in 8°, fol. IV, pp. 51, pl. I; also analysis in Edinb. Journ. of Science, Vol. VI, pp. 362-370, 1826*.

FERRARA F. — Storia di Catania sino alla fine del XVIII secolo.— *Catania, 1829*.

FERRARA F. — Sopra l'eruzione dell'Etna, segnata da Orosio nell'anno 122 B. C. — *Atti di Acc. Gioenia, Vol. X, pp. 141-158, Catania, 1833*.

FERRARA F. — Storia generale della Sicilia. — *Palermo, 1838*.

FICHERA S. — Acqua minerale di S. Venera di Acireale. — *Relaz. d. Acc. d. Sc. Lett. ed Arti, di zelanti di Acireale, Pag. 37-38. Palermo, 1836*.

FILOTHEO (DE HOMODEIS) A. — Siculi Aetnae topografia atque eius incendiorum historia. — *Thes. Sic. Vol. IX. Venetiis, 1591, in 4°, fol. III, pp. 56*.

FILOTHEO (DEGLI OMODEI) A. — Descrizione della Sicilia. — *Manuscript in the Biblioteca communale di Palermo, printed in the Biblioteca de De Marzo, Vol. XXIV and XXV*.

FILOTHEO (DEGLI OMODEI) A. — Descrizione del sito del Mongibello, trad. in Italiano. (See Orlandini) — *Palermo, 1811, in 4°, pp. 85*.

FLEURIAN DE BELLEVUE. — *See Bellevue*.

FODERA DR. — Extrait de la description de l'Etna de M. F. Ferrara. — *J. de Phys. T. LXXXVIII, pp. 283-289, 364-372. 1819 (C. A.)*.

FOUQUÉ F. — Lava del 1865. — Analisi data nel lavoro di Silvestri. — *Also in Zeitschr. d. deutsch. geol. Ges. Bd. XVII, Sett. 606, Berlin, 1865.*

FOUQUÉ F. — Rapport sur l'éruption de l'Etna en 1865. — *Archives des Miss. Sc.* 2me *Sér. T. II. pp. 321-359, Paris, 1865.* (C. A.).

FOUQUÉ F. — Sur l'éruption de l'Etna du 1r février 1865; lettres à M. Elie de Beaumont et Ste. Cl. Deville.—*Compt. Rend. de l'Acad. d. Sc. de Paris. 20 Mars 1865. T. LX et LXI, pp. 23-25, 31-34, 215, 421-424, pp. 548-555 et 555-556, also pp. 1135-1140 and 1140-1142, also pp. 1185-1189, also pp. 1331-1335, pp. 564-567, Tom. LXII, p. 1366 (Rapport). — Ann. d. Missions Sc. et Lett. 5e Sér., Tom. II, p. 321. Paris, 1866.*

FOUQUÉ F. — Phénomènes chimiques de l'éruption de l'Etna en 1865. — *Compt. Rend. d. S. Acad. d. Sc. Tom. LXII, pp. 616 et 1366. — Ann. d. Missions Scient. et Lett. 2e Sér., Tom. III, p. 165. Paris, 1866.*

FOUQUÉ F. — Chlorure de Sodium fondu dans les fentes de la lave de 1879 — *Compt. Rend. d. l'Acad. d. Sc. d. Paris, 7 Juillet, 1879.*

FOUQUÉ F. — Santorin et ses éruptions.—*Cap. XX. Paris, 1879.*

FOUQUÉ F. — Sur la récente éruption de l'Etna (1879). Lettre à M. le Secrétaire perpétuel.— *Compt. Rend. d. l'Acad. d. Sc. Tom. LXXXIX, p. 33. Paris, Juillet-Décembre 1879.*

FOUQUÉ F. — Reproduction artificielle des roches et des minéraux par la voie ignée. — *Compt. Rend. d. l'Acad. d. Sc. Bull. d. l. Soc. géol. d. France. Paris 1879-81.*

FREDA G. — Sulla crisocolla dei monti Rossi all'Etna. — *Gaz. Chim. Ital. Vol. XIV, 1884, p. 3.*

FRESENIUS W. — Phillipsit von Acicastello. — *Analyse u. opt. Prüfung. Groths Zeitschr. für Krystallogr. Bd. III. p. 44. Leipzig, 1880.*

FROMONDUS LIBERTUS. — Meteorologia, corum Libr. VI. I, I, cap. II. (Cit. by Massa, Etna). — *Antwerpen, 1627.*

FUCHS C. W. C. u. GRAEBE.—Die Lava der Aetna— Eruption des Jahres 1865. — *N. Jahrb. f. Min. Geol. u. Pal. pp. 711-715, Stuttgart, 1866.*

FUCHS K. — Vulkane u. Erdbeben.—*Leipzig, 1875. (Aetna, Krater 1869, Seit. 274). French edition :* Les Volcans et les tremblements de Terre, *Vol. I, Paris, 1866. (Etna, pag. 217.*

FUCHS K. — Aetna Ausbruch 27 nov. 1868 — *N. Jahrb. f. Min. Geol. u. Pal. pp. 694, Stuttgart, 1869.*

FUCHS K.—Die vulkanischen Ereignisse des Jahres 1870 Aetna. —

Miner. und Petrogr. Mitth. vom G. Tschermak, Seil, 37, Wien, 1880.

GAETANI (DE) G. — Sopra l'acqua solforosa del pozzo di S. Venera. — *Atti d. Acc. Gioenia. Ser. 1ª, Vol. XIV, Catania, 1839.*

GAETANI (DE) G. — Sopra l'acqua acidula di S. Giacomo. — *Atti d. Acc. Gioenia, Ser. 1ª, Vol. XVI, Catania, 1840.*

GAETANI (DE) G. — Intorno alle acque solforose del Pozzo di S. Venera. Nuove Osservazioni. — *Atti d. Acc. Gioenia, Ser. 1ª, Tom. XX, Catania, 1843.*

GALANTI G. M. (Luigi?). — Descrizione geografica delle due Sicilie. — *Tom. I-IV, Napoli, 1787. German by Jagemann. Bd. I-IV, Leipzig. 1790.*

GALLIANO D. — Liste des éruptions de l'Etna. — *Ext. Voyage Pittoresque des Isles de Sicile, de Malta et de Lipari par J. Houel. T. XI, pp. 115-120.*

GALVAGNI G. A. — Fauna Etnea. — *Atti d. Acc. Gioenia. Ser. 1ª, Vol. XI, XII, XIII, XIV, e XX, Catania, 1834-1843.*

GALVAGNI G. A. — Sopra un nuovo fenomeno sonoro accaduto nella sommità dell'Etna. — *Atti d. Acc. Gioenia. Ser. 1ª, Vol. XII, Catania, 1835, pp. 8.*

GEMMA F. — L'incendio di Mongibello del 1669. Poema in cento stanze. — *Catania, 1674, in 8°, pp. 50.*

GEMMELLARO CARLO. — Sopra alcuni pezzi di granito e di lava antica trovati presso la cima dell'Etna. — *Catania, 1823, and Atti d. Acc. Gioenia, N. 11, pp. 190-223.*

GEMMELLARO C. — Prospetto di una topografia fisica dell'Etna. — *Atti d. Acc. Gioenia, Ser. 1ª, Vol. I, Catania, 1825, pp. 19-34.*

GEMMELLARO C. — Sopra il Basalto e gli effetti della sua decomposizione naturale. — *Atti d. Acc. Gioenia, Ser. 1ª, Vol. II, Catania, 1825, pp. 49-66.*

GEMMELLARO C. — Sopra le condizioni geologiche del tratto terrestre dell'Etna. — *Atti d. Acc. Gioenia, Ser. 1ª, Vol. I, Catania, 1825, pp. 183-211.*

GEMMELLARO C. — Sopra il confine marittimo dell'Etna. — *Atti d. Acc. Gioenia, Ser. 1ª, Vol. IV, Catania, 1827, pp. 179-193.*

GEMMELLARO C. — Sopra la fisionomia delle montagne di Sicilia. — *Atti d. Acc. Gioenia. Ser. 1ª, Vol. IV, Catania, 1827, pp. 73-93.*

GEMMELLARO C. — 1828. — *See Beffa.*

GEMMELLARO C. — Sopra il clima di Catania.—*Att. d. Acc. Gioenia. Ser. 1ª, Vol. VI, Catania, 1829.*

GEMMELLARO C. — Cenno sulla vegetazione dell'Etna. — *Atti d. Acc. Gioenia, Ser. 1ª, Vol. IV, Catania, 1830, pp. 77-86.*

GEMMELLARO C. — Saggio sopra il clima di Catania, abbozzato dietro un decennio di osservazioni meteorologiche. —*Atti d. Acc. Gioenia. 1830, pp. 133-175 (C. A.).*

GEMMELLARO C. — Ueber Basalt. u. basaltische Lava. — *Brief. Not. N. Jahrb. f. Min. Geol. u. Pal. S. 246, Stuttgart, 1830.*

GEMMELLARO C. — Relazione accademica per l'anno VII dell'Accademia Gioenia letta nella seduta ordinaria dei 12 maggio 1831. — *Catania, 1831, in 8º, p. 29. (Etna , eruption of 1494).*

GEMMELLARO C. — Sopra un masso di lava corroso dalle acque marine. — *Atti d. Acc. Gioenia, Ser. 1ª, Vol. VI, Catania, 1832, pp. 71-83.*

GEMMELLARO C. — Aetna. Eruption 31 oct. 1832. — *Neu. Jahrb. f. Min. Geol. u. Pal. Seit. 182, u. 641. Stuttgart, 1833.*

GEMMELLARO C. — Alcuni fenomeni osservati all'eruzione del 31 ottobre 1832. — *Atti d. Acc. Gioenia, Ser. 1ª, Vol. IX, Catania, 1833.*

GEMMELLARO C. — De vallis Bovis in Monte Aetna geognostica constitutione. —*Atti d. Acc. Gioenia, Ser. 1ª, Vol. XII, 1834.*

GEMMELLARO C. —Sulla costituzione fisica della valle del Bove.— *Atti d. Acc. Gioenia, Ser. 1ª, Vol. XII, 1835.*

GEMMELLARO C, — Cenno geologico sul terreno della piana di Catania. — *Atti d. Acc. Gioenia, Ser. 1ª, Vol. XIII, p. 117, Catania, 1836.*

GEMMELLARO C. — Sulla causa geognostica della fertilità del suolo di Sicilia. — *Atti d. Acc. Gioenia, Ser. 1ª, Vol. XIV, Catania, 1837.*

GEMMELLARO C. — Aetna-Eruption 1838. — *Neu. Jahrb. f. Min. Geol. Pal. Seit. 531, Stuttgart, 1838.*

GEMMELLARO C. — Cenno sull'attuale eruzione dell'Etna. — *Catania, 1838, in 8º, pp. 37.*

GEMMELLARO C. — Elementi di Geologia. — *Catania, 1840.*

GEMMELLARO C. — Cenno storico sulla eruzione Etnea del 27 novembre 1842. — *Atti d. Acc. Gioenia, Ser. 1ª, Vol. XIX, Catania, 1842, pp. 18.*

GEMMELLARO C. — Sulla varietà di superficie nelle correnti vulcaniche. — *Atti d. Acc. Gioenia, Ser. 1ª, Vol. XIX, Catania, 1842.*

GEMMELLARO C. — Actna-Eruption 1843. — *Neu. Jahrb. f. Min. Geol. u. Pal. Seit. 180, Stuttgart, 1844.*

GEMMELLARO C.—Memoria sull'eruzione del 17 novembre 1843.— *Atti d. Acc. Gioenia, Ser. 1ª, Vol. XX , Catania , 1844 , pp. 35.*

GEMMELLARO C. — Sulla costa meridionale del Golfo di Catania.— *Atti d. Acc. Gioenia, Ser. 2ª, Vol. II, Catania, 1845.*

GEMMELLARO C. — Sui crateri di sollevamento e di cruzione. — *Atti d. Acc. Gioenia, Ser. 2ª, Vol. III. Catania, 1846.*

GEMMELLARO C. — Sul basalto decomposto dell'isola dei Ciclopi.— *Atti d. Acc. Gioenia, Ser. 2ª, Vol. II. Catania, 1846.*

GEMMELLARO C. — Saggio sulla costituzione fisica dell' Etna. — *Catania, 1847, in 4.°*

GEMMELLARO C. — Saggio di storia fisica di Catania. — *Atti d. Acc. Gioenia, Ser. 2ª, Vol. V. Catania, 1848.*

GEMMELLARO C. — Breve ragguaglio dell' cruzione dell' Etna del 21 agosto 1852. — *Atti d. Acc. Gioenia , Ser. 2ª , Vol. IX. Catania, 1852, pp. 30, pl. 3.*

GEMMELLARO C. — Una corsa intorno all' Etna. — *Atti d. Acc. Gioenia, Catania, 1853.*

GEMMELLARO C. — Sulla struttura del cono dei monti Rossi e dei suoi materiali. — *Atti d. Acc. Gioenia, Ser. 2ª, Tom. XI, Catania, 1854.*

GEMMELLARO C. — Sul profondamento del cono dell'Etna avvenuto il 6 settembre dell'anno 1857. — *Atti d. Acc. Gioenia, Ser. 2ª, Vol. XIV. Catania, 1857.*

GEMMELLARO C. — Vulcanologia dell'Etna. — *Atti d. Acc. Gioenia; Vol. XXXIV, p. 176. Catania, 1857.*

GEMMELLARO C. — La Vulcanologia dell' Etna che comprende la topografia, la geologia, la storia delle sue eruzioni, non che la descrizione e lo esame dei fenomeni vulcanici. — *Catania, 1858, in 4°, pp. XIV + 266, I map. pl. II.* (C. A.).

GEMMELLARO C. — Ulteriori considerazioni sul Basalto in appendice alla volcanologia dell'Etna. — *Atti d. Acc. Gioenia, Ser. 2ª, Vol. XVI, Catania, 1860, pp. 24.*

GEMMELLARO C. — Dell'eruzione dell'Etna 1863. — *Catania, 1863.*

GEMMELLARO C. — Sulla cima dell'Etna considerata sotto il rapporto dell' utile che appresta al viaggiatore istruito e allo scienziato. — *Atti d. Acc. Gioenia, Ser. 2ª, Vol. XIX. Catania, 1863.*

GEMMELLARO C. — Breve ragguaglio della eruzione dell'Etna negli ultimi di Gennaio 1865. — *Catania, 1865.*

GEMMELLARO C. — Un'addio al maggior vulcano di Europa. — Catania 1866. 2ª ediz. con aggiunte, Catania, 1866.

GEMMELLARO C. — Descrizione di una nuova carta geologica di Sicilia. — Giorn. d. Sc. Lett. d. Palermo, N. 134.

GEMMELLARO C. — Sopra il terreno giurassico. — Memoria prima sul terreno giurassico di Tauromina, in 4°. p. 23.

GEMMELLARO GAETANO GIORGIO. — Sul ferro oligisto di monte Corvo. — Atti d. Acc. Gioenia, Ser. 2ª, Vol. XIV. Catania, 1858.

GEMMELLARO G. G. — Sui modelli esterni della Quercia in contrada Pinitella sull'Etna. Lettera al Prof. Guiscardi. — Giorn. d. Gab. Lett. d. Acc. Gioenia. Fasc. 6°, Nov.-Dec. 1858, in 8°, pp. 6. (C. A.).

GEMMELLARO G. G. — On the volcanic cones of Paternò and Motta (S. Anastasia). — Proceed. of the Geol. Society of London. 29 Nov. 1861.

CEMMELLARO GIUSEPPE. — Quadro istorico topografico delle eruzioni dell'Etna. — Catania, 1824. London, 1828, fol. I.

GEMMELLARO G. — Sunto del giornale dell'eruzione dell'Etna nel 1852. — Atti d. Acc. Gioenia. Ser. 2.ª, Vol. IX. Catania, 1853.

GEMMELLARO MARIO. — Memoria dell'eruzione dell'Etna nel 1809. — Messina 1809, in 8°, 2ª ediz. Catania, 1820, pp. 41, pl. 1.

GEMMELLARO M. — Giornale dell'eruzione dell'Etna avvenuta a 27 ottobre 1811. — Manuscript, with appendix?

GEMMELLARO M. — Giornale dell'eruzione dell'Etna avvenuta alli 27 Maggio 1819. — Catania, 1819.

GEMMELLARO M. — Extrait d'un Journal tenu à Catane pendant quatorze ans. — M. S. copy, fol. II. (C. A.),

GEMMELLARO M. — Registro di osservazioni. — (See: Vulcanologia di C. Gemmellaro.

GEMMELLARO RAIMONDO. — Manoscritto sulla eruzione del 1766 con appendice.

GENTILE-CUSA B. — Sulla eruzione dell'Etna di Maggio-Giugno 1886. — Catania, 1886, in 8°, fol. II. pp. 210, VII heliotypes, 6 figs. (C. A.).

GHIGI I. B. — Nuova carta della Sicilia. Fogli I-IV. — Roma, 1779.

GIACOMO (DI) A. — Idrologia generale dell'Etna. Discorso d'introduzione. — Atti di Acc. Gioenia. Ser. 1.ª, Vol. IX. pp. 23-40, Catania, 1832.

GIGAULT DE LA SALLE A. E. — Voyage pittoresque en Sicile. — Vol. I-II, in fol. pl. 92, Paris, 1822.

GIMMA. — Fisica sotterranea. — *Vol, I-II. Napoli, 1730.*

GIOENI G. (1747-1822). — Relation d' une pluie colorée de sang tombée sur le versant méridional de l'Etna (in English). — *Philos. Trans. London, 1782.*

GIOENI G. — Relazione dell' eruzione dell' Etna avvenuta nel luglio del 1787. Scritta d. C. G. G. abitante della prima regione del monte. — *Catania, 1787.*

GIOENI S. — Alcune lettere di uomini illustri nella storia naturale dirette al Cav. G. Gioeni. Aggiuntavi la descrizione data dall' ab. Spallauzani del Gabinetto di St. Nat. Siciliana in casa del medesimo Gioeni.—*Catania, 1815. in 4°, fol. III. pp. 41.* (C. A.).

GIUFFRIDA A. — Quaesita medica. — *With notes on the mineral waters of Palermo. Catania 1753.*

GIUSTI G. D. — Lettera intorno all' ultima eruzione (1819) dell' Etna. — *Giorn. Encicl. d. Napoli. Fasc. VII. 1819.*

GOLZIUS H. — De vita rebusque gestis Regum Siciliae. — *Pars. II. (Thes. Sic. VIII. p. 1144.)*

GONTOULAS. — Erupt. d. 1321, et 1323. — *Hist. Profan. dec. IV. sect. 14.*

GORINI P. — Sull' origine dei Vulcani. — *Napoli, 1872, in 8.°*

GOSSELET G. — Observations géologiques faites en Italie. — *Mém. d. Soc. Imp. des sciences d. l'.Agriculture et des Arts de Lille, IIIe série, vol. 6, Lille, 1869. See pp. 25 and follow.*

GOURBILLON (DE) A. — Voyage critique à l' Etna. — *II. Paris, 1820, 2 vols. in 8°, Vol. I, pp. 541, Vol. II, pp. 403, pl. 31.*

GOUTOUL. — Historia profana. — *Dec. 6, See. XII.*

GRAEBE D. — 1865. — *See Fuchs, C. W. C.*

GRAEVII. — Thesaurus scriptorum Siciliae, Sardiniae, Corsicae Cura P. Burmanni. (Also cited several times in books and in the present Bibliographics as Thesaurus antiquitatum et Historiarum Siciliae, or as : Thesaurus Siculus.) — *Lugd. Batav. 1723-25. Vol. I-XV. See also under names of respective authors.*

GRANDJEAN. — 1850. — *See Deville (Sainte-Claire).*

GRASSI M. — Relazione storica dell' eruzione Etnea del 1865. — *Catania, 1865, in 8°, pp. 92.*

GRASSI M. — Relation historique de l'éruption de l'Etna en 1865.— *Bull. d. l. Soc. d. Géog. Juillet, 1866, pp. 5-29. (C. A.).*

GRASSI M. — Relazione dell' eruzione dell' Etna nel novembre e dicembre 1868. — *Il Nuovo Cimento, Ser. 2ª. Tom. I, pp. 186-191. 1869. (Translated by J. Roth ; See Roth.).*

GRASSI M. — Sull' eruzione dell'.Etna del 1870. — *Cenni Giorn.*
d. Stella. N.° 154. Palermo, 7 giugno 1870.

GRAVINA C. (PRINCIPE DI VALSAVOYA). — Poesie. — *Sonetto per*
la Eruzione dell'Etna del 1832, pp. 72, Catania, in 12. (p. 10).

GRAVINA M. B. — Notes sur les terrains tertiaires et quaternaires
des environs de Catane. — *Bull. d. l. Soc. géol. d. France,*
2.e Sér., Tom. XV. pp. 391-421. Paris, 1859.

GREGORI A. — 1870. — *See: Mantovani.*

GREGORI SANCTI PAPAE — I. Dialog. — *Lib. IV. Cap. 30, n.° 35.*

GREGORIO (DI) G. — 1703-1771. Lettera sulle acque acidole di Paternò. — *Opusc. di Aut. Siciliani. III. 269. Palermo, 1788.*

GROSSO (DE) D. J. B. — Catanense decachordum sive novissima
sacrae Catanensis Ecclesiae notitia. — *.Thes. sic. X. Cata-*
niae, 1654.

GROSSO (DE) D. J. B. — Catana sacra[(Erupt. 1654.). — *Cataniae,*
1642.

GROSSO (DE) D. J. B. — Agatha catanensis. (Erupt. 1656. — *Ca-*
taniae, 1656.

GSELL FELS, TH. — Sicilien aus Meyers Reischandbüchern. Cap.
XVIII. Etna. — *Leipzig, 1877. Short and detailed descript.*
of Etna and its eruptions, with a history of the last.).

GUALTERIUS G. — Siciliae objacent, insularum et Brutiorum antiq.
tabulae. — *Messanae, 1664.*

GUARNERII G. B. — Le zolle narrationi storiche Catanesi. — *Tom.*
III. (Cit. by Massa, Etna.) Catania, 1651.

GUARNERII G. B. — Dissertationes historicae Catanenses. — *Thes.*
Sic. XI.

GUGLIELMINI D. (1655-1710). — Catania distrutta dal terremoto
nel 1693. — *Palermo, 1695.*

GUIDO V. A. DI PATERNÒ. — Breve istorica descrizione del por-
tentoso miracolo della gloriosa vergine e martire Santa Bar-
bara, principale patrona della fertilissima città di Paternò,
operato al 27 maggio dell'anno 1780. Della liberazione del
feudo di Villabona o sia Ragalna dall'incendio di Mongibello.—
Catania, 1785.

GULLI S. — Ricerche sulla profondità dei Vulcani. — *Atti d. Acc·*
Gioenia. Ser. 1.ª, Vol. XI. Catania, 1834.

GÜMBEL C. W. — Ueber das Eruptionsmaterial des Schlammvul-
kanes von Paternò am Aetna und der Schlammvulkane im
Allgemeinen. — *Sitz. Berich. der Bayr. Akad. d. Wiss. Mün-*
chen, 1879, in 8°, pp. 57. — Abstract: Boll. d. R. Com.
geol. d' Italia. Vol. X. pp. 506 e 561, Roma, 1879.

GÜMBEL C. W.—Vulkanische Asche des Aetna von 1870.— *Neues*

Jahrb. für. Min. Geol. u Pal. Bd. I, Sell. 859. Stuttgart,
1879. — Abstract in: Boll. d. R. Com. geol. d' Italia. Vol.
X, pp. 505. Roma, 1879.

GUSTANAVILLA (DE) P. — Notae in Petrum Blesensem. — (*Cit. by*
Massa, Etna.)

HAMILTON W. — Observations on mount Vesuvius, mount Etna
and other Volcanoes of the two Sicilies. — *London*, 1772,
in 8°, pp. IV+179, pl. VI.

HAMILTON W. — Voyage au mont Etna en juin 1769. — *Extract*
from " Voyage en Sicile et dans la Grande Grèce "... ad-
dressed to Winchelmann and translated from the German,
Lausanne, 1773, in 12°, M. S. Copy. (C. A.).

HAMILTON W. — Campi Flegraei. — *I. II. Napoli, 1776. Supple-*
ment 1779.

HAMILTON W. — Voyage au Mont Etna en 1869, (trad. en français par M. de Villebois). — *1780.*

HAMILTON W. — Oeuvres complètes, commentées par l' Abbé Giraud-Soulavie. — *Paris, 1781, pl. 1.*

HAMILTON W. — Account of the Earthquakes in Italy. — *London*
1783. Tradotto ed illustrato da G. Sella. — Firenze 1783.

HAMILTON W. — Waarneemingen over den Vuurbergen in Italie,
Sicilie, en omstreiks den Rhyn als mede over de Aardbeevingen, voorgevallen in Italie, 1783. — *Amsterdam, 1784, in*
8°, pp. 552. (C. A.).

HAUER (VON) K. — Analyse der Lava von 1852. — *Wien, Akad.*
Ber. 11, 89, 1853.

HERBERGER J. E. — Chemische Analyse der körnigen Lava vom
Aetna. — *Brandes' Archiv. d. Apoth. Vereins, 1830. — N.*
Jahrb. f. Min. Geol. u. Pal. Seil. 426. Stuttgart, 1832.

HERBINII J. — Dissertationes de admirandis mundi Cataractis supra
et subterraneis, carumque principio, elementorum, etc. etc. —
Amstelodami, 1678, in 4°, pp. 14-267-17, fig.

HOFF K. E. A. (VON). — Geschichte der natürlichen Veränderungen
d. Erdoberfläche. — *I-III. Gotha, 1822,1824, u. 1834.* (*Etna*
und Sicilien. Bd. II. § 10. Seit. 221-232. Die Liparischen
Inseln. Bd. II §, Seitz. 252-267.

HOFF K. E. A. (VON) — Cronik der Erdbeben u. Vulkan-Ausbrü-
che. — *2 Theile. Gotha 1840-41. — Also in the Poggd. Ann.*
Bd. VII, 1841.

HOFFMANN F. — Verhalten der in den letzten 40 Jahren zu Palermo beobachteten Erdstösse. — *Poggd. Ann. Bd. XXIV,*
1832.

HOFFMANN F. — Mémoire sur les terrains volcaniques de Naples, de la Sicile etc. — *Bull. d. l. Soc. géol. de France. Vol. III, 1833, pp. 170-180.*

HOFFMANN F. — Physikalische Geographie, Erhebungskratere, Vegetation Siciliens, Actna u. a. — *Berlin, 1837.*

HOFFMANN F. — Geschichte der Geognosie u. Schilderung der vulkanischen Erscheinungen. — *(Aetna, pp. 273 u. ff.) Berlin, 1838.*

HOFFMANN F. — Geogn. Beobachtungen auf einer Reise durch Italien u. Sicilien in Jahre 1830-32 — *(d. v. Dechen). Karstens Archiv. Bd. XIII, 1839.* — *(Auch separat erschienen). Berlin, 1839.*

HOFFMANN F. — Karte von Sicilien. — *(Aetna mit seinen Lavaströmen). Karstens Archiv. Bd. XIII, 1860.*

HOFFMANN F. — Lettera al signor E. Repetti. *(Sull' altezza dell'Etna. — Giorn. Letter. d. Sicilia, Tom. XXXV, pp. 54.*

HOFFMANNUS G. — Lexicon topographicum Siculum. (*Cit. by Massa, Etna.*) ı

HOLM. A. — Geochichte Siciliens im Alterthum — *2 Bde. Bd. I, 1869, Bd. II, 1874, Leipzig.*

HOLM A. — Das alte Catania, mit Plan. — *Lübeck, 1873.*

HOLM A. — Geschichte Siciliens —?.

HOMODEIS (FILOTEO DE) A. — *See Filoteo.*

HOROZCII COVARRUVIAS DE LEYVA J. — Episcopi Agrigentini. Emblemata moralia memoriae Sanc. D. D. Didaci Covarriuivas de Leyva, etc. etc. — *Agrigenti, 1601, 3 books in 1 vol. in 8°. with 100 pl. (book I, pp. 4-256, Book II, pp. 110. Book III, pp. 100.).*

HOUEL J. — Voyage pittoresque dans les îles de Sicile, de Malte et de Lipari. — *Tom. I-IV, in fol. Paris, 1782-87.*

HOUEL J. — Reisen durch Sicilien u. Malta, etc. — *Uberselzt v. J. L. Heerl, mit Kupft. Vol. I-IV, 8.° Gotha, 1797-1809.*

HUMBOLT (VON) A. — Ueber Bau u. Wirkungsart der Vulkane. — *Berl. Akad. d. Wiss. Seit. 137, Berlin, 1825.*

HUMBOLT (VON) A. — Ansichten der Natur. — *Bd. II, Tübingen 1839. (Notices on Etna, without historical date).*

IDACII. — Chronica. — *(Erupt. of 72, A. D.).*

ILMONI I. — Misceller an Vulcanen Actna. — *Föredr för Vet. Soc. d. 4 Mars 1839 och d. 26 Apr. 1841, in 4°, pp. 4. (C. A.).*

INGO V. — See: Alfieri G.

ITERLANDI E SIRUGO. — Osservazioni geologiche e geognostiche sopra i terreni di Avolo. — *Atti d. Acc. Gioenia. Ser. 1ª, Vol. XII, Catania, 1835.*

134 ETNA

INTERLANDI E SIRUGO.— Sopra i basalti globulari del Morgo.— *Atti d. Acc. Gioenia, Ser. 1.ª, Vol. XIV, Catania, 1837.*

INTERLANDI E SIRUGO. — Sopra i terreni di Lognina, Aci-Trezza e Castello. — *Atti d. Acc. Gioenia, Ser. 1.ª, Vol. XV. Catania, 1838.*

ISSEL A. — Saggio di una teoria sui vulcani. — *N. Antologia di Firenze , 1875. In abstract: Boll. d. Vulc. Ital. Fasc. I, II, III, pp. 13, Roma, 1875.*

ISSEL A. — Sullo stato sferoidale dell'acqua nelle lave incandescenti. — *Nota, Boll. Vulc. Ital. Fasc. IV e V, pp. 57, Roma, 1876.*

ITTAR S. — Viaggio pittorico all' Etna contenente le vedute più interessanti di questo Monte, e gli oggetti più rimarchevoli che nelle sue Regioni esistono, con una Carta Topografica, che indica con numeri i punti dai quali il Disegnatore le ha espresse, ed una breve descrizione di esso e del suo itinerario. —? *in 4°, obl. pl. XXV. (B. N.).*

JERVIS G. — Tesori sotterranei dell' Italia. — *4 Vols. in 8.°, Torino, 1874-1888. Numerous plates.*

JOHNSTON-LAVIS H. J. — Note on the occurence of Leucite at Etna. — *Reports British Association for Advancement of Science. London, 1888.*

JOHNSTON LAVIS H. J. — Su una roccia contenente leucite trovata sull' Etna. — *Boll. d. Soc. d. Microscopisti It., Vol. I, 1889. Fasc. 1-2, pp. 26, with one photo-engraving.*

JOHNSTON-LAVIS H. J. — Viaggio scientifico alle regioni vulcaniche italiane nella ricorrenza del centenario del « Viaggio alle due Sicilie » di Lazzaro Spallanzani.—*(This is the programme of the excursion of the English geologists that visited the south Italian volcanoes under the direction of the author. (It is here included as it contains various new and unpublished observations. — Naples, 1889, in 8.° pp. 1-10.*

JOHNSTON-LAVIS H. J. — The State of the active Sicilian Volcanoes in September 1889. — *Scottish-Geograph. Mag. Vol. VI. N.° 3, March 1890. pp. 115-150.*

JOY. — Analyse der Lava des Stromes d. J. 122 — *See C. G. Rammelsberg: Handwörterbuch. Supplem. 5, Seite 157.*

JUDD W. J. — Contributions to the Study of Volcanoes. — *Geol. Mag. Vol. II. London, 1876.*

JULLIEN JOHN J. — L'Etna et le troisième congrès du Club Alpine Italien à Catane. — *Echo d. Alpes, publicat. du Club Alpin Suisse, N.° 4, pp. 263, Genève, 1889.*

JUSTINUS. — Hist. Lib. IV. cap. 1, 5, 14. —

KARACZAY F. — Manuel du Voyageur en Sicile, avec une carte. — *Stuttgart, 1826*.

KEPHALIDES M. A. W. — Voyage à l'Etna. — *Nouv. Ann. des Vo. 2° série, t. 4, Juin 1827, in 8.°, pp. 289-306*. (G. A.).

KIRCHER A. — Mundus subterranens in XII libres digestus. — *Amsterodami, Waesberg 1678.* (*Vol. I. Lib. IV. Pyrographicus, Cap. VIII. Aetnae Descriptio, Cap. IX. Crateris Aetnae descr.*).

KLUGE E. — Ueber Synchronimus u. Antagonimus von vulkanischen Eruptionen. — *Leipzig, 1863*.

KUDERNATSCH J. — Chemische Untersuchung des Auget vom Aetna. — *Poggd. Ann. Bd. XXXVII, Seit. 577; — auch Neu. Jahrb. f. Min. Geogn. u. Geol. Seit. 597, Stuttgart, 1836*.

LALLEMONT. — Lettre à Dolomieu sur l'éruption de l'Etna en 1792. — *Journ. de Phys. 1792, T. 40, pp. 481-482 and t. 41, pp. 120-122*. (C. A.).

LANDGREBE G. — Naturgeschichte der Vulkane. — *2 Bd. Gotha, 1855*.

LANZA P. — Eruzione del 1646. — (*Cit. by Recupero*).

LASAULX (VON) A. — Etnabesteigung am 2 oct., 1878. Brief. Schlesische Ztg. v. 15 Oct. 1878.

LASAULX (VON) A. — Der Aetna und seine neueste Eruption von 1879. — *Deutsche Rev. 1879. Septemberheft*.

LASAULX (VON) A. — Sicilien, ein geogr. Charakterbild. — *Bonn, 1879.* (*Aetna Seit. 13-26*).

LASAULX (VON) A. — Szaboit von Biancavilla am Etna, Eisenglanz ebendaher. — *Groths Zeitschr. f. Krystallohr. Bd. II, Seit. 288. Leipzig, 1879. — Abstract: Boll. d. R. Com. Geol. d'Italia, Vol. X, pp. 372, Roma, 1879*.

LASAULX (VON) A. — Ueber die Salinellen von Paternò and ihre neueste Erupt. — *Zeitschr. d. deutsch. geol. Gesell. Berlin, 1879*.

LASAULX (VON) A. — Der Aetna, 1880. — *See: Sartorius von Waltershausen*.

LAVINI, CONTE G. — Rime filosofiche colle sue annotazioni alle medesime. — *Milano, 1750, in 4°, pp. XXXII-232. (pp. 74-85)*.

LAVIS. — See Johnston-Lavis.

LAZZARO N. — Da Napoli all'Etna. — *Due corrispondenze a'l'Illustraz. Ital. di Trèves. Sem. 1°, pp. 388. Sem. 2°, pp. 6, con una incisione. Milano, 1879*.

LEANTI A. — Stato presente della Sicilia. — *Vol. II. Palermo, 1761, in 8°,* (*With a few notes on Etna*).

LEBLANC. — 1858. — *See Deville.*

LENTINI (DA) S. — Chronicon o Chronaca in Rosario di Gregorio. — *Biblioth. script. Arag. Panormi, 1691. (Cit. by Ferrara and Caruso).*

LENTINI (DA) S. — Historia M. S. del conte Rogero. — *(Cit. by Massa, Etna). Biblioth. script. Arag. Vol. II, Panormi, 1691.*

LEONHARD (VON) C. C. — Die Basalt Gebilde. — *2 Bde. Stuttgart 1832.*

LEONHARD (VON) C. C. — Geologisches Atlas. — *Stuttgart, 1841. (2 Karten des Etna, die der Lavaströme nach Gius. Gemmellaros Quadro).*

LEONHARD (VON) C. C. — Geologie oder Naturgesch. der Erde.— *Bd. I-V, Stuttgart, 1844. (Der Aetna 72te Vorlessüng. Bd., V, Seit. 189).*

LIGHT MAJOR. — Sicilian scenery. — *London, 1828, in 4.º*

LOEWE. — Analyse der Lava von 1669. - *Poggd. Ann. Bd. XXXVIII, pp. 160, 1836.*

LONGO A. — Colpo d'occhio geologico sul terreno di Caltagirone.— *1864.*

LONGO A. — Ad un addio del Prof. Cav. Uffìciale C. Gemmellaro. Parole di Risposta del vecchio Mongibello. — *Catania, 1866.*

LONGO A. — Un apostrofe all'Etna oggi Mongibello. — *Catania, (in versi), 1869.*

LONGO A. — Sulle cagioni probabili delle accensioni vulcaniche subaeree. — *Atti d. Acc. Gioenia. Ser. 3ª, Vol. IV, 1869.*

LONGO A. — Sulle interpretazioni dei fenomeni chimici in rapporto alle leggi della natura. — *Atti d. Acc. Gioenia. Ser. 3ª, Vol. V. Catania, 1870.*

LONGO A. — Osservazioni sopra alquanti squárci della memoria del Sig. Mallet « Volcanic Energy ». — *Atti d. Acc. Gioenia, Ser. 3ª, Vol. XIV, Catania, 1879.*

LONGO A. — L'Etna al cospetto della Scienza. — *Catania, 1886, in 4º; pp. 63, Map. 1. (C. A.).*

LONGOBARDO A. — Extrait d'une lettre à M. Ch. Sainte-Claire Deville (éruption de l'Etna 7 Juillet 1863). — *Compt. Rend. Acad. Sc. Paris, 1863, T. 57, p. 157. (C. A.).*

LONGOBARDO A. — Extrait d'une lettre à M. Ch. Sainte-Claire Deville (Eruption de l'Etna, 31 janv. 1865). — *Compt. Rend. Acad. Sc. Paris, 1865, T. 60, p. 354. (C. A.).*

LONGUS J. — Continuation du Maurolycus. — *See Maurolycus.*

LUCA P. DE. — Lettera sul miserando caso della esplosione avvenuta addì 25 Nov. 1843 durante la eruzione dell'Etna.—*Rend. R. Accad. Sc. Fis. Mat. Napoli, 1843-46, T. III, pp. 177-185.*

‹ Luc J. A. DE. — Formation des Montagnes Volcaniques. — *Observations au Vésuve et à l'Etna. La Haye. Paris, 1780, in 8.°, pp. 19.*

Luca Pl. DE. — Eruzione dell'Etna in Novembre del 1843, e suoi effetti nell'industria dei Brontesi.—*Napoli, 1844, in 8°, pp. 27, pl. 1.* (B. N.). *See: Museo di Scienze e Letteratura. pp. 145-169.*

Lucilius Junior.—Ancienne traduction de A. C. Schmid.—*1769.*

Lucilius Junior. — Meinecke. — *Quedlinburg, 1818.*

Lucilius Junior. — Recensuit notasque Jos. Scaligeri, Arid. Lindenbruchii et suas addidit F. Jacobs. (With translation in verse) — *Lipsiae, 1826, in 8°.*

Lucilius Junior. — Traduction française par J. Chenu. — *Paris, 1843.*

Lucilius Junior. — Etna; — (*See:* « *Wernsdorffs excurs. ad Luc. Aetn.* in *Poëtae latini minores* »). —

Lucretius. — *Lib. VI, 639-702.*

Ludovicus Aurelius. — Erupt. of 1169. — *Ex Baronio in epist. Lib. XI.*

Ludovicus Cremonensis. — *See Cremonensis.*

Luebeck (von) A.—Cronica Slavorum. — (*Written in 1209*) *Vol. XIX, Pertz. Monument. Historic. Germanic. Scriptorum.* (*Cap. XXI, pag. 159. Epistola Conradi cancellarii episcopi electi Hildescimensis. Anno 1195.*) *Notice on Etna and an eruption of the period.*

Lycurgos. — Contra Leokrates 95. — *Erupt. des Fratelli pii.*

Lyell C. —· Principles of Geology. — *London. Numerous editions. Principes de Géologie. Traduction française sur la sixième édition anglaise, Lyon, 1846. (Etna. Part. 1*, pag. 137-205. Stromboli. Part. 3*, pag. 357).*

Lyell C. — On Lavas of Mount Etna formed on steep Slopes and on Craters of elevation. — *Philosoph. Trans. Part. II, 1858, Abstract in French. See : Archiv. d. Sc. d. l. Biblioth. Uni. ver. Genève, novembre 1859. Abstract in German: Zeitscrift, d. deut.geol. Gesell. Berlin, 1859, and Journ. Sc. 2nd Ser. Vol. XVI, pp. 214-219, etc.*

Lyell C. — Ueber die auf steil geneigter Unterlage erstarrten Laven des Etna und über Erhebungskratere. — *Taf. I-IV, in 8°, Auch. Roth. J. Zeitsch. d. Deutsch. geol. Ges. Band. XI, Seit. 149, Berlin, 1859*

6*

LYELL. C.—Principles of Geology. — *2.ᵃ Edit., Chap. Aetna, London, 1872*

MACRI V. (DI NICOLOSI). — (Erupt. of XVII century)—*Manuscript. cit. by Recupero: abstract printed in the appendix to his work.*

MACROBIUS. — Saturn. — *Vol. V, pag. 17.*

MAGGIORE D. G. — 1842. — *See, Aradas A.*

MAGINI G. M, — Descrizione della Sicilia. — (*Cit. by Di Gregorio M. sur les eaux minérales de Palerno).*

MAGNETI V. — Notitie istor. di Terramoti.—(*C. by Mongitore).*

MALAGOLI VECCHI M. — Il Mediterraneo illustrato, le sue Isole.— 1841.

MALATERRA G. -- De rebus gestis Roberti Guiscardi ducis Calabriae et Rogeri comitis Siciliae libri quatuor.—*Lib. II, cap. XXX, Erupt. of XI century). Thes. Sic. Vol. VIII, 1725.*

MALHERBE A. — Ascension à l'Etna.—*Ext. des Mém. de l'acad. 1841, pp. 32, (C. A.).*

MALLET R.—On the mechanism of production of volcanic dykes.—*Quart. Journ. of geol. Soc. of London, Vol. XXXII, pag. 472. London, 1876.*

MALVICA F. — Gita alle Madonie. — *Effemeridi Scientifiche e Letteraria per la Sicilia, Fasc. 35, Palermo, 1835, in 8°,* (B. N.).

MANCINI C. — Narrativa del fuoco uscito dal Mongibello il dì undici marzo 1809. — *Messina, 1669.*

MANCUSI P. ANT. — Istoria di Sᵗᵃ. Rosalia etc. — *Palermo, 1721, in 4°, Vol. I, pp. 17-456. Vol. II, pp. 19-221, (Notizia del Monte Pellegrino (Etna o Mongibello) (Vol. I,ᴵp. 117-137).*

MANNERT H. — Geographie der Griechen und Römer. — *(Bd. IX, Seit. 2 Geogr. v. Sicilien). 8.° Leipzig, 1832. (Ancient notices of Etna).*

MANTOVANI P, — Sulla formazione basaltica delle Isole dei Ciclopi presso Catania. — *Roma, 1870.*

MANTOVANI P. E GREGORI A. – La eruzione dell' Etna 1879. — *Bull d. Club Alpino ital. N. 37. Torino, 15 giugno, 1879.*

MARANA. — Des Montagnes de Sicile et de Naples, qui jettent des feux continuels: de la nature de leurs effets. —*Lettre XLIII, de l'Espion Turc, T. I, pp. 153-157 (C. A.).*

MARASCHI G. — Lettera sulla costruzione della Gratissima e della casa inglese detta altrimenti di Mario Gemmellaro fabbricata sopra l' Etna sin dal 1804. — *Palermo, 1829, in 8°, pp. 13* (C. A.).

MARAVIGNA C. — Memorie compendiose dell'ultima eruzione dell' Etna, accaduta nel mese di novembre 1802. — *Catania, 1803, in 4°, pp. 19.*

MARAVIGNA C. — Tavole sinottiche dell' Etna. — *Catania, 1811, Paris, 1838.*

MARAVIGNA C. — Istoria pell'incendio dell'Etna del mese di maggio 1810. — *Catania, 1819, in 4°, pp. 102, pl. 2. Also in the Bibliolh. Ital. Tom. XVIII, pag. 108.*

MARAVIGNA C. — Della causa dei vulcani, dei loro fenomeni e delle sostanze eruttate. — *Giorn. d. Sc. Lett. ed Arti, Tom. I, pp. 223, e Tom. II, pp. 3, Palermo, 1823.*

MARAVIGNA C. — Su i miglioramenti che le recenti scoperte chimiche hanno apportato alla soluzione di alcuni fenomeni geologici e particolarmente alla teoria dei vulcani. — *Atti d. Acc. Gioenia, Ser. 1ª, Vol. VII. Catania, 1830.*

MARAVIGNA C. — Alcune idee sull'eruzione del fuoco nella produzione di alcuni membri della serie geognostica sui rapporti del terreno trachitico e basaltico con quello dei vulcani estinti ed attivi. — *Atti d. Acc. Gioenia, Ser. 1ª, Vol. VIII, Catania, 1831, pp. 25.*

MARAVIGNA C. — Materiali per servire alla Orittognosia Etnea.— *Atti d. Acc. Gioenia, Ser. 1ª, Tom. V, pp. 141-161: VI, 205-214, 1832, VIII, 25-51, 1834, IX, 231-295. Catania, 1829-1834.*

MARAVIGNA C. — Memoria sopra la eruzione apparsa nella plaga occidentale dell'Etna nelle notti del 31 ottobre, 1 e 3 novembre dell'anno 1832, per cui fu in pericolo il comune di Bronte.— *Atti d. Acc. Gioenia, Ser. 1ª, Vol. IX, Catania, 1832.*

MARAVIGNA C. — Appendice alla memoria sopra i silicati. Orittognosia etnea. — *Atti d. Acc. Gioenia, Catania, 1834.*

MARAVIGNA C. — Esame di alcune opinioni del sig. N. Boubée contenute nelle sue opere intitolate " Géologie populaire, et tableau de l'état du globe à ces différens âges ". — *IV, édit. 1834, in 4°, pp. 48.*

MARAVIGNA C. — Cenno sul solfato di calce che formasi nell' interno del cratere dell'Etna, sulla genesi di altri sali che ivi rimangono e specialmente di una sostanza molto rassomigliante al Caolino prodotta dalla decomposizione delle lave.— *Atti d. Acc. Gioenia, Ser. 1ª, Vol. XII, Catania, 1835, pp. 13. .*

MARAVIGNA C. — Cenno sul ferro Oligisto ottaedrico del Monte del Corvo vicino a Biancavilla. — *Atti d. Acc. Gioenia, Ser 1ª, Tom. XI, Catania, 1836, pp. 6.*

MARAVIGNA C. — Sulla Jalite del Basalto della Motta, sulla Tremolite dell'Isola dei Ciclopi, sull'Idroclorato di ammoniaca della eruzione di Bronte nel 1832. — *Atti d. Acc. Gioenia, Ser. 1ᵃ, Tom. XII, Catania, 1835, pp. 81-88.*

MARAVIGNA C. — Memorie di Orittognosia Etnea e dei Vulcani estinti della Sicilia. — *Parigi, 1838, in 8°, pp. 203, pl. 2,* (C. A.). *An abridged form of the above appears in the reports of Congrès Sc. de France à Clermont en 1838, in 8°, pp. 331-349* (C. A.).

MARAVIGNA C. — Mémoires pour servir à l'histoire naturelle de la Sicile. — *Paris, 1838, in 8°, pp. 86, pl. 6.*

MARAVIGNA C. — Su i rapporti che passano fra le rocce dell'Etna e sul modo di loro emissione — VII *Congresso degli Scienziati Italiani in Napoli. Napoli, 1845, in 4°, pp. 40* (C. A.).

MARAVIGNA C. — See Zuccarello.

MARGALLE. — 1866. —· *See Zurcher.*

MARINI N. — De formidabilissimo terrae motu, etc. Poema. — *Panormi, 1729, in 8.°*

MARMONT (DUKE OF RAGUSA). — Voyage en Sicile. — 3ʳᵈ *Edition, Paris, 1839, in 8°, pp. 372* (C. A.).

MARMONT. — Carte des Eruptions du Mont Etna. Appended to the work of Marmont. " Voyage en Sicile " — (*The original M. S. in library* (C. A.).

MARMOR PARIUM (PSEUDON?). — Marmora Oxoniensia ex Arundellianis. Seldenianis aliisque conflata. — *Oxonii, 1676. (Erupt. of 475. B. C.)*

MARTINES A. M. — De situ Siciliae et insularum adjacentius. — *1580.*

MARZO (DE) G. — Biblioteca storica e letteraria di Sicilia. — *Vol. I-XXVI. Palermo, 1869-76.*

MASCULUS J. B. — De incendio Vesuvii XVII, Kal. Januar, an. 1631, Libri X, Cum chronologia superiorum incendiorum et ephemeride ultimi. — (*Erupt. 1631). Neapolis, 1633, in 4°, pp. 312, 37, pl. 2.*

MASSA (PADRE) G. A. — Della Sicilia grand'isola del Mediterraneo in prospettiva e il monte Etna o il Mongibello esposto in veduta da un religioso della compagnia di Gesù. — *Palermo, 1708. (Cap. XVIII, cronologia degli incendi, in 4°, pp. 126. Also, Palermo, 1709, 2 Vols. in 4°, Pl. I, pp. 12 + 359. Pl. II, pp. 503.*

MAUGINO F. — L'eruzione dell'Etna 1879. Lettera al prof. Ragona. — *Ann. d. Soc. Met. Ital. Vol. II, N.° 41-44, pp. 306.*

MAURO CIRINO. — Lentini abbattuta dai terremoti. — *Messina, 1700.*

MAUROLYCUS F. (1404-1575). — Cosmographia dial. 3. — (*cit. by Massa, Etna) Veneliis, 1543.*

MAUROLYCUS F. — Sicaniarum rerum compendium. — *Thes. Sic. IV, Messanae, 1562. Better edition by Giacomo Longo. Messanae, 1716.*

MAZZA. — Storia di Adernò. — *Catania, 1820.*

MAZZARA G. — Poema del Mongibello. — (*rep. by Ventimiglia). Poeti siciliani, cap. 30 (Cit. by Massa, Etna).*

MAZZETTI L. — 1879-1880. — *See Ba'dacci.*

MELLONI M. — Sulla Polarità magnetica delle lave e rocce affini. — *Atti d. Acc. d. Sc. di Napoli, An. 1853.*

MERCALLI G. — Sull'eruzione Etnea del 22 marzo 1883. — *Atti d. Soc. It. d. Sc. Nat. Vol. XXVI, Milano, 1883. in 4°, pp. 11.* (C. A.).

MERCALLI G. — Le ultimi eruzioni dell' Etna del 22 marzo 1883 e del 18 maggio 1886. — *Firenze, 1887, in 4°, pp. 8* (C. A.).

MERCURIO G. A. — Sulla salsa di Fondachello nel comune di Mascali, del profondamento parziale del cono argilloso e dell'apparizione di un'acqua minerale gasosa. — *Catania, 1847.*

MERCURIO G. A. — Relazione della grandiosa eruzione Etnea della notte dal 20 al 21 agosto 1852. — *Catania, 1853. Palermo, · 1853, pl. 1. A short abstract in: v. Rath's Aetna, pag. 32. Also in Baltzer, pag. 38.*

METAPHRASTUS S. — Editio Migre. — (*Erupt. 251, A. D.) Vol. I, pag. 346, Paris, 1864.*

METAPHRASTUS S. — Vitae Sanctorum, in Neander's Historia Ecclesiac. — (*Erupt. 253, A. D.). Vol. I.*

MINASI G. — Relazione de' Tremuoti di Sicilia — *Messina, 1783 Supplemento, Messina, 1785.*

MIRONE G. — Sopra un'acqua minerale (Acqua Santa) nelle vicinanze di Catania. — *Catania, 1786.*

MIRONE G. AND PASQUALI G. — Descrizione dei fenomeni osservati nell'eruzione dell'Etna accaduta in quest'anno 1787 e alcuni prodotti vulcanici che v'appartengono. — *Catania, 1788, in 12°, pp. 29. (There is also the descript. of the Aurora borealis of 13 July 1787 during the erupt). French translation also.*

MOLL H. — Map of Italy, containing a representation of Etna, during the eruption of 1669. — *London, 1714.*

MOMPILERI. — Relazione M. S. dell' eruzione del 1536. — (*cit. by Massa*).

MONACO F. — Cataclysumus aetnaeus, sive inundatio ignea Aetnae montis anni 1669. — *Venetits , 1669 , in 4°, fol. 6, pp. 60, wtth figs.* (C. A.).

MONGITORE A. — La Sicilia ricercata nelle cose più memorabili.— *Vol. I-II, Palermo, 1742-43, in 4° (pag. 286, Tom. II, Etna ovvero Mongibello: pag. 345 Istoria cronologica dei terremoti in Sicilia). etc.*

MONGITORE A. — Diario Palermitano 1640-1743. — *Biblioteca de Marzo, Vol. VII, IX e XII.*

MONGITORE A. — Cronologia de tremuoti di Sicilia. — *Palermo, 1743, in 4.°*

MONTICELLI T. AND COVELLI N. — Analisi del fango dell'Etna.— *" Biblioteca Analitica di Sc. Lett. ed Arti, Napoli, August 1823, pp. 143-148.* (C. A.).

MONTICELLI T. E COVELLI N. — Esame chimico d'una pioggia di polvere caduta il 21 di Giugno 1822. — *Giorn. di Sicilia, N. 5, 1823.*

MONTLÉMENT A.—Des volcans en général et plus spécialement du Vésuve et de l'Etna. — *Bull. Soc. Géol., t. XVI, Sept. 1841, pp. 137-158.* (C. A.).

MORICAUD S. — Eruption de l'Etna en 1819. — *"Nouv. Ann. des Voyages" 1819, t. III, pp. 455-462.* (C. A.).

MORIS A. — Account of the eruption of 1822. — *Rodwell's Etna, pag. 106.*

MORTILLARO B. V. — Discorso su la vita, e su le opere dell'Abbate Domenico Scinà. — *Palermo, 1837, in 8°, p. 61. (Etna in 1811).*

MUENSTER S. (1489-1552). — Cosmographia universalis. V. lib. II, 257 et lib. I, cap. VII. Meber: " De igne in terrae visceribus flagrante. — *(Description of the changes in crater of Etna since the time of Strabo, to the eruption of 1537).*

MURABITO F. — Catania liberata dall'incendio dell'Etna del 1669 in X canti. — *Catania, 1675. (Cit. by Ferrara: Storia di Catania).*

MURATORI. — Annales. — *Tom. V, pag. 743; Tom. VII, p. 342, Tom. X, p. 921, etc.*

MUSUMECI M. — Sopra l'eruzione dell'Etna dell'anno 1832. — *Atti d. Acc. Gioenia, T. IX. pp. 207-218.*

MUSUMECI M. — Sopra una colonnella nella lava. — *Opere Archeol. ed Art. Vol. I, p. 59 (Nota).*

NARBONE. — Bibliographia Sicula. — *Palermo, 1854. (Vol. III, p. 139 descript. of Auria during the eruption of Etna, 1669).*

NATALIS COMITIS. — Universae historiae sui temporis libri triginta ab anno salutis 1545 usque ad annum 1581. — *Acta*

Sanct. II, p. 650. Venetiis, 1581, lib. XVII, p. 370. (Eruption, 1566).

NAUDÉ G. UND GIULIANI G. B. — Ueber den Vesuv und Aetna.—? *1632.*

NEGRI F. — Pianta del Monte Etna. — *(Cit. by Massa: Etna).*

NIGER M. — Siciliae insulae descriptio. — *Scriptores Rer. Sicul. III.*

NOUGARET J. — Lettres écrites de la Sicile, à l'occasion de l'éruption de l'Etna. — *Moniteur, 18 mars, 26 avril 1865, pp. 32.* (C. A.).

OBSEQUENTIS J. — Prodigiorum liber, ed O. Jhan.—*(Pp. 118 and follow. Erupt. 141, 135, 126, 122. B. C.).*

ODELEBEN (VON) E. G. — Beiträge zur geol. Kenntniss v. Italien.— *2 Th. Freiberg, 1819.*

OLDENBOURG. — Cronologia dell'eruzioni del Monte Etna. — *Compendio delle transazioni filosofiche. Anno 1869, Venezia, 1793, pp. 1-4.* (C. A.).

OMODEI F. G. — See Filoteo.

ORLANDINI L. — La descrizione latina del sito di Mongibello di Ant. Filoteo degli Homodei tradotta in lingua italiana. — *Palermo. 1611, in 4°, fol. 4, pp. 87, pl. 1.*

OROSIUS. — Lib. II, cap. XIV-XVIII; lib. V, cap. VI, X and XIII.— *(Eruptions 125, 134, 126, 122 B. C.).*

ORTOLANI G. E. — Prospetto dei Minerali di Sicilia. — *2° ediz. Palermo, 1809, in 8°, pp. 17-18, 30-31.*

ORTOLANI G. E. — Prospectus of the minerals of Sicily. — *1808.*

ORTOLANI G. E. — Nuovo dizionario geografico e biografico della Sicilia antica e moderna. — *Palermo, 1819.*

ORVILLE (COMITIS D') J. PH. — Sicula, quibus Siciliae veteris rudera illustrantur. — *In fol. Amstelodam, 1764.*

OTTAVIO (PADRE) G. — Isag. ad hist. Sic. (Cap. XIII, N. 15. Earthquake at Etna.) — *Repeated under Cajetani Syra cusani Pactris Octavii). (Cit. by Mongitore).*

OTTAVIO (PADRE) G. — Sicul. in animad. — *T. I, fol. 22. Earthquake of 1619. (Cit. by Mongitore).*

OVIDIUS. — Metamorph. lib. XV, 340-55.

PACICHELLI G. B. — Memorie di viaggi per l'Europa Christiana scritte a diversi in occasione dei suoi Ministeri. — *Napoli, 1685, Vols V in 12°. Parte I, pp. 40-743-53. Parte II, pp. 8-827 40 ; Parte III, pp. 8-761-27. Parte IV, vol. I, pp. 4-541-20. Parte V, vol. II, pp. 4-438-18. (Parte IV, vol. II, pp. 66 and follow. Del Mongibello.*

PACICHELLI G. B. — Lettere familiari istoriche ed erudite. — *Na poli, 1695. Vols. II, in 12°. (Vol. I, pp. 12-490; Vol. II, pp. 20-432-31).*

PAGLIA B. — Epigrammata in XII. Suetoni Caesaris. — *Neapoli, 1693, in 8°, p. 200. — Messana fugit, Aetna teritus, pp. 65.*

PALGRAVE (SIR) F. — An account of the eruption of Mount Etna in the year 1535 from an original contemporary document communicated in a letter to J. G. Children, Esq. Secretary of the Royal Society. — *Proceed. of the Roy. Soc. of London, Jan. 15, 1835.*

PALGRAVE F. — Sac. Dr. Children communique une lettre de Sir Francis Palgrave contenant le récit d'une éruption du mont Etna dans l'année 1535 d'après les documents originaux contemporains, trouvés parmi les papiers qui renferment la correspondance de Henry VIII avec les princes d'Italie, dans les archives de Westminster. — *L' Institut. Journ. général. N. 5, 1835.*

PALMIERI L. — Un fatto che merita di essere registrato. — *Rend. R. Acc. Sc. Fis. Mat. An. XXV, Napoli, 1886, pp. 125.*

PAPIN S. — Theb. XII, 274; Silv. III, I. ·130.

PARTHEY. — Wanderungen durch Sicilien u. die Levante. — *2 Th. Berlin, 1834.*

PARUTA F. — La Sicilia descritta. — *In fol. (Cit. by Carrera). Lione, 1617.*

PASCALE V. — Descrizione storico-topografico-fisica delle isole del regno di Napoli. — *Napoli, 1796, in 8°, pp. 138, pl. 1.* (C. A.).

PATERNIO J. — Matricula monasteriorum S. Mariae, S. Leonis.— *Catanae, 1693, (Repeated under the name of Bartolomeo (don) a Palernione).*

PATERNÒ T. — 1669. — *See Tedeschi.*

PATERNÒ A. — Cronaca di Sicilia. — *(Cit. by Auria).*

PATERNÒ I. (Principe di Biscari). — Viaggio per tutte le antichità della Sicilia. — *Napoli, 1781, in 4°, pp. 200. pl. 2.* (C. A.).

PENK A. — Ueber Palagonite u. Balsalttuffe. — *Zeitschr. d. Deutsch. Geol. Ges. Seit. 504. Berlin, 1879. (Also tuffs and Palagonite of Etna).*

PERERIA A. — Die Aetna (Eruption 1879)—*Verhhl d. K. K. geol. Reichsanstalt. N.° 10, 231. Wien, 1879.*

PEROU (DU). — Notice sur l'éruption de l'Etna de février 1865.— *Bull. d. l. Soc. Géol. de France. Paris, 1866.*

PETAVIUS D. — Uranologia I, VII, c 10. — *(Cit. by Massa, Etna). Parisiis, 1630.*

PETRARCHA F. — De Rom. pontif. et Imp. (*Eruption 1169*).

PHILOSTRATUS. — Vita Apoll. — *V, cap. 16 and 17.*

PHILOTHEUS. — See Filoteo.

PIAZZA (DI) F. M. — Cronaca m. s. — (*Earthquake 1176. (Cit. by Massa, Etna*).

PILLA L. — Parallelo tra i tre Volcani ardenti delle Sicilie. — *Atti d. Acc. Gioenia, 1837. — Cit. Jahrb. f. Min. p. 347. Stuttgart, 1836.*

PILLA L. — Sopra la produzione delle fiamme nei vulcani. — *Pisa, 1837, in 4° — Also: Bull. d. l. Soc. Géol. de France, VIII, p. 262; 1837, — J. Roth, d. Vesuv. S. 350. Berlin, 1837.*

PILLA L. — Studii di Geologia. — *Napoli, 1841.*

PILLA L. — Aggiunte al discorso sopra la produzione delle fiamme nei vulcani. — *Nuovo Cimento, Pisa, 1844.*

PILLA L. — Sur les phénomènes volcaniques de l'Italie Méridionale. — *Mém. d. l. Soc. géol. d. France, 2e Sér., Vol. I, page 179, 1844.*

PILLA L. — Orittognosia e Geognosia in Italia.—*Progresso delle Scienze, Lettere ed Arti, art. of Vol. II, 3, 5;*

PINDARUS. — Pythia. — *Ode 29 and follow.*

PIRRO D. R. -- Catanensis Ecclesiae Notitiae.— *1 lib. III, Thes. Sic.*

PIRRO D. R. — Chronologia regnum Siciliae. — *Thes. Sic. L. V. Panormi, 1643.*

PIRRO D. R. — Sicilia Sacra. — *3ª Edition Vol. II, fol. Lugg. Bat. 1722. — Panormi 1733. — Also: Thes. Sic. I, and II; (Annales Panormi sub annis D. Ferdinandi de Andrada archiepiscopi Panormitani, ab anno 1646). Bibliolh. De Marzo. Vol. IV, pp. 58-252.*

PISTOIA C. F. — Carta dell'Etna in rilievo per uso dell'istituto topografico militare a Firenze (colorata geol.) Scala verticale 1:25000 ; scala orizzontale 1:50000. — (*Constructed on the Base of the map of Sartorius von Wallershausen*).

PLATANIA GIOV.—Les tremblements de terre de Nicolosi (Sicile).— *La Nature, 1885, II, p. 350.*

PLATANIA GIOV. — La récente éruption de l'Etna. — *La Nature, II, 1886, pp. 97-99, map. 1, figs.*

PLATANIA S. — Sul carbonato di Soda nativo nelle Lave dell'Etna. — *Atti d. Acc. Gioenia, Vol. VIII, Ser. 1ª, pp. 153-176.*

PLATTNER. — Analyse des Gesteins von Serra Giannicola in Val del Bove. — *Mitgeth in Fr. Hoffmann's Beobacht, Karsten's Archiv. XIII, Seit. 702, 1839.*

PLINIUS. — Historia natur. — *II, 103, 106, and III, 8.*

7*

POMPONIUS MELA. — De situ orbis II , 7. — (*Eruption of first cent. B. C.*).

POLEMONE. — De admirabilibus Siciliae Fluminibus. — (*Cit. by Alfiio Ferrara. Mem. sulle acque d. Sicilia, pag. 7, Londra, 1811.*

PORTAL. P. — Osservazioni sopra il ferro speculare vulcanico trovato nell'Etna. — *Without locality or date , in 8°, pp. 10,* (C. A.).

PORTIUS S. — Physiologicum opus. De Aetnae ignibus acorumque causis. — *Messinae, 1618.*

POWER JEANNETTE. — Itinerario delle Due Sicilie riguardante tutt'i rami di storia naturale, e parecchi d'antichità che essa contiene. — *Messina, 1839, in 4°, p. VIII, 249, pl. (Etna pp. 63-82).*

PREVOST C. — Rapport fait à l' Academie royale des sciences sur le voyage à l'île Sicilia en 1831-32. — *Paris, 1832.*

PREVOST C. — Sur un projet d'exploration de l' Etna et des formations volcaniques d' Italie. — *Compt. Rend. Acad. Sc. t. XXXV, Paris. 1852, pp. 409-413* (C. A.).

PREVOST C. — Etude des phénoménes volcaniques du Vésuve et de l'Etna. — *Compt. Rend. Acad. Sc. t. XLI, Paris, 1845, pp. 794-797* (C. A.).

PREVOST C. — Observations géologiques en Sicile. — *Bull. d. l. Soc. Géol. de France, II, pp. 303.*

PRIVITERA F. — Annuario Catanese. — *(Eruptions 1536-37). Catania, 1690.*

PRIVITERA F. — Succinta relazione del tremuoto del 1693. — *Catania, 1694.*

PRIVITERA F. — Dolorosa tragedia, etc. di Catania distrutta nel 1693. — *(earthquake of 1693). Catania, 1695, in 4°, fol. 1, pp. 98.*

PROCOPIUS. — (Eruption, 550.) — *De bello Gothico IV, 35.*

PROTOSPATA L. — Rerum gestarum in Regno Neapolitano ab 860-1102 Chronicon. — *Carusii. Mem. storiche.*

PRRYSTANOWSKY (VON) R. — Ueber den Ursprung der Vulkane in Italien. — *Berlin, 1822.*

QUATREFAGES (DE). — Souvenirs d'un Naturaliste. — *Paris, 1865. Revue des Deux Mondes T. XIX, 1847, pp. 5-36.*

RAFFELSBERGER F. — Gemälde aus dem Naturreiche beider Sicilien. — *Mit. Kpfr. Wien, 1824.*

RAMMELSBERG K. T. — Mineralogische Gemengtheile der Laven, etc. (Actnalava). — *Zeitschr. d. Deutsch. geol. Ges. I, 232, Berlin.*

RAMMELSBERG K. T. — Ausbruch des Aetna 1805. — *Zeitschr. d. deutsch. geol. Ges. XIII, pp. 606, Berlin, 1866.*

RANZANO P. — De auctore et primordiis urbis Panormi. — (Eruption 1444).—*Opusc. di autori siciliani, IX, pag. I. Palermo, 1747.*

RATH (VOM) G. — Der Aetna in den Jahren 1863-66, nach O. Silvestri's "I fenomeni Vulcanici presentati dall'Etna negli anni, etc.". — *Ubertragen im Neu. Jahrb. f. Min. Scit. 51, Stuttgart, 1870.*

RATH (VOM) G. — Der Aetna, Vortrag gehalten am 21 Mai 1872.— *Verhandl d. naturhistor. Vereins für Rheinl. u. Westf. 1872, in 8°, pp. 49-81.*

RATH (VOM) G. — Referat über Silvestri's. Relazione sulla doppia eruzione del 1879. — *Verhandl. d. niederrhein. Ges. f. Nat. u. Heilkunde. Sitzungsberichte, 1879.*

RECLUS E. — La Sicile et l'éruption de l'Etna en 1805. — *Journ. du Monde. 1ro Sem. pp. 353, Paris, 1866. Bibl. di Viaggi X, (La Sicilia). pp. 53, Milano, 1873.*

RECLUS E. — La terre. — 2. *Vols. Paris, 1877. (Vol. I, pp. 575. Erupt. Etna 1856 with coloured geol. map).*

RECUPERO G. — (1720-78). Discorso storico sopra l' acque vomitate da Mongibéllo e suoi ultimi fuochi avvenuti nel mese del Marzo 1755, recitato nell'Accademia degli Etnei. — *Catania, 1755, in 4°, pp. 79, pl. 1.*

RECUPERO G. — Storia naturale e generale dell' Etna. — *Opera postuma con annotazioni del suo nipote Agatino Recupero. Tom. I-II, Catania, 1815, in fol. Vol. I, pp. XX + 244 + LXIV + 15 pl. 3, portrait. Vol. II, pp. 236 + XII+22, map. 1, pl. 4.*

RE L. DEL. — Relazione di una gita in Catania e all' Etna , durante l'eruzione del dicembre 1842 per eseguirvi alcune magnetiche osservazioni. Memoria. — *Atti d. Accad. d. Sc. di Napoli, 1843, in 4°, pp. 46 (C. A.).*

REYER E. — Beitrag zur Physik des Eruptionen und der eruptiv Gesteine. — *Wien, 1877.*

RIBIZZI. — Eruzione del 1646. — (*cit. by Recupero*).

RICCIARDI L. — Ricerche chimiche sulle lave dei dintorni di Catania indicate nella carta geologica di Sciuto Patti. — *Atti d. Acc. Gioenia. d. Sc. Nat. Ser. 3ª , Vol. XV, Catania , 1881.*

RICCIARDI L. — Sopra un'alterazione superficiale osservata sulla selce piromaca dei dintorni di Vizzini. — *Atti d. Acc. Gioenia, Ser. 3ª, Vol. XV. Catania, 1881.*

RICCIARDI L. — Sulla cenere caduta dall'Etna il giorno 23 Gennajo 1882. — *Gaz. Chim. It., t. XII, 1882, pp. 3*. (C. A.).

RICCIARDI L. — Sulla composizione chimica di diversi strati di una stessa corrente di lava eruttata dall'Etna nel 1669. — *Gaz. Chim. Ital. T. XII, 1882, pp. 6*. (C. A.).

RICCIARDI L. — Composition chimique des diverses couches d'un courant de lave de l'Etna. — *Compt. Rend. Acad. Paris, pp. 3*, (C. A.).

RICCIARDI L. — Sulla cómposizione chimica della cenere lanciata dall'Etna il 16 Novembre 1884. — *Atti Accad. Gioenia Sc. Nat. Catania, Ser. 3ª, Vol. XVIII. pp. 5* (C. A.).

RICCIARDI L. — L'Etna e l'eruzione del mese di marzo 1883. — *Atti d. Acc. Gioenia, Sci. Nat. Catania, 1885, ser. III, Vol. XVIII, pp. 195.*

RICCIARDI L. — Recherches chimiques sur les produits de l'éruption de l'Etna aux mois de mai et de juin 1886. — *Compt. Rend. Ac. Sc. Paris, Vol. CII, pp. 1484-1488.*

RICCIARDI L. — Sull'Eruzione dell'Etna del Maggio-Giugno 1886. — *Chieti, 1886, in 4°, pp. 8* (C. A.).

RICCIARDI L. — Sull' allineamento dei vulcani italiani, etc. — *Reggio-Emilia, 1887, in 8°, pp. 10, col. map, 1.*

RICCIARDI L. E SPECIALE S. — Ricerche chimiche sui Basalti della Sicilia. Nota preliminare. — *Atti d. Acc. Gioenia. Ser. 3ª, Vol. XV, 1881.*

RICCIARDI e SPECIALE. — I basalti della Sicilia. — *Gaz. Chim. It. T. XI, 1881, pp. 34* (C. A.).

RICCIARDI L. E SPECIALE S. — Sui Basalti della Sicilia. Ricerche chimiche. — *Atti d. Acc. Gioenia, Ser. 3ª, Vol. XV, Catania, 1881.*

RICCI G. — Rapporto a S. E. il Ministro della Guerra intorno alla misura di una base nella Piana di Catania. — *With a plate of triangulation). Torino, 1867.*

RICCIOLI B. — Chronologia reformata Seti. — (Eruptions from 1321 to 23). — *Bononiae, 1669.*

RICCO A. — Phénoménes atmosphériques observés à Palerme pendant l' éruption de l' Etna. — *Compt. Rend. Ac. Sc. Paris, Vol. CIII, pp. 419-422.*

RICHARDUS (DE) S. G. — Chronicon Siculum ab anno 1189-1243. — *Carusii. Mem. storiche.*

RIEDESEL (VON) J. H. — Reise durch Sicilien und Gross-Griechenland. Zürich. 1771. — *En français. Lausanne, 1773, in 12°, pp. XII+353.*

Rio (DEL) M. — Disquisitiones magic. 1, 2 quest. 10. — (*Ctf. by Massa, Etna*).

Riolo V. — Delle acque minerali di Sicilia.—*Palermo, 1794, in 8.°*

Riso (DE). — Relazione della pioggia di cenere avvenuta in Calabria ulteriore nel dì 27 Marzo 1809.—*Atti d. Acc. Pontan. Napoli, 1809, pp. 23.*

Ritius M. — De regibus Siciliae usque ad 1497. Scriptore Sic.— *Thes. Sicul. V.*

Ritter C. W. — Beschreibung merkwürdiger Vulkane. — *Breslau, 1847.*

Ritter von Hauer K. — Ueber di Beschaffenheit der lava des Etna von der eruption im Jahre 1852. — *Sitz. K. Ak. d. Wissen zu Wien. Math. Nat. T. XI, 1853, pp. 87-92, pp. 8* (C. A.).

Roberto F. de. — L' eruzione dell' Etna. — *Rivista Mensile del Club Alpino It., 1886, N° 6.*

Rodwell G. F. — Etna, a history of the Mountain and of its eruptions, — *London , 1878 , in 8°, pp. XI + 142 , maps. and pl.*

Romualdi. — Salernitani (Archiep.). — *Chronicon, postrema pars ab anno 1159-77; ex Bib. J. B. Carovi. Carusii. Mem. storiche. (Eruption 1169 A. D.). Panormi, 1723.*

Rosenbusch H. — Referat über die eruption des Aetna 1879. — *Neu. Jahrb. f. Min. Geol. u. Pal. I , Seit. 390. Stuttgart. 1880.*

Rossi (DE) M. S. — Terremoti presso l'Etna dal 7 al 20 Gennaio 1875. — *Bull. d. Vulcan. Ital. An. II, fasc. I , II , III. — Bull. d. R. Comit. Geol. d'Italia. Vol. VI, pag. 113. Roma, 1875.*

Rossi (DE) M. S. — Insegnamento di fisico-chimica terrestre nella R. Università di Catania ed Osservatorio Vulcanologico nell'Etna. Lettera di M. S. de Rossi al Prof. Orazio Silvestri.— *Bull. d. Vulcan. Ital. An. VI, fasc. I-III, p. 5, Roma, 22 Gennaio, 1879.*

Roth J. — Lyell's Abhandlung (siehe diese) übersetzt. —*Zeitschr. d. Deutsch. geol. Ges. Bd. XI, Seit. 149, Berlin, 1859.*

Roth J. — O. Silvestri. Ueber die vulkanischen Phänomene des Aetna in den Jahren 1863-66 , mit besonderer Bezugnahme auf den Ausbruch von 1865. — *Atti Accad. Gioenia, Catania. Ser. 3ª, 1867, t. I, pp. 56-285 und, Zeits. d. D. geol. Gesells. 1869, t. XXI, pp. 221-238 (C. A.).*

Roth J. — Ueber die Ausbrüche des Aetna im Nov. u. Dec. 1868, von Mar. Grassi (übersetzt). — *Zeilschr. d. Deulsch. geol, Ges. Bd. XXII, Seil. 189, Berlin, 1870.*

ROTH J. — Ueber Vesuv u. Actnalaven. — *Zeilschr. d. Deulsch. geol. Ges. Bd. XXV, S. 116. Berlin, 1873.*

ROTH J. — Der Ausbruch des Aetna am 26 Mai (nach) Silvestri Baldacci etc. — *Zeilschr. d. Deulsch. geol. Ges. Bd. XXXI, Seil. 398. Berlin, 1879.* Y

ROZET. — Mém. sur les Volcans de l'Auvergne avec un appendice sur les Volcans d'Italie. — *Paris, 1844.*

RUFFO S. — Istoria dell'orrendo terremoto accaduto in Palermo 1 Settembre 1726. — *Palermo, 1726, in 4.°*

RUSSEGGER J. — Reise in der Levante u. in Europa mit besonderer Berücksichtigung der naturwis. — *Verhältnisse der betreff. Ländern. Stuttgart, 1851. (Reise in Sicilien. Seil. 255-363).*

RUSSO A. — Manoscritto che possiede don Ludovico Toscano di Aci Reale.—*(Cit. by Recupero II,p. 58). Eruptions 1651-53.*

RUSSO (GRASSI) G.—Acqua di Santa Venera—*Aci Reale, 1878.— Traduction français par Ingigliardi. Lyon, 1878.*

RUTLEY F. — The mineral constitution and microscopic characters of some of the lavas of Etna. (*Rodwell's Etna pp. 135, London, 1878.*

SACCO F. — Dizionario geografico del Regno di Sicilia. — *Vol. I-II. Palermo, 1790, in 4,°*

SAINTE-NON. — *See De Saint Non.*

SAITTA L. — Sul miserando caso della esplosione avvenuta addi 25 Nov. 1843. — *See, Luca P. de.*

SALIS-MARSCHLINS (VON) K. U. — Beiträge zur natürlichen u. ökonomischen Kenntniss beider Sicilien. — *2 Bad. Zürich, 1790.*

SANCHEZ G. — La Campania sotterranea, e brevi notizie degli edificii scavati entro Roccia nelle Due Sicilie, ed in altre Regioni. — *Napoli, 1833. Vols. II, in 8. pp. 2-656. See pp. 78 und follow. Etna.*

SARTORIUS VON WALTERTHAUSEN W. — Atlas des Aetna. — *Vollst. in 8°, Lief. u. zugehörigem. Text. Imp. fol. m. 57. Kpfrtfn Weimar, 1848-64*

SARTORIUS VON WALTERSHAUSEN W. — Ueber die vulkan. — *Gesteine in Sicilien. u. Island. und ihre submarine Umbildung. Gottingen, 1853.*

SARTORIUS VON WALTEBSHAUSEN W. — Ueber den Aetna u. seine Ausbrüche. — *Leipzig, 1857, in 8°, pp. 23.*

SARTORIUS VON WALTERSHAUSEN W. UND LASAULX (VON) A. — Der

Aetna. — *II, Band. Leipzig, 1880, in fol. Vol. I, pp. XVIII + 371, map. 1, pl. 14, portrait e figs. Vol, II, pp. VIII+ 540, 1 map. pl. 23 and figs.*

SAUSSURE (DE) H. B. — Voyages. — *Vols IV, Genève, 1779-96, in 4.°*

SAUSSURE (DE) H. — Sur la récente éruption de l'Etna. — *Compl. Rend. d. l' Acad. d. Sc. Vol. LXXXIX (1879). p. 35.* — *Journ. de Genéve, Juin 1878.* — *Abstract in: Bull. d. R. Com. Géol. d'Italia, Vol. X, p. 323. Roma, 1879.*

SAUSSURE H. DE. — L'Etna et ses derniéres éruptions. — *Le Globe, 4e sér. Vol. VII, pp. 211, Genève, 1888.*

SAVA R. — Sull'accidentale arsione umana per l'eruzione dell'Etna 1843. — *Rend. d. R. Acc. d. Sc. di Napoli, N. 12, 1843.*

SAVA R. — Lucubrazioni sulla Flora e Fauna dell' Etna e sopra l'o.igine delle spelonche delle lave di questo Vulcano. — *6° Congr. scient. ital. Milano, 1844, in 8°, pp. 36.*

SAVA R. — Sopra alcuni prodotti minerali che si formano in una spelouca dell'Etna. — *Ann. Civ. d. Due Sicilie. Fasc. LX, Vol. XXX, pp. 89-102.*

SAVERIO C. — De Aetna. — (*Poème cit. by Mongitore*).

SAYVE A. — Voyage en Sicile, fait en 1820 et 1821. — *Bibl. Univ. Paris, 1822, t. XX, pp. 131-158, t. XXI, pp. 128-160.* (C. A.).

SCANELLO C. — Descrizione di Sicilia. — (*Eruption 1536*). (*Cit. by Filoteo and Massa*).

SCASSO M. — 1786. — *See Burigny.*

SCHMIDT J. F. — Vulkanstudien. — *Aetna, 1870, Leipzig, 1878, in 8°, fol. 4, pp. 235, map 1, pl. 7.*

SCHOTT G. (1608-66). — Magia universalis naturae et artis IV, — *Herbipoli, 1657, in 4°.*—*In German, Bamberg, 1671, in 4°.*— *Frankfort a. M. 1677.* (*l. I, cap. X et a., cit. by Massa, Etna*).

SCHOUN J. F. — Observations météorologiques sur le mont Etna.— *Bibl. univer. d. Sc. et Arts de Geneve, XII, p. 153.*

SCHOUN J. F. — L'ultima eruzione dell'Etna 1819. — *Giorn. enciclop. di Napoli, 1819. Goetting. Wochenblatt. 18 t. Woche 1819, pp. 71-75.*

SCIACCA E. — Eruzione dell' Etna del 1669.—*Napoli, 1671, in 8.°* (*Cit. by Gemellaro C. in: Origine e progressi delle scienze naturali in Sicilia, Catania 1833*).

SCIGLIANO A. — Posnona Etnea. — *Atti d. Acc. Gioenia VIII. Catania, 1831.*

SCINÀ D. — (Palermitano trovandosi in Catania. Copia di lettera scritta ad un suo amico 2 Nov. 1811. — (*Eruptions October 1811). Catania, 1811.*

SCIUTO-PATTI C. — Della utilità del Drenaggio in talune terre della Pirna di Catania. — *Atti d. Sòc. Econ. d. Prov. di Catania. 1857.*

SCIUTO-PATTI C. — Sull'età probabile della massa subacrea dell'Etna. — *Atti d. Acc. Gioenia. Ser. 3ª, Vol. I, Catania, 1866, pp. 30.*

SCIUTTO-PATTI C. — Carta geologica della città di Catania e dintorni di essa. — *Con 8 tav. Atti d. Acc. Gioenia. Ser. 3ª, Vol. VII. Catania 1880.* — (*Notice in: Atti d. Soc. Ital. d. Sc. Nat. Milano, Agosto 1869).*

SCIUTTO-PATTI C. — Sulla temperatura del mare nel Golfo di Catania. — *Atti d. Acc. Gioenia. Ser. 3ª, Vol. IV, Catania, 1869.*

SCIUTTO-PATTI C. — Carta idrografica della città di Catania e dintorni, — *Atti d. Acc. Gioenia. Ser. 3ª, Vol, XI. Catania, 1878.*

SCIUTTO-PATTI C. — Sul sito dell'antica città di Symactus. — *Atti d. Acc. Gioenia d. Sc. Nat. Ser. 3ª, Vol. XV, Catania, 1880.*

SCROPE G. P. — Volcanoes. The character of their phenomena.— *London, 1862, in 8°, Chap. Sicily, mount Etna and isole Lipari.*—*Translated from English by Endymion, in 8°, Paris 1864.* — *Uebersetz von G. A. von Klöden, Berlin, 1872.*

SCROPE G. P. — On the mode of formation of volcanoes and craters. — *Quarterly Journ. of the Geol. Soc. 1859. London. 1859.* — *Edit. française avec addition, Paris 1860.* — *Ins Deutsche übertragen v. C. L. Griesbach. Berlin, 1873.*

SCUDERI R. — Sopra i segni meteorologici dell' Etna. — *Atti d. Acc. Gioenia, I, 2. Catania, 1824.*

SECCHI P. A. — Lezioni di fisica terrestre. — *Torino e Roma, 1867.*

SEGUENZA A. G. — Di certe rocce vulcaniche interstratificate fra rocce di sedimento. — *Nota ove è illustrata una serie stratigrafica di Salice nella provincia di Messina nella quale s'incontrano materiali vulcanici delle isole Eolie interstratificati nel pliocene antico. (Astiano). Rend. d. R. Acc. d. Sc. Fis. e Mat. di Napoli. Adunanza del 13 maggio 1876.*

SELVAGGIO M. — Descriptio Montis Aetnaei cum horrendis emanationibus ignium a retro seculis usque da tempora nostra — *Venetiis , 1541, in 12°. It is Chap. XLIII, of: De partibus Mundi etc., Venetiis 1542.*

SELVAGGIO M. — Colloqium trium Peregrinorum. — *(Pp. 143. earthquake 1169). (Cit. by Mongitore and Massa, Etna, ibid. Cronaca Siciliana).*

SENECA. — Quast. natural. — *II, I; epist. 51 et 79 ; de benefic. III, 37, 2; VI, 36, 1.*

SERPETRO N. — Mercato delle Meraviglie, ovvero Istoria Naturale. — *Venezia, 1653.*

SERPETRO N. — Trattato della Geografia dell' Etna. — *(Cit. by Massa, Etna). Not Edited.*

SETO. — Opera chronologica.—*(For the "Chronologia reformata auctore Joanne Baptista Riccioli." See Riccioli).*

SEVASTA F. — Istoria dell'orrendo terremuoto di Sciacca nell'anno 1727; colla relazione di altri terremuoti. — *Palermo, 1729, in 8.⁰*

SICKLER. — Actna. — *Allg. Encyclop. d. W. u. k. 2ᴬ , pp. 123-135, pl. 1, representing the panorama of Etna of Gemmellaro (C. A.).*

SIEGERT. — Panorama des Etna und der umliegenden Gegend. — *Breslau, 1822, in 8°, pp. 8, pl. 1. (C. A.).*

SILIUS ITALICUS. — *Lib. XIV, 59 and follow.*

SILLIMAN B. — An Excursion on Etna. — *Am. Journ. of Sc. 2ⁿᵈ Ser. Vol. XIII, N.° 38, 1852, pp. 175-184 (C. A.).*

SILVESTRI O. — Analisi chimica di un prodotto minerale di un vulcano spento della Toscana. Studiato in paragone a un prodotto' analogo dell' Etna. — *Atti d. Acc. Gioenia, Ser. 2ᵃ, Vol. XIX, Catania, 1864.*

SILVESTRI O. — Sulla eruzione dell' Etna nel 1865. Prima relazione al Prefetto della Provincia.—*Catania, 1865. Giornale della Provincia di Catania.*

SILVESTRI O. — Sopra i terremoti dell' Etna nel 1865. Relazione al Prefetto della Provincia.—*Catania 1865.— Compt. rend. d. l'Acad. d. Sc. Paris, 31 Juillet 1865.*

SILVESTRI O. — Sur l' éruption actuelle de l' Etna. Lettres à M. Sainte-Claire Deville. — *Compt. rend. d. l'Acad. d. Sc. Tom. XLI, Paris, Juill. 1865.*

SILVESTRI O. — Découverte du Vanadium dans les laves l'Etna.— *Journ. de Minér. et Géol. de W. Delesse, Paris, 1866.*

SILVESTRI O. — Le salse e la eruzione fangosa di Paternò in Sicilia incominciata a dì 7 febbraio 1866. Ricerche chimicogeologiche. — *Catania, 1866.*

SILVESTRI O. — Relazione scientifica sugli ultimi fenomeni vulcanici presentati dall' Etna fatta al Congresso della Società italiana di scienze naturali , tenuto alla Spezia nell' autunno

8*

1866. — *Atti d. 2ᵃ Riun. Straord. d. Soc. ital. d. Sc. Nat. Vol. IX, Fasc. I, 1866.*

SILVESTRI O. — Sur une récente éruption boueuse des Salses de Paternò en Sicile. — *Compt. rend. d. l' Acad. d. Sc. Paris, 12 mars 1866.*

SILVESTRI O. — Sui fenomeni eruttivi dell' Etna nel 1865. Studi chimici e geologici. — *Nuovo Cimento, Tom. XXI e XXII. Febbraio-marzo, Pisa, 1866-67.*

SILVESTRI O. — Tremblement de terre de la Sicile en 1866. — *Compt. rend. d. l'Acad. d. Sc. Paris, 1866.*

SILVESTRI O. — I fenomeni vulcanici presentati dall'Etna nel 1863 1864-65-66 in rapporto alla grande eruzione del 1865. Studi di chimica geologica. — *Vol. I, Tav. V, e fotografie. Atti d. Acc. Gioenia, Ser. 3ᵃ, Vol. I, Catania, 1867. — Abstract: Zeitschr. d. Deutsch. geol. Gesellschaft. (Von Prof. J. Roth). Berlin, 1869. — Neues Jahrbuch für Min. Geol. u. Pal. (Von Prof. G. vom Rath in Bonn.) Stuttgart 1866; — Verhandl. der K. K. geol. Reichs. (Von Prof. F. v. Hauer.) N. 15, Wien, November 1868. — Zeitschr. der Deut. geol. Gesell. (Von Prof. A. Heim. Zürich), Berlin, Nov. Dec. 1871. Jan. 1872.*

SILVESTRI O. — Proposta di un' Osservatorio sull' Etna in servizio alla Vulcanologia e Meteorologia. Presentata manoscritta al R. Ministero della Pubblica Istruzione. — *See note al pag. 112 of "I fenomeni vulcanici presentati dall' Etna, etc.".*

SILVESTRI O. — Fenomeni eruttivi Etnei in seguito alla eruzione scoppiata il 27 novembre 1868 dal cratere centrale. Relazione. — *Gazz. d. Provincia d. Catania, N.° 147, Dicembre 1868.*

SILVESTRI O. — Sull'eruzione dell'Etna del 27 novembre 1868 dal cratere centrale. Relazione. — *Catania, 30 novembre 1868, Gazz. Piemontese. N. 342. Torino, 10 dicembre 1868.*

SILVESTRI O. — Processi chimici e di dissociazione studiati nella lava fluente e nei fumajoli a elevatissima temperatura sul cratere centrale dell'Etna nel 1868. Lavoro comunicato al Congresso della Società italiana di Scienze naturali tenuto in Catania nell'agosto 1869. — *Atti d. Soc. d. Sc. Nat. Vol. XII, Fasc. III. Milano, 1870.*

SILVESTRI O. — Osservazioni fatte sull' Etna in compagnia della spedizione scientifica inviata in Sicilia dal Governo inglese in occasione dello Ecclisse totale di Sole del 22 dicembre 1870. — *Atti d. Acc. Gioenia, Ser. 3ᵃ, Vol. VI, Catania, 1871.*

Silvestri O. — Notizie sopra un nuovo minerale dell'Etna. — *Miner. Miltheil. V. G. Tschermak. Heft. I, Seil. 54. Wien 1872.*

Silvestri O. — Sopra due sorgenti di acqua minerale salino-solfurea idrocarburata, detta di S. Venera, alla base orientale dell'Etna. Ricerche chimico-geologiche. — *Atti d. Acc. Gioenia, Ser. 3ª, Vol. VIII, Catania, 1872, pl. 2.*

Silvestri O. — Ambrogio Soldani e le sue opere: (con osservazioni del Soldani sulle sabbie dell'Etna). Discorso fatto nella R. Accademia dei Fisiocritici a Siena in occasione del Congresso della Società italiana di Scienze naturali. — *Atti d. Soc. ital. d. Sc. nat. Vol. X, Fasc. IV, Milano, 1873.*

Silvestri O. — Sulle sorgenti idrogassose di S. Venera al Pozzo. — *Catania, 1873.*

Silvestri O. — Emissione di fumo eruttivo straordinario dal cratere centrale dell'Etna. — *Bull. d. Vulc. Ital. Fasc. II, e III, pp. 44, Roma, 1874.*

Silvestri O. — Fenomeni eruttivi dell'Etna nell'interno del cratere centrale. (Col presagio di una prossima grande eruzione laterale). — *Bull. d. R. Com. geol. d'Italia, Pag. 244, An. VI, Roma, 1874. — Bull. d. Vulc. Ital. Fasc. VI, e VII, pp. 73, Roma, 1874.*

Silvestri O. — Eruption dans l'intérieur du cratère central de l'Etna. — *Revue Savoisienne 15 Sept. (Abstract by M. Bollshauser).*

Silvestri O. — Sulla eruzione laterale dell'Etna scoppiata il 29 Agosto 1874. Relazione. — *Bull. d. R. Com. geol. d'Italia, Ann. V, pag. 244, Roma, 1874. — Bol. d. Vulcan. Ital. Fasc. IX e X, pp. 105, Roma, 1874.*

Silvestri O. — Terremoti presso l'Etna e conati eruttivi del medesimo dal 7 al 20 gennaio 1875. — *Boll. d. Vulcan. Ital. Fasc. I, II e III, pp. 19, Roma, 1875.*

Silvestri O. — La scombinazione chimica (dissociazione) applicata alla interpretazione di alcuni fenomeni vulcanici. Sintesi e l analisi di un nuovo composto minerale dell'Etna e di origine comune nei vulcani. — *Gazz. Chim. ital. Tom. V, Palermo, 1875. — Journ. of the Ch. Soc. N. 158, Frebruary 1876. — Atti d. Acc. Gioe. Ser. 3ª, Vol. X, Catania, 1876.*

Silvestri O. — Das Vorkommen des Stickstoffeisens unter den Fumarolen-Produkten des Actna, und künstliche Darstellung dieser Verbindung. — *Uebersetz aus den Atti d'Acc. Gioenia, durch G. vom Rath. — Pogged. Ann. 1876.*

Silvestri O. — Sopra due grandi perdite che ha fatto la vulca-

nologia. Cenni sulla vita scientifica e sulle opere di C. Sainte-
Claire Deville e W· Sartorius v. Waltershausen. — *Boll. d.
Vulcan. ital. Fasc. XI e XII, pp. 179, Roma, 1876.*

SILVESTRI O. — La scienza della terra. Discorso d'inaugurazione
al corso di Chimico-fisica terrestre, Mineralogia e Geologia
nella R. Università di Catania nell'anno 1877. — *(After the
foundation of the chair of terrestrial physics and chemistry
with special application to the vulcanology of Etna). Catania, 1877.*

SILVESTRI O. — Sopra alcune paraffine ed altri carburi d'idrogene
omologhi che trovansi contenuti in una lava dell'Etna. —
*Atti d. Acc. Gioenia, Ser. 3ª, Vol, XII, Catánia, 1876. —
Vorträge und Mittheilungen von G. vom Rath, Bonn, 1877.—
Sistsungsb. der Niederrhein. Gesellsch. f. Natur. und Heilk-
unde in Bonn. Sitz. d. 18 febr. 1877.*

SILVESTRI O. — Importante eruzione di fango comparsa a Paternò
nelle adiacenze dell'Etna ai primi dicembre 1878. — *Boll. d.
Vulcan. ital. Vol. V, pag. 131, Roma, 1878.*

SILVESTRI O. — I Terremoti di Mineo, in Provincia di Catania,
dell'ottobre e novembre 1878 accompagnati da singolari fenomeni di rombi. Relazione presentata al Prefetto della Provincia (in commissione col Prof. Boltshauser. — *Catania, novembre, 1878.*

SILVESTRI O. — Cronaca dei fenomeni etnei del 1878-1879. Osser-
vazioni meteoriche fatte nelle stazioni presso le Alpi e gli
Apennini, pubblicate per cura del Club Alpino Italiano. —
Ann. VIII, Torino, 1878-79.

SILVESTRI O. — Andamento della eruzione fangosa di Paternò nelle
adiacenze dell'Etna in data del 14 gennaio 1879. — *Boll. d.
Vulcan. ital. pag. 30, An. VI, Roma, 1879.*

SILVESTRI O. — Atlante di grandi fotografie sulla eruzione ed effetti dei terremoti dell'Etna nel 1879. — *Diretto e pubblicato
per cura del R. Governo Italiano, Catania, 1879.*

SILVESTRI O. — Continuazione della eruzione fangosa a Paternò
nelle adjacenze dell'Etna e sua fase in data del 20 dicembre
1878.— *Boll. d. Vulcan. ital. An. VI, pp. 28, Roma, 1879.*

SILVESTRI O. — Fenomeni dell'Etna successivi all'ultima eruzione
del maggio-giugno 1879.— *Boll. d. Vulcan. ital. An. VI, pp.
118, Roma, 1879.*

SILVESTRI O. — Il nuovo monte Umberto-Margherita comparso in
5 giorni sull'Etna durante la eruzione del maggio-giugno
1879.— *Illustr. Ital. di Trèves, Sem. 2ª, pp. 309, con incisioni, Milano, 1879.*

SILVESTRI O. — La doppia eruzione dell'Etna scoppiata il 20 maggio 1879. Relazione ai Ministri di Istruzione pubblica, Agricoltura, Industria e Commercio, pubblicata il 30 maggio 1870, con una carta topogr. — *Catania, 1879.* — *Boll. d. Vulcan. ital. 1879 An. VI, Fasc. IV-VII, pag. 67, Roma, 1879.* — *Sitzzungs-Berichlder Niederrg. Gesell. für Nat. und Heilkunde zu Bonn, 1879.* — *Vorlräge und Mitth. von G. vom Rath, 1880.*

SILVESTRI O. — L'attuale eruzione di fango, termale, salato, petrolifero dell'Etna presso Paternò. — *Illustrazione Ital. di Trèves N. 8, Descrizione con figura 23 febbraio, Milano, 1879.*

SILVESTRI O. — Sulla doppia eruzione e i terremoti dell'Etna nel 1879 2ª ediz., ampliata del primo rapporto presentato al R. Governo. — *Catania, 1879, p! 1.* — *Abstract: Boll. d. R. Com. geol. d' Italia, Vo!. X, pp. 590, Roma, 1879,* — *Sitsber. der Niederrh. Gesell. für Nat. und Heilk. zu Bonn, 1879.* — *Vorlräge und Mitth. von G. vom Rath, 1880.*

SILVESTRI O. — Sulla eruzione dell'Etna del 1879. — Tre incisioni sopra disegni originali fatti sul teatro eruttivo e che rappresentano : 1° Un gruppo di bocche eruttive formatosi a 1950 metri di altitudine sul livello del mare alla base del Monte Nero. — 2° Eruzione dell'Etna sul fianco nord-nord-est osservato da Randazzo la notte del 28 maggio a ore 3 ant. — 3.° Fenditure ed avvallamenti di suolo che fanno capo alle bocche eruttive situate tra il Monte Timparossa e il Monte Nero. — *Illustr. Ital. Sem. 2°, pp. 5, Milano, 1879.*

SILVESTRI O. — Sulla eruzione di fango a Paternò nelle adiacenze dell'Etna dal suo principio fino alla data del 25 maggio 1879. — *Relazioni al Giorn. d. Sicilia, N. 304, 25 dicembre 1878.* — *N. 314, 25 dicembre 1878.* — *N. 18, 15 gennaio 1879.* — *N. 56, 8 aprile 1879.* — *N. 146, 20 maggio 1879.*

SILVESTRI O. — Un viaggio all'Etna. Vol. I, con la descrizione storica, topografica, geologica, altimetrica e pittoresca del grande vulcano, con una carta topografica e un' appendice con le norme e tariffe per i viaggiatori all'Etna, stabilite dalla Sezione catanese del Club Alpino Italiano. — *Torino-Roma-Firenze, 1879, in 16°, map 1.*

SILVESTRI O. — Continuazione della eruzione fangosa di Paternò e sulle condizioni attuali dell'Etna. (25 dicembre 1879). Lettera diretta da O. Silvestri al Prof. Luigi Palmieri, direttore dell'Osservatorio Vesuviano. — *Boll. d. Vulc. ital. An. VII, pp. 9, 1880.*

SILVESTRI O. — Fenomeni vulcanici dell' Etna avvenuti dal gennaio a tutto aprile 1880. — *Boll. d. Vulcan. ital. An. VII, pp. 80-83. Roma, 1880.*

SILVESTRI O. — Fenomeni vulcanici dell'Etna nel maggio e giugno 1880. — *Boll. d. Vulcan ital. Vol. VII, p. 86. Roma, 1880.*

SILVESTRI O. — Programma per il XIII congresso degli Alpinisti italiani da tenersi a Catania il 15 settembre 1880 con ascensione all'Etna. — *Catania, 1880, Boll. d. Club Alp. ital. Torino, 1880.*

SILVESTRI O. — Sullo sfeno trovato per la prima volta tra i prodotti minerali dell'Etna. — *Riv. scient. e indusir. di G. Vimercati, N. 12, Firenze, giugno 1880.*

SILVESTRI O. — Cronaca della eruzione di fango a Paternò e dei fenomeni vulcanici generali dell' Etna durante l'anno 1880. An. IX, 1879-80. — *Osservazioni meteorologiche fatte nelle stazioni presso le Alpi e gli Appennini e pubblicate per cura del Club Alpino Italiano, Torino, 1880. — Boll. d. Vulcan. ital. Fasc. III. Roma, 1881.*

SILVESTRI O. — Continuazione del periodo eruttivo (con eruzione di fango) presso Paternò e cronaca dei fenomeni vulcanici generali dell'Etna durante l'anno 1881. — *Boll. dec. dell'assoc. meteorol. Ital. An. X. Torino, 1880-81. — Boll. d. Vulcan. ital. Roma, 1881.*

SILVESTRI O. — I fenomeni vulcanici presentati dall'Etna dal 1866 al 1881. Studi di geologia chimica, (Sequel to « I fenomeni presentati dall'Etna 1863 al 1866.» which appeared in 1867). — *Accademia Gioenia di scienze naturali, seduta del 20 marzo 1881, Vol. XVII.*

SILVESTRI O. — Nota preliminare sopra un lavoro in corso di esecuzione riguardante la Petrografia e Mineralogia micrografica delle roccie eruttive dell'Etna e in generale della Sicilia. — *Boll. d. R. Com. geol. d'Italia. Roma 1881.*

SILVESTRI O. — Progetto di una rete sismica estesa dal centro alla periferia dell' Etna con a capo l'Osservatorio centrale a 3000 metri di elevazione e l'Istituto vulcanologico di Catania a 10 metri sopra il mare. Presentato al R. Ministero di Agricoltura, Industria e Commercio nel 1881. — *Giorn. d. Sicilia. N. 137, Palermo, 20 maggio 1881.*

SILVESTRI O. — Ricerche chimiche sulla composizione dell' acqua minerale acidulo-alcalina, magnesiaco-ferruginosa (conosciuta col nome volgare di acqua grassa) delle sorgenti idrogassosa di Paternò alla base occidentale dell' Etna, 3ª memoria per servire ad un'opera completa di Idrologia generale dell'Etna sotto il punto di vista della chimica geologica, -- *Atti d. Acc. Gioenia, Vol. XVI, pp. 89, pl. 2, map. 1, table 1, Catania, 1881.*

SILVESTRI O. — Sopra una singolare lava Basaltica di Paternò nelle adiacenze dell'Etna con piccole geodi ripiene di paraffina cristallizzata. (In appendice alla Memoria « Sopra alcune paraffine ed altri carburi, etc. »). — *Boll. d. R. Com. geol. d'Italia, Vol. II, Roma, 1881.*

SILVESTRI O. -- Album fotografico di 12 fotografie che riproducono i fatti più caratteristici della eruzione suddetta. — *Catania, 1883.*

SILVESTRI O. — Sulla eruzione dell'Etna scoppiata il 22 marzo 1883. Rapporto al R. Governo. — *Catania, 1883, map 1.*

SILVESTRI O. — Sopra una particolare specie di quartzite semivetrosa, contenuta nell'interno di alcune bombe projettate dall'Etna nella éruzione del 22 marzo 1883.—*Atti Acc. Gioenia, Ser. 3, Vol. XVII, Catania, 1884.*

SILVESTRI O. — Sulla esplosione eccentrica dell'Etna avvenuta il 22 marzo 1883 e sul contemporaneo parossismo geodinamico-eruttivo. — *Catania, 1884 in 4°, pp. 195, pl. VI, map 1.* (C. A.).

SILVESTRI O. — Fenomeni Etnei. — *Boll. Soc. Meteor. It., Vol. V, Torino, 1885.*

SILVESTRI O. — I terremoti di Nicolosi avvenuti nel settembre e ottobre 1885. — *Boll. Soc. Meteor. It. Vol. V, Torino, 1885.*

SILVESTRI O. — Sulla esplosione Etnea del 22 marzo 1883, in relazione ai fenomeni vulcanici presentati dall'Etna durante il quadriennio compreso dal genn. 1880 al dec. 1883. — *Atti Acc. Gioenia Sc. Nat. Catania, 1885, ser. III, Vol. XVIII, pp. 237.*

SILVESTRI O. — Der letzte Ausbruch des Aetna. Brief des Prof. O. Silvestri in Catania au Prof. E. Suess in. Wien. — *Neue Freie Presse, Wien, 10 juli, 1886.*

SILVESTRI O. — Fenomeni geodinamici e vulcani osservati nella regione dell'Etna e nel rimanente del suolo Siciliano durante l'anno 1885. — *Annuario Soc. Meteor. It. Torino, 1886.*

SILVESTRI O, — La recente eruzione e i danneggiati dell'Etna. — *Nuova Antologia, fasc. XIII, Roma, 1886.*

SILVESTRI O. — Observations sur les phénomènes éruptifs de l'Etna depuis le 18 mai jusqu'au 7 juin 1886. — *Bull. Soc. Sc. Flammarion, Marseille, 1886. 2me An. p. 97.*

SILVESTRI O. — Sulle acque che circolano e scaturiscono nella

regione dell'Etna. Ricerche di chimica-geologica. Monografia
IV. Acqua potabile detta Reitana presso Acireale. — *Atti
Accad. Gioenia, Ser. 3, Vol. XIX, Catania, 1886.*

SILVESTRI O. — Sulle eruzioni centrale ed eccentrica dell'Etna del
maggio-giugno, 1°. Rapporto al R. Governo. — *Catania, 22
maggio 1886, with map.*— IDEM. — 2. Rapporto, 12 giugno
1886. 2ª Edizione dei due detti Rapporti. — *Catania.*

SILVESTRI O. — Sunti di fatti più rimarchevoli dell'eruzione del-
l'Etna del maggio-giugno 1886. — *Boll. Soc. Geograf. It. ser.
II, Vol. XI, Roma, 1886. Also in Annuar. Meteor. It. 1886.*

SILVESTRI O. —Sur l'éruption de l'Etna de mai et juin 1886. Lettre
à M. Daubrée. — *Compl. Rend. Ac. Sc. Paris, 1886. Vol.
CII, pp. 1589-1592.*

SILVESTRI O. — Etna, Sicilia ed isole vulcaniche adiacenti, sotto
il punto di vista dei fenomeni eruttivi e geodinamici presentati durante l'anno 1888. — *Atti Acc. Gioenia, Sc. Nat. S.
IV, Vol. I, Catania, 1888.*

SILVESTRI O. — Etna, Sicilia nel 1887 sotto il punto di vista dei
fenomeni eruttivi e geodinamici — *Annuario Met. It. Anno
III, Torino, 1888.*

SILVESTRI O. — La recente eruzione dell'Etna. — *Firenze?*

SILVESTRI O. BLASERNA P. E GEMELLARO G. G. — Sulla eruzione
dell' Etna del 26 maggio 1879 e successivi terremoti. Relazione della Commissione nominata dai Ministri di Agricoltura
Industria e Commercio e della Pubblica Istruzione per lo studio della eruzione dell' Etna del 26 maggio 1879; pubblica-
zione fatta dal R. Governo. — *Roma, luglio 1879. — See
also: Boll. d. R. Com. geol. d'Italia, Vol. X, pag. 309. Roma, 1879.*

SIMON L. — A Tour in Italy and Sicily. — *London, 1828. (Eruption of Etna). — Neu Jahrb. f. Min. Geogn. u. Geol. pp.
358, 1833.*

SINCELLO G. -- Chorographia. — *Tip. Reg. pag. 257, Paris, 1652,
(Eruption of Etna).*

SIRUGO. — 1835-1837-1838. — *See Interlandi.*

SMYTH W. H. — Carta generale dell' isola di Sicilia, compilata,
disegnata ed incisa nell' uffleio topografico di Napoli sui mi-
gliori materiali esistenti e sulle recenti operazioni fatte dal
Cav. G. E. Smyth. — *Napoli, 1814.*

SMYTH W. H. — A descriptive memoir of the resources, inhabi-
tants and hydrography of Sicily. — *London, 1824, in 4.° (with
plate).*

SOLDANI A. —Testaccographia ac Zoophytographia parva et microscopica. — *Tom. I, and II, in fol. Senis, 1789-98. (Tom. II, Cap. II, Volcanic sand of Etna.)*

SOLINUS J. — Polyhistoria. — *Cap. XI, und Collectanea Rerum memorabilium, recog. Th. Mommsen, pag. 54.*

SOMMA AGAT. (DI) — Historico racconto dei Terremoti della Calabria dell'anno 1638-41. — *Napoli, 1641.*

SOMMA ANT. — Osservazioni vulcanologiche sulle fenditure esistenti in Mascalucia volgarmente chiamate Cavòli. — *Atti d. Acc. Gioenia, Ser. 1ª, Vol. XVI, 1839.*

SOMMA ANT. — Sul luogo e tempo in cui avvenne l'eruzione dell'Etna appellata dei Fratelli Pii. — *Atti d. Acc. Gioenia, XX, pp. 59, Catania, 1843.*

SOMMA ANT. — Sopra le stratificazioni alluviali del Fasano. — *Catania, 1845.*

SPALLANZANI L. — Viaggi alle due Sicilie e in alcune parti dell'Appennino. — *Vol. I-IV, Pavia, 1792. In German Bd. I-VIII. Leipzig, 1794-96.*

SPALLANZANI L. — Travels in the two Sicilies and some parts of the Apennines. — *Translated from the Original Italian, 4. vols. with 11 plates, London, 1798.*

SPECIALE S. — All'Etna! Escursione del 6 agosto 1876. Relazione. — *Musumeci-Papale, Catania, 1876.*

SPECIALE S. — 1881. — *See Ricciardi.*

SPECIALIS N. — Rerum Syculorum Libri octo (1282-1337. — *Thes. Sic. Vol. V.*

SQUILLACI P. — Progressi portentosi dell'incendio di Mongibello. — *Catania, 1669, in 8°, fol. 8.*

STAMPINATO B. — Osservazione sui tremuoti in occasione del tremuoto che scosse orribilmente la città di Catania la sera del 20 febbrajo 1818. — *Catania, 1818, in 4°, pp. 64, pl. 1 (C. A.).*

STOBAEUS. — Flor. 70,38. — *(Cit. by Aelian).*

STOPPANI A. — Corso di Geologia. — *Vol. I-III. — Bernardoni e Brigola. Milano, 1873. (Etna. Vol. I. § 585, 596, 602, 603, 611, 617, 618, 650, 651, 683; Vol. II, § 599; Vol. III. § 137.*

STOPPANI A. — Sull' opuscolo " Esperimenti vulcanici del Prof. Gorini di Arturo Issel." Nota. — *Rend. d. R. Ist. Lomb. Ser. II, Vol. VI, Fasc. VIII, Milano, 1873.*

STRABO. — Geogr. 6, 2; De natura rerum. — *Lib. VI.*

SUESS F. — Die Erdbeben des südlichen Italien. — *Denkschr. d. K. Akad. d. Wissensch. Wien, 1874. — Abstract: Boll. d. R. Com. geol. d'Italia, Fasc. Aprile 1875. — Boll. d. Vul. ital. Fasc. I, II e III, pag. 42. Roma, 1875.*

SURITA. — Annales Rerum Aragonensium. — *I, III, c. 86.*

SVETONIUS C. — Caligula c, 51. — (*Eruption 38-40 A. D.*)

SWINBURNE. — Travels in the Two Sicilies. — *Vol. IV.* (*Ascension and description of Etna*). — *London, 1795.* — *German by I. R. Forster, II. Th. Hamburg, 1785.*

TARCAGNOTTA. — Istoria del Mondo. (Earthquake of 1169).—(*Cit. by Mongitore*).

TEDESCHI (DI) E. V. — Eruption boueuse à la base de l' Etna. — *La Nature. Fasc. I. Paris, Mai, 1880*

TEDESCHI V.—A propos des recherches chimiques faites par M. M. Ricciardi et Speciale sur les laves des environs de Catane et sur les basaltes de la Sicile. — *Acad. des Sciences de Paris, 1882, in 4°, pp. 7* (C. A.).

TEDESCHI V. — La récente éruption de l'Etna (22 mars 1883). — *La Nature (1883) I, 305-306.*

TEDESCHI E PATERNÒ T. — Breve ragguaglio dell'incendio di Mongibello, avvenuto in quest'anno 1669, con tre piante: una di Catania antica in tempo della gentilità, altra della medesima prima degli incendii e la terza dell'istessa già deformata dal fuoco. — *Napoli, 1669, in 4°, pp. 70, pl. 3, portrait.*

TENORE M. — Ragguagli di alcune peregrinazioni effettuate in diversi luoghi delle provincie di Napoli, e di Terra di Lavoro 1832. — *?, in 8°.*

THUCYDIDES. — III. 116. — *Eruptions 475, 425* (B. C.).

TOLOMEO (DI LUCCA). — Bibl. Part. Tom. XXV. (*Earthquake of 1669*). (*Cit. by Mongitore*).

TORNABENE PAD. — Lettera sull'attuale eruzione dell'Etna di Nov. 1843, etc. — *Rend. R. Accad. Sc. Fis. Mat. Napoli, 1843-1846, T. II, pp. 441-447.*

TORNABENE F.—Sulla Eruzione presente dell'Etna 1.ᵃ e 2.ᵃ parte.— *Rend. d. R. Acc. d. Sc. di Napoli, 1852, pp. 113-120, 146-154.*

TORNABENE F. — Flora fossile Etnea. — *Atti d. Acc. Gioenia. Ser. 2ᵃ, Vol, XVI, Catania, 1859.*

TORNARENE F. — Come si rendano coltivabili le lave dell'Etna.— *Rend. d. Acc. d. Sc. di Napoli, 1864.*

TORNABENE F.—Sull'arginazione del Simeto.— *Giorn. di Agricolt. Ind. e Comm. del Regno d'Italia, 30 novembre 1864.*

TORNABENE F. — Condizioni della Provincia di Catania in rapporto alle acque potabili. — *Giorn. di Agricolt. Ind. e Comm. d. Regno d'Italia, 31 ottobre, 1865.*

TRAVAGLIA R. — 1870. — *See Baldacci.*

TRITHEMIUS. — Chronologia monasteriorum. — (*Cit. by Massa, Etna, and by Mongitore*).

UGHELLI. — Cronica pisana. (Earthquake of 1169.) — *Cit. by Mongitore*).

VAGLIASINDI P. DI RANDAZZO.—Memoria sull'eruzione accaduta nella pioggia accidentale dell'Etna al primo novembre 1832. — *Palermo, 1833.*

VALERIUS Flaccus. — Argon. — *lib. II, 24, 33.*

VALGUARNERA. — Origine di Palermo. — (*Cit. by Massa, Etna*).

VARENIUS B.—Geographia generalis, in qua affectiones generales telluris explicatur. (Lib. I, cap. X, sur l'ile de Vulcano et sur l'Etna). Elzev. — *Amstelodami, 1664.*

VECCHI (DE) E. — Notizia su di alcune altitudine determinati geodeticamente nella regione dell' Etna. — *G. G. Casson, Torino, 1866.*

VENTIMIGLIA D. C. — Pianta del monte Etna. — (*Cit. by Massa, Etna*).

VETRANI A. — Sebethi vindiciae, sive dissertatio de Sebethi antiquitate, nomine, fama, culto, origine, prisca magnetudine, decremento, atque alveis, adversus Jacobus Martorellium. — *Neapoli, 1767, in 8°, pp. 8 + 213, pl. II..*

VIGO L. — La eruzione etnea del 1852. Testimonianza. — *Atti dell'An. d. Sc. e Lett. di Palermo, N. S., t. II, 1853, in 4°, pp. 28* (C. A.).

VIGO L. — Poesie e prose. — *Palermo, 1823, in 4°, pp. V. + 171+3.*

VIOTTI G. — Cenni sulla eruzione del Gennaro 1865. — *Gemmellaro Carlo, Ragguaglio etc. pag. 13.*

VIRGILIUS MARO P. — Aeneis. — *III, 571.*

VIRGILIUS MARO P. — Georgica. — *I, 471.*

WALKIDI. — (*Cit. by Amari*).

WAGLER P. R. — De Aetna poemata questiones criticae. — *Berolini, 1884, in 8°, pp. 107* (C. A.).

WHITE J. — Eruption of Mount Etna. — *Nature , Vol. XXXIV, pp. 82, 108.*

WENTHERN POLYCARPUS. — Brieg der Elementen , wider das baimmerus-würdige Sicilia ader Beschereibung des erschruklichen Bebeus und Erschüt,tern der Erder gransamen Ubelauffs und Sturn des meers auch höchstensselichen Toben und siedenden Wültur des Jener-auspeinden Bergs Aetna.—?, *1693, in 4°, pp. 52* (C. A.).

WINCHELSEA (EARL OF.) — A true and exact relation of the late prodigious earthquake and eruption of mount Etna or Mongibello , as it came in a letter written to his Majesty from Naples. Together with a more particular narrative of the

same , as it is collected out of several relations sent from
Catania. — *Published by Authority. Printed by Newcomb in
the Savoy, 1669, in 4°, pp. 30, with a sketch of the eruption*).
ZURCHER ET MARGALLE. — Volcans et tremblements de terre (Etna
et Stromboli). — *Paris, 1866.*

ESUVIUS

ABATI A. — Il Forno, Poesia heroica burlesca e latina sopra il Monte Vesuvio etc. -- *Napoli, 1631, in 8°, (B. N.).*

ABBATI B. — Epitome meteorologica di tremuoti con la cronologia di tutti quelli che sono accorsi in Roma dalla creatione del mondo sino agli ultimi successi sotto il pontificato del regnante pontefice Clemente XI il dì 14 Gennaro giorno di Domenica su le due della notte meno un quarto, e 2 di Febbraio del corrente anno 1703. — ? (C. A.).

ABICH H. — Sur la formation de l'Hydrochlorate d'Ammoniaque à la suite des éruptions volcaniques et en particulier de celle du Vésuve en 1834. — *Bull. d. l. Soc. géolog. d. France, 1.re Sér. Tom. III. Paris, 1835.*

ABICH H. — Sur les phénomènes volcaniques du Vésuve et de l'Etna. — *Bull. d. l. Soc. géol. de France, 1.re Sér. Tom. III. Paris, 1835.*

ABICH H. — Vues illustratives de quelques phénomènes géologiques, prises sur le Vésuve et l'Etna pendant les années 1833-34. — *Paris, 1836 pp. 8, pl. 10.*

ABICH H. — Erläuternde Abbildungen der geologischen Erscheinungen am Vesur und Aetna in den Jahren 1833 u. 1834. — *Berlin, 1837.*

ABICH H. — Ueber Lichterscheinungen auf dem Kraterplateau des
Vesuv im Juli 1857. — *Zeits. d. Deutsch. geol. Gesell. Berlin, 1857.*

ABICH H. — Ueber die Erscheinung brennenden Gases im Krater des Vesuv im Juli 1857, und die periodischen Veränderungen; welche derselbe erleidet (1857). — *Bull. d, l, Class.
Phys. Math. d. l. Acad. Imp. d. Sc. d. S.ᵗ Petersburg, Vol.
XVI, 1858.*

ACCADEMIA DI NAPOLI. — Istoria dell' incendio del Vesuvio accaduto nel maggio 1738, 2.ᵃ ediz. — *Napoli, 1740.*

ACCADEMIA PONTANIANA. — Relazione intorno all'incendio del Vesuvio cominciato il dì 9 dicembre 1861. — *Napoli, 1862, pp.
36, 3 plates.* (C. A.).

ACCADEMICO IMANTO.—Incendio del Vesuvio, pubblicato per cura di
Vincenzo Bone.—*Napoli, 1632, in small 8°, fol. X.* (C. A.).

ACERBI F. — De Vusuviano incendio anno 1631. In « Polypodium
Apollineum » — *Napoli, 1674, in 8.°*

ADAMI P. — Napoli liberata dalle stragi del Vesuvio. — *Napoli,
1633, in 8,° (O. V.).*

ADAMO F. M. D' — L'avampante ed avampato Vesuvio, in ottava
rima. — *Napoli, 1632, in 12°, fol. XII. (C. A.).*

AFELTRO O. DE — De Monte Vesuvio ac ejus eruptione. — *M. S.
in the Biblioteca Brancacciana. Copy in* (C. A.).

AGNELLO DI SANTA MARIA. — Trattato scientifico delle cause che
concorsero al fuoco e terremoto del Monte Vesuvio. — *Napoli, 1632, in 8.°, pp. 100.* (C. A.).

AGRESTI A. — Pochi versi sulla Torre del Greco nel 1861. —*Napoli, 1862, in 8.°, pp. 12.* (C. A.).

AGRESTI G. D. — il Monte Vesuvio (Song) In Vol. delle Rime d'illustri ingegni napoletani.—*Venezia, 1633, in 12.°, pp. 37-48.*
(C. A.).

ALBINUS F. — Dialogus de Vesuvij incendio.—*V. Falcone.* (C. A.).

ALEXANDER C. — Practical remarks on the lavas of Vesuvius,
Etna, and the Lipari Islands. — *Proceed. Scient. Soc. of London. Vol. I. London, 1839.*

ALOIA. — Eruzione del Vesuvio nella notte degli 8 agosto 1779.—
One plate. (C. A.).

ALVINO F. — Il Vesuvio. Cenno brevissimo sugli antichi suoi nomi,
sue dimensioni, istorie di tutte l'eruzioni, ragioni fisiche di
tal fenomeno, ed uno sguardo sul cratere (eruzione del 1794).—
Napoli, 1841, in 8.°. pp. 18, with a coloured figure. (C. A.).

ALZARIO DELLA CROCE V. — Vesuvius ardens idest motum et incendium. — Vesuvii Montes in Campania XVI. Mensis De-

cemb. Anno 1631. — *Romae, 1632, in 4.°, fol. IV, pp. 318, fol. 1.*

AMATO P. G. (D') — Giudizio filosofico intorno ai fenomeni del Vesuvio. — *Napoli, 1755, in 4.°, pp. 38.*

AMATO P. G. (D') — Divisamento critico sulle correnti opinioni intorno ai fenomeni del Vesuvio e degli altri vulcani. — *Napoli, 1756, in 8.°, pp. 90, pl. 1.*

AMITRANO A. — Encomium sacri sanguinis gloriosi martyris et pontificis Januarii. — *Neapoli, 1632, in 8°, fol. IV, (O. V.).*

AMODIO G. — Breve trattato del terremoto, scritto in occasione dell' incendio successo nel monte Vesuvio nel giorno 16 dicembre 1631 etc., — *Napoli, 1632, in 8.°, pp. 60, fig. 1.*

ANCORA G. (D') — Prospetto storico fisico degli scavi d'Ercolano e di Pompei e dell'antico e presente stato del Vesuvio. *Napoli, 1803, in 8.°, pp. 137, pl. II.*

ANDERSON. — Volcanic vapours of Mount Vesuvius. — *Proceed. of the Phil. Soc. of Glasgow, Vol. III, N. 2, 1872-73.*

ANDERSON T. — The Volcanoes of the two Sicilies. — *Geol. Mag., Dec. III, Vol. V. p. 473.*

ANDOSILLA LARRAMEN DI JUAN. — A Vesuvio. A sonnet. — *See Quiñones. (A. C.).*

ANDRINI. — La grande éruption du Vésuve, Naples 17 Décemb.— *Press. Scient. des Deux Mondes, N. 2, 1862, T. 1.ʳ , Paris, in 8.°, pp. 114-119. (C. A.).*

A. N. M. — Un Papiro, ossia i gladiatori nella caverna del Vesuvio. — *Venezia, 1826, in 4.°, pp. 197. (C. A.).*

ANNA A. (D') — Eruzione Vesuviane dal 1779 al 1794 — *Napoli Pl. 1 in fol. (O. V.).*

ANONYMOUS. — Descripcion del Monte Vesuvio, y relacion del incendio, y terremotos que empezaron a 16 di diziembre 1631—?, *in small fol. pp. 7. (C. A.).*

ANONYMOUS. — Nackte Beschrijvinge van de... Aerdbevinge ende Brandt van den bergh Soma... twee miglen van Napels... 15-17 December 1631.— *Leyden, 1632, in 4.°, pp. 8. (C. A.).*

ANONYMOUS. — Relacion del incendio de la montaña de Soma—(?) *1631, in small fol. pp. 8.*

ANONYMOUS. — Vedute della Eruzione del Vesuvio del 1631. — *Amsterdam, Fol. 1. (C. A.).*

ANONYMOUS. — Vero ritratto dell'incendio nella Montagna di Somma, altrimenti detto Mons Vesuvi, distante da Napoli 6 miglia, successo alli 16 decembre 1631. — *Loose sheet? with explanation, 1631. (C. A.)*

ANONYMOUS. — Avvisi e notizie sull'Eruzione del Vesuvio del 1631, provenienti da Roma e da Napoli dal dì 27 dicembre 1631 al dì 21 febbraio 1632. — *In the Cancelleria Ducale Estense.*

ANONYMOUS. — Copia eines Schreibens aufz Neapolis darinnen berichtet werden etliche Erschreckliche Wunderzeitungen welche sich imend desz nechstabgelauffennen 1631.—*Neapolis? 1632, in 4°. fol. 4.* (C. A.).

ANONYMOUS, — Devotione per il terremoto. — *Napoli, 1632, in 8.°* (*loose sheet*).

ANONYMOUS.—Discours von dem brennenden Berg Vesuvio, oder Monte di Somma etc. — *Loc.? 1632, fol. VII.* (C. A.)

ANONYMOUS. — Novissima relatione dell'incendio successo nel Monte di Somma a dì 16 Decembre 1631, con un avviso di quello successo nell'istesso dì nella Città di Cattaro nelli parti d'Albania.—*Venetia. Reprinted in Napoli, 1632, in 8.°, p. 16.* (C.A.).

ANONYMOUS. — Vesuviani incendii elogium. — *Napoli, 1632, in fol. 2.* (C. A.)

ANONYMOUS.—Extrait d' une lettre ècrite de Naples à l'auteur du journal (des Sçavans) touchant l' embrasement du Mont Vésuve, arrivé au commencement du mois de Janvier dernier 1682. — *" Journal des Sçavans ", 1683, in 12.°, pp. 61-62.* (C. A.).

ANONYMOUS. — Relazione dell'incendio del Vesuvio seguito l'anno 1682 delli 14 di Agosto fino alli 26 del medesimo. — *Roma, 1682, in 4°, fol. 2, figured.* (C. A.).

ANONYMOUS. — Feureyferige zorn die Ruthe Gottes auff dem Brennenden Berg Vesuvio in Campania über Italien und alle Jündtliche Königreiche weit und breit aussgéstrecket: nach ihren Eigenschafften etc. etc. — (*? loc.) 1633, in 4° (fol. 28).*

ANONYMOUS. — Distinta relatione dei portentosi effetti cagionati dalla maravigliosa eruzione fatta dal Monte Vesuvio detto di Somma, di pietre infuocate, e di fiumi di acceso bitume con mistione di minerali di tutte le sorti — *Napoli, 1694, in 4.° fol. 4* (C. A.).

ANONYMOUS. — Succinta relazione dell'incendio del Vesuvio accaduto alla fine di luglio e progresso di agosto 1606. — *Napoli, 1696.*

ANONYMOUS. — Distinta relazione del grande incendio e meravigliosa Eruzione fatta dal Monte Vesuvio detto volgarmente la montagna di Somma, nella quale si dà distintissimo ragguaglio di quanto ha eruttato dalli 29 di aprile per infino alli 10 del corrente giugno 1698 et il danno, spavento, e fuga, che ha apportato a' popoli. — *Napoli and Roma, 1698, in 4,° fol, 2.* (C. A.).

ANONYMOUS. — Diario della portentosa eruzione del Vesuvio nei mesi di luglio e agosto 1707. — (?).

ANONYMOUS. — Relazione dei meravigliosi effetti cagionati dalla portentosa eruzione del monte Vesuvio detto di Somma, di pietre infocate, gorghi di fuoco, tuoni saetti e pioggia infinita di arenosa cenere seguita dal dì 20 del caduto luglio per tutti li due del corrente agosto 1707.—*Napoli, 1707, in 4,° fol. 2.* (C. A.).

ANONYMOUS. — Touchant le mont Vésuve et tremblement à Naples le 5 juin 1688. Lettre écrite le 12. — *Tiré d'un voyage en Italie en 1688, 4·me Edit. t. III, La Haye, 1717, in 4.°, pp. 391-418.* (C. A).

ANONYMOUS. — Dissertatione della grande Eruttione fatta dal Vesuvio nel maggio del 1737. — *M. S. S. in Library of St. Martino Museum. Napoli, in 4°, pp. 62.* (C. A.).

ANONYMOUS.—Neapolitanae scientiarum Academiae de Vesuvii conflagratione commentarius. — *Neapolis, 1738.*

ANONYMOUS. — An account of the eruption of mount Vesuvius 18 May, 1737.—*Philos. Trans. of the R. Soc. of London, 1739, N.° 4555,p. 352, fol. 3.*

ANONYMOUS.— Historia dell'incendio del Vesuvio accaduto nel mese di maggio dell'anno 1737. — *Napoli, 1740.*

ANONYMOUS. — Histoire du Mont Vésuve, avec l' explication des phénomènes qui ont coûtume d'accompagner ses embrasements. — *Paris, 1741.*

ANONYMOUS.— De Monte Vesuvio, disquisitionis.—*Acta Helvetica, t. I, 1751, pp. 97-104.*

ANONYMOUS. — An account of the eruption of Mount Vesuvius in Octob. 1751. *Philos. Trans. of the R. Soc. of London, 1751-52 Vol. XL, pp. 409-412.*

ANONYMOUS. — Laves qui sortaient des flancs du Vésuve à la suite de l'Eruption de 1754. — *A. plate.* (C. A.).

ANONYMOUS.—Veduta interiore del Vesuvio nel 1755. — *A ·Plate?* (C. A.).

ANONYMOUS. — Sur l'éruption du Vésuve en août 1756. — *Journ. Etrang. mars, 1757.*

ANONYMOUS. — Vue générale du Vésuve en 1757. — *A plate in fol.?* (C. A.).

ANONYMOUS. — Ragionamento historico intorno a nuovi vulcani.— *Napoli, 1761.*

ANONYMOUS. — Eruzione del Monte Vesuvio nell' anno 1767 veduta da Portici. — *A. plate.* (C. A.).

ANONYMOUS. — An account of the eruption of Mount Vesuvius in 1707. — *Trans. of the American Philos Soc. Philadelphia, Vol. I. 1771.*

ANONYMOUS. — Geschichte des Vesuv. In Vol. I, pag. 92-114 of «*Vermischte Beiträge zur physikalischen Erdbeschreibung,*» *6 vols. Brandenbourg, 1774-1787.*

ANONYMOUS. — Dei Vulcani o monti ignovomi piu noti, e distintamente del Vesuvio.—*Osservazioni fisiche e notizie istoriche di uomini insigni di varî tempi. Livorno, 1779, in 8.° Vol. 1, pp. LXX + 149, Vol. II, pp. VIII + 228.* (C. A.).

ANONYMOUS. — Piano del Volcano di Napoli denominato il Vesuvio; colle vieppiù rimarchevoli eruzioni seguite in più tempi. Dedicated to the Princess Jablonouka born Princess Sapieha, Palatina di Braclau. *Filip Morg scp.— Published about 1779. Loc. unknown. Explanation in Italian, My own collection.*

ANONYMOUS. — Ragguagglio di una nuova eruzione fatta dal monte Vesuvio nei primi giorni del corrente agosto 1779. — *Roma, 1779, in 4,° fol. 2.* (C. A.).

ANONYMOUS. — Relazione o sia descrizione della spaventevole eruzione del monte Vesuvio distante alcune miglia da Napoli verso Levante, seguita la sera delli 8 del corrente mese d' agosto (1779) avendo la stessa cagionati grandissimi danni a tutti que' luoghi, a cui si è estesa.—*Bologna, 1779, in 4,° fol. 2.*

ANONYMOUS. — Raccolta di lettere scientifiche ed erudite dirette dall'abate Genovesi a diversi suoi amici. (Letter VII. An account of the Vesuvian eruption of 1779 and in the letter VIII of Padre Ant. de Sanctis, that of 1631. — *Napoli, 1780, in 8,° fol. 3, pp. 247.* (C. A.).

ANONYMOUS.—Carteggio di due amanti alle faldi del Vesuvio.— *Pompei, 1783, in 8°, pp. 45.* (C. A.).

ANONYMOUS.— Untergang der Stadt Messina. Ingleichen eine kurze Beschreibung von den beiden Feuerspeyenden Bergen Vesuv und Aetna — ?. *1783.*

ANONYMOUS. — Dettaglio su l'antico stato ed eruzioni del Vesuvio ed eruzione nel 1794. — ?

ANONYMOUS. — Dialoghi sul Vesuvio in occasione dell' eruzione del 15 giugno 1794. — *Napoli, 1794.*

ANONYMOUS. — Lettera ragionata ad un amico, nella quale si dà un esatto ragguaglio dell'eruzione del Vesuvio , accaduto ai 15 giugno 1794.— *Napoli, 1794, in 8,° pp. 24.* (C. A.).

ANONYMOUS. — Relazione ragionata della eruzione del Vesuvio di Napoli, accaduta ai 15 giugno 1794, con la storia di tutte le eruzioni memorabili fino al presente avvenute. — *Without loc. or date, in 8,° pp. 52, pl. 1.* (C. A.).

ANONYMOUS. — Riscontro di un avvocato napoletano ad un suo amico di provincia della cruttazione del Vesuvio dei 15 giugno 1794. — ? *In 8,° pp. 40.* (C. A.).

ANONYMOUS. — Seconda lettera di un legista napoletano ad un suo fratello in provincia, in cui gli da distinto ragguaglio di quanto e avvenuto in Napoli in occasione dell'orribile eruzione del Vesuvio avvenuta ai 15 giugno 1794. — *Napoli? in 8,° pp. 16.* (C. A.).

ANONYMOUS. — Account of a descent into the crater of Mount Vesuvius by eight Frenchmen on the night between the 18th and 19th of july 1801. — *The Philos. Magaz. Vol. XI, London, 1801.*

ANONYMOUS. — Account of the late eruption of Mount Vesuvius may 31.st — *Moniteur, june 22nd 1806. Nicholson's Journ. of Nat. Phil. N° 58 Aug. 1806, pp. 345-350.*

ANONYMOUS. — Le Vésuve. — *Journ. d. l'Empire 7 et 10 nov. Paris, 1807.*

ANONYMOUS. — Cenno storico dell'eruzioue del Vesuvio in ottobre 1822. — *Napoli, 1822, in 8.°, pp. 29.* (C. A.).

ANONYMOUS. — Vue de l'église de Resina et de l'éruption du Vésuve du 23 oct. 1822. — *A plate ?* (C. A.).

ANONYMOUS. — Deux lettres sur l' éruption du Vésuve , 22 oct. 1822. — *Bibl. Univ. Vol. XXI, nov. 1822, pp. 190-191, 226-228, and Vol. XXII, feb. 1823, pp. 138-139.* (C. A.).

ANONYMOUS. — Remarques sur le Vésuve. — *Bull. Sc. Nat. et de Géol. Paris, in 8,° p. 39. Also Edinb. Journ. of Sc. July, 1827, p. 11.* (,C. A.).

ANONYMOUS. — Notice sur Herculaneum , Pompei et Stabiae. — *Edinb. Journ. of Sc. N.° XIX. Bibl. Univ. Vol. XL, pp. 411-426, 1829.* (C. A.).

ANONYMOUS. — Eruzione del Vesuvio. — *Giorn. d. Farmacia, Vol. XVIII, Milano, 1833.*

ANONYMOUS. — Raccolta di osservazioni chimiche sull'uso dell'acqua termo-minerale vesuviana Nunziante. — *Fascicolo Primo, Napoli, 1833, in 8,° pp. 76.* (C. A.).

ANONYMOUS. — Sur l'éruption du Vésuve en juillet et août 1832 (Translation.). — *Biblioth. Univ. d. Sc. etc., Vol. LII. Genève, 1833, pp. 376-388.* (C. A.).

ANONYMOUS. — Eruption of Vésuvius , January , 3 rd 1839. — *A. plate.* (C. A.).

ANONYMOUS. — Notes on Vésuvius. — *Am. Journ. Sc., Newhaven, 2nd ser. Vol. XIII, N° 37, 1852, pp. 131-133.* (C. A.).

ANONYMOUS. — Eruption du Vésuve le 1er mai 1855. — *Cuttings from newspapers*. (C. A.).

ANONYMOUS. — Neuer Ausbruch des Vesuvs.—*Mulhaüsen, 1855*.

ANONYMOUS. — Phénomènes observés au Vésuve. — *La Science pour tous, 1re An., N° 15, août 1856, pp. 119-120.* (C. A.).

ANONYMOUS. — Eruzione del Vesuvio del maggio 1858. — *21 numbers of "Giornale del Regno delle due Sicilie, 1858.* (C. A.).

ANONYMOUS. — Carta della regione perturbata dai fenomeni vesuviani, cominciati il dì 8 decembre 1861. — *Napoli, 1862.*

ANONYMOUS.—Conto reso dalla Commissione centrale pei danneggiati di Torre del Greco dal dì 16 Dicembre 1861 al 27 aprile 1802. — *Napoli, 1862, in 8.°, pp. 28.* (C. A.).

ANONYMOUS. — Intorno all'incendio del Vesuvio cominciato il dì 8 dicembre 1861; relazione per cura dell' Accademia Pontaniana, — *Napoli, 1862.*

ANONYMOUS. — Il Vesuvio — Strenna pel 1869, pubblicata a pro dei dannegiati dell'eruzione del 1868. — *Napoli, 1869, in 8.°.* (C. A.).

ANONYMOUS. — Eruption du Vésuve des 25, 26 et 27 avril 1872.— *Figures in La Presse Illustrée, 4 mai, 1872. Paris.* (C.A.).

ANONYMOUS. — Eruzione del Vesuvio dell'anno 1872. — *M. S. documents in the archives of the municipality of Torre del Greco. Fol. 17 Copy.* (C. A.)

ANONYMOUS. — Relazione della Giunta comunale di Napoli al Consiglio su' provvidimenti adottati per la eruzione del Vesuvio del 1872 ed atti relativi. — *Napoli, 1872, in 4.° pp. 31.* (C. A.)

ANONYMOUS. — Statuto e regolamento organico dell'Associazione Vesuviane di mutuo soccorso per assicurare le proprietà dai danni delle lave vulcaniche. —. *Napoli, 1873, in 4.°, pp. 42.* (C. A.).

ANONYMOUS.—Il Vesuvio.—Un Sonetto ed un Ode.—? *fol. 2.* (C.A.).

ANONYMOUS. — Sommet du Vésuve; autre vue du même sommet durant une petite éruption. — *Plate in fol. ?...* (C. A.).

ANONYMOUS. — Partenope terraemotu vexata. — *M. S. Copy?* (C. A.).

ANONYMOUS. — Veduta del Monte Vesuvio e parte della città di Napoli. — *Napoli?.* (O. A.).

ANONYMOUS. — Veduta di parte delle lave di bitume che nelle eruzione vomitate dal Vesuvio coprirono l'antichissima Città di Ercolano; prima della nostra era posta al di sotto di questi luoghi, tra il presente Portici e la Torre del Greco. — *Napoli?* (C. A.).

ANONYMOUS. — Vesuvius. — *Encycl. Britannica, Vol. XVIII, pl. II, in 4.°, pp. 728-734.*

ANONYMO US. — Vesuvius morum magister. – *M. S.*, *fol. 3.* (C. A.).

ANTICI S. — Sonetto (1031). — *V. G. Urbano.* (C. A.) .

APOLLONI. G. — Il Vesuvio ardente. — *Napoli, 1632, in 12° fol. 15, 2.ᵈ Edit.*

ARAGO F. — Liste des Volcans actuellement enflammés. — *Annu. d. Bur. d. Longit. année 1824, pp. 167-189.* (C. A.).

ARCONATI, VISCONTI G. M. — Appunti sull' eruzione del Vesuvio del 1867-68. — *Giornale Politecnico di Milano, Vol. V. fasc. III, March, 1868, in 8°, pp. 237-253.* (C. A.).

ARDINGHELLI M. – Eruption du Vésuve en 1767. — *Compt. rend. d. l'Accad. d. Sc. Paris, 1767.* .

ARMFIELD H. T. — At the crater of Vesuvius in eruption. A word picture. — *Salisbury, 1872.*

ARMINIO I. D'. — De terraemotibus et incendiis eorumque causis, et signis naturalibus et supranaturalibus. Item de flagrationi Vesuvii ejusque mirabilibus eventis et auspiciis. — *Neapolis, 1632, in 8°, pp. 16.*

ARTHENAY (d') — Journal d'observations dans les différents voyages qui ont été faits pour voir l'éruption du Vésuve. — *Mém. d. Matem. et d. Phys d. l'Acad. d. Sc. Vol. IV, Paris, 1773, pp. 247-280.*

ASCIONE C. — Breve compendio della descrizione della Torre del Greco antica e moderna, delle sue chiese esistenti prima e dopo 1631. — *Napoli, 1836, in 4°, pp. 120.* (C. A.).

ASTERIO P. — Discorso aristotelico intorno al terremoto, etc. — *Napoli, 1632, in 4° (in Biblioteca Vit. Eman. Roma).*

ASTORE F. A. — Eruzione del Vesuvio del 1794. — *Napoli, 1794.*

ATLANTE. Di vedute de principali incendi del Monte Vesuvio etc. *V. Duca della Torre.* (C. A.).

ATTUMONELLI M. — Della eruzione del Vesuvio accaduta nel mese d'agosto dell'anno 1779. Ragionamento istorico-fisico. — *Napoli, 1779. In 8°, fol. IV, pp. 147, pl. I.* (C. A.).

AUBRYET H. — Pompei et les pompeïens', par Marc Monnier. — *Compt. Rend., Acad. Sc., Paris, 1864, in 8°, fol. V.* (C. A.).

AUDOT. — Quattro vedute del Vesuvio riguardanti le eruzioni del 1751, 1804, 1822. *Iz 8.° — Napoli* (?)

AUGEROT A. (d') — Le Vésuve; description du volcan et ses environs. — *Limoges, 1877.*

AULDJO J. — Vue du Vésuve, avec un précis de ses éruptions principales depuis le commencement de l'Ere Chrétienne jusqu'à nos jours. — *Spettatore del Vesuvio. Fasc. I, Napoli, 1832, in 8°, pp. 102, pl. XVII, map. 1.*

AULDJO J. — Veduta del Capo Uncino presso Torre dell'Annun-

ziata, della cosi detta sorgente del Vesuvio, e degli avanzi
di un cipresso giacente nel tufo a quaranta palmi di profondità. — *Spettat. d. Vesuvio. Fasc. II, Napoli, 1832.*

AULDJO J. — Sketches of Vesuvius with a short account of its principal eruptions. — *Naples, 1832, fol. III, pp. 96, map. 1, pl. XVII, London, 1833.*

AULDJO J. — Source jaillissante d', eau minérale, découverte, en
1831, prés du Cap Uncino, dans le Royaume de Naples. —
*Biblioth. Univ. d. Sc. Bell. Lettr. et Arts, faisant suite à
la Biblioth.] Britann. red. à Genéve. Partie d. Sc. Ire Sér.
Vol. LII, Genéve, 1833. The American Journ. of Sc., and
Arts; by B. Silliman etc. First Ser. Vol. XXV. New Haven,
1834.*

AULISIO G. D. (D') — Divotissime Orationi ecc. — *Mentioned in
the catalog. of Vinc. Bove. V. Mormile.*

AUTORI VARII. — Dei vulcani o monti ignivomi più noti, e distintamente del Vesuvio. — *Livorno, 1779.*

AYALA S. (d') — Copiosissima y verdadera relacion del incendio
del Monte Vesuvio, donde se da cuenta de veinte incendios
que ha habido sin este último. — *Napoles, 1632, in 4°, pp. 28.*

AYELLO F. A. (DE) — De ingenti ac repentino in hoc tempore
Vesuvii Montis lamentabile incendio. Epistola. — *Neapolis,
1632, fol. 4.*

AYROLA F. L. — L'Arco celeste, overo il trionfo di Maria dell'arco
e suoi miracoli. — *Napoli, 1688, in 4.° fol. XII, pp. 328, fol. 12.*
(C. A.)

BADILY W. — Estratto di una lettera intorno alla pioggia di ceneri nell'Arcipelago, nell'incendio del Vesuvio 1631. — *Giornale dei Letterati per l'anno 1674, pp. 146-147, Roma, in 4.°*
(C. A.). *(In French) Mem. de Phys de toutes les Acad. d.
Sc., Lausanne, 1754.*

BAILLEUL. — Remarques sur l'éruption de 1850. — *Compt, rend,
Hebd, des Séanc. d. l'Acad. d. Sc. Vol. XXXI. Paris, 1850.*

BALDUCCI F. — Gli incendi del Vesuvio. Discorso accademico. At
the end of the rhymes of the same author. — *A vol. in 18°
with a portrait. Venetia, 1642, pp. 459-750. (C. A.).*

BALZANO F. — L'antica Ercolano, ovvero la Torre del Greco tolta
all'oblio. — *Napoli, 1688. 3 books in one vol. in 4.° (Eruption of Vesuvius, 1680, pp. 81-85. (C. A.).*

BANIER ABBÉ. — Des embrasements du mont Vesuve. — *Ac. des
Inscript. et Bell. Lett. Vol. IX. pp. 14-22, 1736, in 12.°
M. S. copy. (C. A.).*

BARBA A. — Ragionamento fisico-chimico sull'eruzione ultima del
Vesuvio accaduta ai 15 giugno 1794. — *Napoli, 1794.*

BARBAROTTA L.—Il Vesuvio: a song. (Il Fausto ritorno da Vienna di Ferdinando IV) — *Napoli, 1791, pp. 13-19.* (C. A.).

BARBERIUS F. — De prognostico cinerum quos Vesuvius Mons dum conflagrabatur eructavit. — *Neapoli, 1632, in 4.°, pp. 64.*

BARBERIUS F. — Manifestum eorum quae omnino verificata fuerunt jam antea ab ipso praedicta in prognostico cinerum quos Mons Vesevus emisit dum comburebatur. — *Neapoli , 1635, in 4°, pp. 14. (*O. V.*).*

BARONIUS ET MANFREDI F. — Vesuvii montis incendium. — *Neapolis, 1632, in 4.°, pp. 8.*

BARRA C. — Partenope languente per l'accaduto terremoto al 5 giugno 1688. — *Napoli, 1688, in 12.° .(* C. A.).

BARRIER (Abbé) A. — Des embrasements du mont Vésuve.—*Acad. d. Inscr. et Bell. Lettr. Tom. IX. Paris, 1736.*

BARTALONI D. — Osservazioni sopra il Vesuvio. — *Atti d. Acc. d. Sc. d. Siena. Vol. XXV. Siena, 1776.*

BARTOLI P. — Continuazione dei successi del prossimo incendio del Vesuvio cogli effetti della cenere etc. — *Napoli, 1662.*

BARUFFALDI G. — Vesuvio. Baccanale. — *In 12.° pp. 32, without other indications. (* C. A.).

BASILE G. B. — Tre sonetti nelle Rime d'illustri ingegni Napoletani. — *Venezia, 1633, in 12°, pp. 133-136.* (C. A.).

BASSI B. — Two songs referring to the vesuvian eruption of June 15th 1794. — *Cont. in a vol. entitled Opuscoli Varj. Napoli, 1794, in 8.° pp. 41-42.* (C. A.).

BASSI U. — (under pseudonym Plangeneto). La lacrima di Monte Vesuvio, volgarmente Lacrima Christi. A bacchanalian song.— *Napoli, 1841, in 16.° pp. 67.* (C. A.).

BEALE N. — Analisi qualitative della cenere del Vesuvio eruttata nella notte del 27 al 28 aprile p. p. — *Ann. d. Chim. Vol. LIV. Milano. 1872.*

BEAUMONT J. B. ELIE DE—Remarques sur une note de M. Constant Prévost relative à une.communication de M. L. Pilla , tendant à prouver que le còne du Vésuve a été primitivement formé par soulèvement. — *Compt. rend. Hebdom. d. Séanc. d. l'Acad. d. Sc. Vol. IV. Paris, 1837.*

BELLANI A. — Salita al Vesuvio — Milano, 1835.

BELLICARD. — Observations upon the 'Antiquities of the town of Herculaneum, discovered at the foot of Mount Vesuvius. — *London, 1753, in 8.°, pp. VII, plates 42. Another edition in 1756.* (C. A.).

BELTRANO O. — Vesuvio centone. — *Napoli, 1633, in 8°, fol. 1, pp. 30 (*O. V.*).*

BENIGNI D. — Sonetti tre (1631). — *V. G. Urbano.* (C. A.).

BENIGNI D. — La strage di Vesuvio. Lettera all'abate Perretti.— *Napoli, 1632, in 4.°, fol. 6·*

BERGAZZANO G. B. — Bacco arraggiato co'Vorcano, discurzo ntrà de lloro. — *Napoli, 1632, in 8°. small, fol. 8,* (C. A.).

BERGAZZANO G. B. — Il Vesuvio fulminante. Poema. — *Napoli, 1632, in 8.° (B. N.).*

BERGAZZANO G. B. — I prieghi di Partenope. Idillio. — *Napoli, 1632, in small 8.°, fol. 8.* (C. A.).

BERGAZZANO G. B. — Vesuvio Infernal. Scenico avvenimento. — *Napoli, 1632, in 12° (V. Quadrio, Storia e ragione d'ogni poesia) (III, Par. I, p. 88).* (C. A.).

BERKELEY E. — Eruption du Vésuve en 1717. — *Phil. Transact. of the R. Soc. of London. Vol. VI. 1717.*

BERNARDINO F.—Discorso istorico intorno all'eruzione del Monte Vesuvio 15 giugno 1794.—*Napoli, 1794, in 4.° pp. 22.*

BERNARDO F. — L'incendio del Monte Vesuvio etc.—*Napoli, 1632, in 4.°, pp. 32.*

BETOCCHI A. — Sulla cenere lanciata dal Vesuvio alla fine della passata straordinaria eruzione (24 aprile). — *Atti d. R. Acc. d. Lincei Vol. XXV, 1872.*

BEULÉ. — Le drame du Vésuve.—*Paris, 1872, in 12°, pp. 336, map 1.* (C. A.).

BINNET-HENTSH J. L.—Une excursion au Vésuve. — *Echo d'Alpes. N. 2, 1876. (?)*

BITTINI G. — Sonetto (1631). — *V. G. Urbano.* (C. A.).

BITTNER A. — Beobachtungen am Vésuv. — *Verh. d. K. K. géol. Reichs Anst. N. 12, Wien, 1874.*

BLACK J. M.—An account of the eruption of Mount Vesuvius of April 1872. — *Proc. Geol. Assoc. Vol. III, N. 6. 1874.*

BLAKE J. F. — A Visit to the Volcanoes of Italy. — *Proceed, Geol. Assoc, London, 1889, Vol. IX, pp. 145-176,*

BLAEV J. A. — Theatrum civitatum nec non admirandorum Neapoli et Siciliae regnorum. — *In fol. pp. 78-30 with 30 plates* (C. A.).

BLANC LE. — 1863. — *V. Fouqué.*

BLASIIS G. DE — La seconda congiura di Campanella. — *Giornale Napoletano, Vol. I, Napoli, 1875, in 8°.*

BOCANGEL Y UNÇUETA G. — Epitafio al Vesuvio y sus incendios, Sonnet. — *V. Quinones.* (C. A.).

BOCCAGE MME DU. — Sur le Vésuve. Deux lettres en date des 8 et 15 oct. 1757. — *(Ext. du Recueil de ses œuvres), Lyon 1770, III Vols. in 8°, Vol. III p. 265-286.* (C. A.).

Boccosi Ferdinand (Biondi Francesco). — Delle centurie poetiche. — *Napoli, in 12°, pp. 203, VII, in 2 parts.* (C. A.).

Bomare de M. — Sopra il Vesuvio ed altri vulcani. Dei Vulc. o monti ignivomi. — *V. Anonymous, Livorno, 1775.*

Borkowsky (Dunin) S. — Sur la sodalite du Vésuve. — *Ann. d. Mines, Vol. I, Paris, 1816. — Journ, d. Phys. d. Ch. et d. l'Hist. nat. Vol. LXXXIII, Paris, 1816. — Ann. of Philos. or Magaz. of Chem. Miner. Mech. and the Arts. 1 Ser. Vol. X, London, 1817. Ann. der Physik von L. W. Gilbert. Bd. LXIII. Halle und Leipzig, 1819.*

Bornemann J. G. — Sur l'état des volcans d'Italie pendant l'été de 1856. — *Translation of De Perrey from Tageblett. der 32 Versam. Deutch. Naturf. und Aertze in Wien, 1856, p. 114-141. Original M. S. pp. 4. (C. A.).*

Bornemann J. G. — Ueber Erscheinungen am Vesuv und geognostisches aus den Alpen. — *Zeitschr. d. Deuts. geol. Gesell. Bd. IX. Berlin, 1857.*

Borzi. — 1814 (?). — *V. Granville.*

Bottis G. (de) — Ragionamento istorico intorno ai nuovi vulcani comparsi nella fine dell'anno scorso 1760 nel territorio della Torre del Greco.—*Napoli, 1761; in 4°, pp. 67, pl. 2. (C.A.)*

Bottis G. (de) — Ragionamento storico sull'incendio del Vesuvio nell'ottobre del 1767. — *l'apoli, 1768.*

Bottis (de) G. — Ragionamento istorico dell'incendio del Monte Vesuvio, che cominciò nell'anno 1770 e delle varie eruzioni che ha cagionate. — *Napoli, 1776, in 4.° pp. 84 + III, pl. 4.* (C. A.).

Bottis G. de — Ragionamento istorico intorno all' eruzione del Vesuvio che cominciò il di 29 luglio dell'anno 1779 e continuò fino al giorno 15 del seguente mese di agosto.—*Napoli, 1779, in R. 4°, pp. 117 + III, pl. 4. (C. A.).*

Bottis G. (de)— Istoria di varii incendi del Monte Vésuvio. — *Napoli, 1786.*

Bottoni D. L. — Pyrologia Topographica id est de igne dissertatio juxta loca cum eorum descriptionibus. — *Neapolis, 1692. Messanae, 1721.*

Bourke E. (de). — Le Mont Somma et le Vésuve. — *Notice sur le ruines les plus remarquables de Naples et ses environs, Paris, 1823, in 8., pp. 167-174. (A. A.).*

Bourlot J. — Etude sur le Vésuve , son histoire jusqu'à nos jours. — *Paris, 1867.*

Bove V. — Il Vesuvio accesso. — *Napoli, 1632, in 12.°, fol 12.*

Bove V. — Nuove osservazioni fatte sopra gli effetti dell'incendio del Monte Vesuvio aggiunta alla decima Relatione etc. — *Napoli, 1632, in 8.°, pp. 31.*

Bove V. — Decima relazione nella quale più dell' altre si dà breve et succinto ragguaglio dell'incendio risvegliato nel monte Vesuvio o di Somma, etc. — *Napoli, 1632, in 4.°, pp. 11.*

Bovio G. — La Geologia dell' Italia meridionale rispetto all' indole degli abitatori. — *Napoli, Ernesto Anfossi, 1883, in 8°, pp. 31. (C. A.).*

Braccini (Ab.) G. C. — Relazione dell' incendio fattosi nel Vesuvio ai 16 dicembre 1631, e delle sue cause ed effetti etc. — *Napoli, 1632, in 4.° fol. 2, pp. 104.*

Braccini (Ab.) G. C. — Relazione dell'incendio del Vesuvio ai 16 dicembre 1631. — *Napoli, 1632, in 8°, pp. 40.* (This rare account was referred to by P. G. M. della Torre under the title «Colonna Girolamo. Cardinale.» Lettera sopra l'incendio del Monte Vesuvio del 1631.

Brard. — Une Eruption du Vesuve. — *Feuilleton de l' Echo de Numidie. 15-29 mai and 5 juin 1861.* (C. A.).

Breislak S. — Institutions géologiques. — *Vol. I-III. Milan, 1818.*

Breislak S. e Winspeare A. — Memoria sull' eruzione del Vesuvio accaduta la sera del 15 giugno 1794. — *Napoli, 1794, fol. 12.* (C. A.)

Breislak S. und Winspeare A. — Fortgesetzte Berichte vom Ausbruche des Vesuvs am 14 junius 1794. — *Dresden, 1795.*

Brocchi G. B. — Sull'eruzione del Vesuvio del 1812. — *Bibl. Ital. ossia Giorn. di Letter. Sc. etc. Vol. VI. Milano, 1817.*

Brooke H. J. — On the Comptonite of Vesuvius, the Brewsterite of Scotland, the Stilbite, and the Heulandite. — *The Edinburg. Philos. Jour. exhib. a view of the Progr. of. Discov. in Nat. Philos. Vol. VI. Edinburgh, 1822.*

Bromeis T. — Analyse eines glimmers vom Vesuv. — *Ann. der Phys. und Ch. von I. C. Poggendorff. Bd, LV. Leipzig, 1842.*

Bruni A. — Canzone sull' incendio del Vesuvio (1631). — *V. G. Urbano.* (C. A.).

Brydone P. — Lettre sur l'éruption du Vésuve en 1770. — (?)

Buch L. (von) — « Bocche nuove ». — *Fragment aus einer Reihe von Briefen über den Vesuv. Jahrb. der Berg und Hüttenkunde, von Karl E. von Moll. Bd. II, Salzburg, 1801.*

Buch L. (von) — Lettre sur la dernière éruption du Vésuve et sur une expérience galvanique nouvelle. — *Bibl. Britannique, ou Rec. d. extr. d. Ouvr. Anglais period.et autres: part. d. Sc. et Arts. Vol. XXX, Genève, 1805.*

BULIFON. A. — Lettere memorabili. Lettera a Mabillon del 1689 e fig. — *Napoli, 2 Vols. 1693, in 12,° (C. A.).*

BULIFON A. — Ragguaglio dello spaventevole moto del Vesuvio succeduto il mese di dicembre 1689. Lettera al R. P. Mabillon. — *Napoli, 1693.*

BULIFON A. — Incendio del Vesuvio dell'aprile 1694. *Lettera.* — *Napoli, 1694. in 12, pp. 88, pl. 1. Another edition, 1696.*

BULIFON A. — Compendio istorico degl'incendi del monte Vesuvio fino all'ultima eruzione accaduta nel giugno 1698. — *Napoli, 1701, in 12°, pp. 106+III, pl. II.*

BULWER-LYTTON. — The Last Days of Pempeii. — *Numerous editions.*

BURIOLI P. — Vera relatione del terremoto, e Voragine occorso nel Monte Vesuvio il 16 Dicembre 1631 a ore 12 etc. — *Bologna, 1632, in 4°, fol. 4.* (C. A).

BUSCA D. — Sonetto (1631) -- *V. G. Urbano.* (C. A.).

CACCABO G. B. — Januarius poema sacrum. — *Napoli, 1635, in 4°, fol. 4, pp. 46. (O. V.).*

CAGNAZZI LUCA. — Discorsi meterologici degli anni 1792-93-94. — *Extract from Giornale Letterario di Napoli. In 8°, pp. 46, 29.* (C. A.).

CAGNAZZI DE S. LUCA — Lettera sull'elettricismo della cenere lanciata dal Vesuvio, diretta al P. Em. Taddei. — *Extracted from « Giornale Enciclopedico ». — Napoli, June 10th, 1806, pp. 8.* (C. A.).

CAMERLENGHI G. B. — Incendio del Vesuvio. Poema. — *Napoli, 1632. in 4°, fol. 2, pp. 190, frontisp.*

CAMOLA G. P. — Sonetto (1631). — *V. G. Urbano.* (C. A.).

CAMPO D. — Histoire des phénomènes du Vésuve. — *Naples, 1771.*

CAMPONESCHI F. — De Vesevo monte. Epigramma. — *V. G. Urbano.* (C. A.).

CANEVA S. (Hermit priest). — Lettera dell'Eremita del S. S. Salvatore sito alle falde del Vesuvio per dare ad un amico suo un succinto ragguaglio dell'accaduta Eruzione la sera del 15 Giugno 1794. — *Without authors name or locality. fol. 2.* (C. A.).

CANGIANO L. — Sur la hauteur du Vésuve. — *Compt. rend. hebdom. d. Séanc. d. l'Acad. d. Sc. Vol. XXII. Paris, 1846.*

CANTALUPO G. — Réminiscenze Vesuviane di un profugo. — *Napoli, 1872, in 8°, pp. 66.* (C. A.).

CAPACCIO G. C. — Incendio del Vesuvio ; in • Forastiero, dialoghi, etc. • — *Napoli, 1634, in 4° pp. 86.*

CAPECELATRO F. — Historia di Napoli — *Vol. I-II. Napoli., 1724.*

CAPECE P. A. —.Lettere scritte al P. A. Capece della Comp. di Gesù a Roma. (Erupt. 1631). — *M. S. in library of the Faculty of Medecine of Montpellier. Copy* (C.A). *V. Riccio L.*

CAPOCCI E.—Su di un raro fenomeno vulcanico che il Vesuvio ha offerto nel mese scorso. — *Rend. d. Adun. ed. lavori d. R. Acc. d. Sc. Fis. e Mal. Vol. V. Napoli, 1846.*

CAPOCCI E. — Relazione del fenomeno delle corone di fumo e di cenere presentato dal Vesuvio nell'eruzione del dicembre del 1846, e nei mesi seguenti. — *Rend. d. adun. ed. lavori d. R. Acc. d. Sc. Fis. e Matém. Vol. V. Napoli, 1846.*

CAPOCCI E.—Sulla eruzione del Vesuvio degli 8 Dicembre 1861.— *pp. 3. At pages 23, 24, 25 of « Raccolta di Scritti varii per cura di Rinaldo C. de Sterlich » — Napoli, 1863, in 8°* (C. A.).

CAPPA R. — Delle proprietà fisiche, chimiche e terapeutiche dell'acqua termo-minerale Vesuviana Nunziante.—*Napoli, 1847, in 8°, pp. 12.* (C. A.).

CAPRADOSSO A. — Il lagrimevole avvenimento dell'incendio del monte Vesuvio per la città di Napoli e luoghi adiacenti etc.— *Napoli, 1632, in 4°, fol. 4. (also an edit. 1631).*

CAPUANO G. A. — 1779. — *V. Sanfelice.* (C. A.).

CARAFA G. — De novissima Vesuvii conflagratione. Epistola isagogica. — *Neapolis. 1632, in 8°, fol. 64, pl. 1. (also 2nd edit. 1632, in 4°, pp. 93, fol. 4, pl. 1.*

CARDASSI S. — Relazione dell'irato Vesuvio, dei sui fulminanti furori ed avvenimenti compassionevoli.—*Bari, 1632, in 12°, pp. 46.*

CARDONE A. — Saggio di poetici componimenti (Sull'ultima eruzione del Vesuvio. Poem.)—*Napoli, 1828, in 8°, fol. I, pp. 25* (C. A.).

CARDOSO F. — Al Vesuvio, Sonnet. — *V. Quinones.* (C. A.).

CARNEVALE G. A.—Brevi e distinti ragguagli dell'incendio del Vesuvio del 1631.—*Napoli, 1632. (Quoted by V. Bove, Soria and Bucca).*

CARPANO G.—Giornale dell'incendio del Vesuvio nell'anno 1660—(?)

CARREY E. — Le Vésuve. — Feuilleton du « Moniteur » des 16, 17 et 18 Octobre, 1861. (C. A.).

CARUSI G. M. — Tre passeggiate al Vesuvio nei dì 3 e 21 Giugno e 27 Settembre 1858. — *Napoli, in 4°, pp. 44.* (C. A.).

CASORIA E. — L'acqua della fontana pubblica di Torre del Greco ed il predominio della potassa nelle acque vesuviane.—*Idrologia e Climatologia Medica, Anno VII, N.° 9, Firenze 1885, in 8°, pp. 12.* (C. A.).

CASORIA E. — Sopra due varietà di calcari magnesiferi del Somma. — *Boll. d. Soc. d. Naturalisti in Napoli. Ser. 1.ª, Vol. I, Anno I, fac. 1.°, 1887, pp. 2.*

CASORIA E. — Composizione chimica e mineralizazione delle acque potabili vesuviane.—*Idrologia e Climatologia Medica. Anno IX, N. 3, Firenze, 1887, pp. 11.* (C. A.).

CASORIA E.—Composizione chimica di alcuni calcari magnesiferi del Monte Somma.— *Boll. d. Soc. d. Naturalisti in Napoli Ser. 1ª, Vol. II, Anno 2, fasc. 2. 1888 in 8°, pp. 7.*

CASORIA. E. — Mutamenti chimici nelle lave vesuviane per effetto degli agenti esterni e delle vegetazioni. — *Boll. d. Soc. Naturalisti in Napoli. Ser. 1ª, Vol. II. an. 2.°, 1888, fasc. 2, pp. 18.*

CASORIA E.—Sulla presenza del calcare nei terreni vesuviani. — (*Boll. Soc. Naturalisti, Vol. II, fasc. 2°*). *Napoli, 1888.*

CASSANO (Aragona Prince of). — An account of the eruption of Vesuvius in March 1737.—*Phil. Trans. of the R. Soc. of London, 1739.*

CASSOLA. — 1832 — *V. Pilla.*

CASTELLI P. — Incendio del monte Vesuvio etc. — *Roma, 1632. in 4.°, fol. IV., pp. 92. fol. 4.*

CASTRUCCI G.—Breve cenno della eruzione vesuviana del maggio 1855.—*Napoli, 1855, in 4,°, pl. 1.* (O. V.).

CATANI A. — Lettera critica filosofica su della vesuviana cruttazione accaduta il 19 ottobre 1767. — *Catania, 1768, in 4.°, fol. 3, pp. 42.*

CATANTI.—Lettera all' Abbate Mecatti sul Vesuvio. — (?)

CAVALLERI G. M.—Considerazioni sul vapore e conseguente calore che manda attualmente (8 ottobre 1836) il vulcano di Napoli. — *Atti d. Acc. d. Fis. e Mat. di Napoli Vol. XII. 1856-57.*

CAVALLI A. — Il Vesuvio, poemetto storico-fisico con annotazioni.—*Milano, 1749, in 8.°, pp. 157, pl. 2.* (C. A.).

CAVA P. (LA) — Sulla efflorescenza della soda clorurata che trovasi in taluni fumaiuoli attivi del Vesuvio. —*Napoli, 1820.*

CAVAZZA G. — Sonetto che l'incendio del Vesuvio è stato per salute dell' anime nostra. *Napoli, 1632, loose sheet (B. N.).*

CAVOLINI F. — Cenno storico dell'eruzione del Vesuvio dell'ottobre 1822. — *Napoli, in 8°, pp. 29.* (O. V.).

CAVOLINI F.—Piano del Vulcano di Napoli denominato il Vesuvio colle più rimarchevoli eruzioni seguite in più tempi.—*Pl. in fol. (published about 1854).* (O. V.).

3

CERASO F.—? L' opre stupende e meravigliosi eccessi dalla natura prodotti nel monte Vesuvio etc.—*Napoli, 1632, in 8.°, fol. 18.*

C. F. T.—L'Eruzione del Vesuvio della notte del 15 Giugno 1794 poeticamente descritta.—? *In 4.°, pp. 30.* (C. A.).

CHATEAUBRIAND.—Le Vésuve. Voyage en Italie, 1804—(?)

CHAVANNE De La — Histoire du Vésuve. (In: Audot, Voyage de Naples)—(?)

CHAVANNES DE.—Le Vésuve.—*Tours 1859 and 1867, pp. 127, pl. 1.* (C. A.).

CHIARINI G. e PALMIERI L. — Il Cratere del Vesuvio nell'8 novembre 1875. — *Napoli, 1876.*

CHODNEW A. — Untersuchungen eines schwärzlich grunden glimmers vom Vesuv. — *Ann. d. Phys. und. d. Ch. Bd. LXI. Leipzig, 1841.*

CHRISTIAN F. (Princ. d. Danimarca). — Osservazioni sulla lava del Vesuvio del 26 gennaio 1820 — *Napoli, 1820.*

CHRISTIAN F. (Fürst v. Dänemark). — Beobachtungen am Vesuv angestellt in Jahre 1820. Taschenb. für die ges. Miner., von Karl Cäsar Leonhard. Bd. XVI. Frankfurt. am Main 1822.— *Atti d, R. Acc. d. Sc. e Lett.; Sez. d, Soc. R, Borbonica. Vol. II, Napoli, 1825.*

CHRISTIANI P. — De Vesevo Monte, epigramma.—*See G. Urbano.*

CICADA H. — De Vesevi conflagratione. In the work entitled "Carmina ". — *Lycii, 1647, in 8°, pp. 206 (O. V.)*

CICCONI M. —Il Vesuvio, canti anacreontici — *Napoli, 1779, in 8.° pp. 96.* (C. A.).

CILLUNZIO N. — Versi per la eruzione del Vesuvio, accaduta a' 12 Agosto 1804. —? *In 8.° fol. VIII.* (C. A.).

CINITANO G. — Sonetto. (1631). — *V. G. Urbano.* (C. A.).

CIOFI A. — Dimostrazione scenografica e iconografica di tutti gli effetti prodotti dall' eruzione del Vesuvio de' 15 giugno 1794. *Pl. in fol. with description.*

CIRILLI. — Vita S. Januarii. — *Venetiis, 1776, in 4°, pp. 48.*

CLARO F.—Humanae calamitatis considerationes.—*Neapoli, 1632, in 4ª, fol. 14, pp. 87. (O. V.).*

CLASSENS DE LONGSTE A. — Souvenirs d'une promenade au Mont Vésuve. — *Naples, 1841, in 8.°, pp. 61.* (C. A.).

COCHIN ET BELLICARD. — Observations sur les antiquités d' Herculaneum, avec quelqus reflexions, etc.—*2ⁿᵈedit. Paris, 1757, in 8°, pp. XLI+ 84, pl. 40. See also Bellicard.*

COLLINI M. — Considerations sur les Montagnes Volcaniques, etc. — *Manheim, 1781, in 4°, pp. VIII, + 61, pl. 1.*

COLOMBO A. — Osservazioni sulla conformazione sottomarina del

Golfo di Napoli.—*Rivista Marittima, ottobre-dicembre 1887 with col. maps.*

COLONNA C. — Lettera sopra l'incendio del Vesuvio nel 1631. — *Napoli, 1632. V Braccini.*

COLUMBRO G.—Rime e prose.—*Napoli, 1817, Vols. II, in 8.° (Vol. 1. p. 21., 72-75 a poem « Il Vesuvio » speaks of the eruption of 1794. Vol. 2. p. 254, p. 202-211 letter to a friend at Resina speaking of Vesuvius.* (C. A.).

COMES O.—Le lave, il terreno vesuviano e la loro vegetazione.— *(Lo spettatore del Vesuvio e dei Campi Flegrei, Nuova serie, vol. 1.°) Napoli, 1887.*

COMPTE A. C.—Lettera critico-filosofica sull'eruzione del Vesuvio del 1767. — *Catania, 1868.*

CONNOR B.—De Montis Vesuvii incendio. Dissertationes medphis. Auxonisi, 1695. — *Acta Eruditorum. Lipsiae, 1696.*

CONNOR O'. — Mirabilis viventium interitus in Charonea Neapolitana crypta. Novissimum Vesuvii montis incendium anni acre salutis 1694. —, *Coloniae Agrippinae, 1694, in 12°, pp. 68 (O. V.).*

CONTE DI CORUNA. — Sonetto (1631). — *V. Quintones.* (C. A.).

COOMANS J. — Sur l'éruption du Vésuve en 1858. —« *Moniteur — du 10 Juin 1858, in 8.°, pp. 6.* (C. A.).

COPPOLA M. — Contribuzione alla storia chimica dello « Stercocaulon Vesuvianum ». Notizie preliminari.—*Rend. d. Accad. Sc. fis. e Mat. di Napoli. Fasc. 10°, Oct. 1879. in 4.°, pp. 4.* (C. A).

COPPOLA M. — Produzione artificiale dell'oligisto sulla lava vesuviana. — *Gaz. Chim. It: Vol. IX., 1879, in 8.° pp. 4.* (C. A.).

CORAFÀ G. — Veduta del Vesuvio dalla parte di mezzogiorno 25 ottobre, 1751. — *Plate in fol. with descript. (O. V.).*

CORAFÀ G. — Dissertazione istorico-fisica delle cause e degli effetti delle eruttazioni del Monte Vesuvio negli anni 1751-52— *Napoli, 1752, in 4.° fol. II, pp, 86.*

CORBONE. — Veduta dell'Eruzione del Vesuvio nel 1631 con illustrazione stampata dei danni prodotti— *Without date or loc.* (C. A.).

CORNELIUS T.—De sensibus.—*Neapoli, 1638, in 8°, fol. 7, pp. 119.*

CORRADO M. — Descrizione del fenomeno cagionato dal Monte Vesuvio nella sera del dì 15 di Giugno dell'anno 1794 de' fatti occorsi in seguito e della somma religiosità de' cittadini Napolitani. Verses.—*Napoli, 1794, in 8.° pp. 7.* (C. A.).

COSSA A. — Sulla predazzite periclasifera del Monte Somma. *Atti*

di R. Acc. d. Lincei. Ser. 2.ª, Tom. III. Roma, 1876, in
4.° p. 8.

COSTA O. — Fauna Vesuviana, ossia descrizione degl'insetti che
vivono nei fumajuoli del Cratere del Vesuvio. — *Ann d' Ac-*
cad. Sc. Vol. IV, per 1826, Napoli, 1839 , in 4.°, pp. 32,
pl. II. (C. A.). See also « *Giambattista Vico* «, *Vol. I, pp.*
39-44. Napoli, 1857.

COSTA O. G. — Rapporto sulle escursioni fatte al Vesuvio in ago-
sto-dicembre 1827 — *Atti di R. Acc. di Sc. e Lett. Sez. di*
Soc. R. Borbonica. Vol. IV per 1826. Napoli, 1839.

COVELLI N. — V. Monticelli. *

COVELLI N. — Cenno sullo stato del Vesuvio dalla grande eruzione
del 1822 in poi (1824) — *?, in 8°, pp. 9. (O. V.).*

COVELLI N. — Débit des minéraux du Vésuve. Catalogue pour 1826.
—*Naples, 1826, in 8°, pp. 16.*

COVELLI N. — Sur le bisulfure de cuivre qui se forme actuellement
dans le Vésuve. (1826). — *Ann. d. Ch. ou Recueil d. Mém.*
concern. l. Ch. et l. Art. qui en dépend. Vol. XXVII. Paris
1827. — *Bull. d. Sc. nat. et d. Geol. par le Bar. de Ferus-*
sac. Vol. XI. Paris 1827. — *Ann. der Phys. und. d. Ch.*
v. J. C. Poggendorff. Bd. X. Leipzig, 1827. — *Quart. Journ.*
of. Sc. Lit. and Arts, Vol. II. London, 1827. Atti d. Accad-
d. Sc. Napoli, 1826, pp. 7.

COVELLI N. — Memoria sulla costituzione geognostica della Campania. — *Atti R. Accad. Sc. fis. nat. Napoli (An. 1827)ʻ*
pp. 37.

COVELLI N. — Relazione di due escursioni fatte sul Vesuvio e di
una nuova specie di solfuro di ferro, che attualmente produ-
cesi in quel vulcano. *Atti d. Accad. Sc. Napoli, 1827. Na-*
poli 1839, in 4°pp. 15. (C. A.).

COVELLI N. — Mineralogia Vesuviana. — *" Il Pontano" Napoli*
1828-29, p. 19 and 145 (C. A.).

COVELLI N. — Sulla Beudantina, nuova specie minerale del Ve
suvio (1836). — *Atti d. R. Acc. d. Sc. e Lett. Sez. d. Soc.*
R. Borbonica, Vol. IV, Napoli 1839.

COVELLI N. — Sulla natura dei fumaioli e delle termantiti del
Vesuvio dove vivono e si moltiplicano varie specie d'insetti
(1826). — *Atti d. R. Acc. d. Sc. e Lett. d. Soc. R. Borbo-*
nica, Vol. IV, Napoli, 1839.

COZZA P. — Erupt. of Vesuvius A. D. 787. — *V. Archivio Storico*
per le Province Napoletane. Anno XV, fasc. III, Napoli,
1890, pp. 642-646.

Cozzolino V.—Cataloghi di minerali vesuviani. — *Three sheets.* 1844-1846.

Criscoli P. A. — Vesevi montis clogica inscriptio. — *Neapoli, 1632, 1 leaf in fol. fig. Vesuvius (O. V.)*

Crisconio P. — Il Vesevo: (Ode). — *Napoli, 1828, in 12°, pp. 12. (C. A.).*

Cristiano Fred. Prince of Denmark. — Memoria sulla eruzione del Vesuvio del 1820. — *Att. d. Accad. d. Sc. Napoli, 1820 in 4°, pp. 5, (C. A.).*

Crivella Antonio, detto il "Monaciello" improvissante. — Il fumicante Vesevo, ovvero il Monte di Somma bruggiato. Con diverse Terre, Casali, e luoghi situati nella sua falda. Con esservi anco un minuto ragguaglio di quanto in quello è successo. Composto in ottava rima. — *Napoli, 1632, in 12°, fol. 6. (C. A.).*

Curtis L. M. de. — Saggio sull'elettricità naturale diretto a spiegare i movimenti e gli effetti dei Vulcani. — *Napoli, 1780, in 8°, pp. 88 (O. V.)*

Cyrillo M. — Eruption du Vésuve en 1730. — *Philos, Trans. of the R. Soc. of London. Vol. XXII. London, 1732, and Vol. XXIII, 1733.*

Damiano P. — Breve narratione de' meravigliosi esempi occorsi nell' Incendio del Monte Vesuvio circa l'anno 1038, cavata dall'opera del B.° Pietro Damiano. — *Napoli, 1632, in 12°, fol. 4 (C. A.).*

Damiano P. — Il Vesuvio considerato qual bocca dell' Inferno. (Refers to an eruption A. D. 993. — *Opera Parissiis, 1663. 4 vols bound together in fol. Opusc, XIX, 6, 9, and 10, pp. 191-192. (C. A.).*

Damour A. A. — Analyse de la Périclase de la Somma. — *Bull. d. l. Soc. géol. d. France, 2.e Série, Vol. VI. Paris, 1849.*

Damour A. A. — Relation de la dernière éruption du Vésuve en 1850. — ?

Dana J. D. — On the condition of Vesuvius in Italy 1834. — *The American Journ. of Sc. and Arts; By B. Silloman. Ire Ser. Vol. XXVII. Newhaven, 1835.*

Dana J. D.—Abstract of a paper on the Leucite of Monte Somma by A. Scacchi, with observations.—*The American Journ. of Sc. and Arts; by B. Silliman. 2 Ser., Vol III. New-haven, 1847.*

Dana J. D. — Abstract of a paper on the humite of Monte Somma; by A. Scacchi, with observations. — *Am. Journ. Sc. 2nd Ser. pp. 175-182, Newhaven, 1852. (C. A.).*

Danza E. — Breve discorso dell' incendio succeduto a 16 dicembre 1631 del Vesuvio e luoghi circonvicini e dei terremoti della città di Napoli. — *Trani, 1632, in 8°, (O. V.).*

DARBIE F. — Istoria dell' incendio del Vesuvio dell' anno 1737. (In the work " Dei Vulcani e Monti Ignivomi " V. Anonymous). — *Livorno, 1779.*

DARTHENAY. — Mémoire sur la ville souterraine découverte au pied du mont Vésuve. — *Paris, 1748, in 8°, fol. II, pp. 52. (C. A.).*

DAUBENY C. — 1825. — *V. Raffles.*

DAUBENY C. — Some account of the eruption of Vesuvius which occured in the month of August 1834. Extracted from the manuscript notes of cav. Monticelli, foreign member of the Geological Society and from other sources; together with a statement of the products of the eruption and of the condition of the volcano subsequently to it. — *Philos. Trans. of the R. Soc. of London, London, 1835, pp. 153-159.*

DAUBENY C. — Remarks on the recent eruption of Vesuvius in december 1861. — *The Edinburgh Philos. Journ. Vol. XVII. Edinburgh, 1863.*

DAU L. — Lettere al Barone Durini intorno ad una nuova teoria spiegatrice dei fenomeni dei Vulcani. — *?, 1835, in 8°, pp. 32. (O. V.).*

DELAIRE. — Osservazioni sul Vesuvio negli anni 1745-1752. In the work of Mecatti. " Discorso storico filosofico del Vesuvio".— *Napoli 1752.*

DELVAUX. — Vue du Vésuve et d'une partie du Golfe de Naples—? (C. A.).

DEMARD E. — Extinction des Volcans. Etude sur les volcans en général et principalement sur les monts Vésuve et Etna. — *Rouen, 1873.*

DEQUEVAUVILLER. — Vue du Vésuve prise sur le bord de la mer et de côté de Portici — ?. (C. A.).

DE QUINONES J. — El Monte Vesuvio aora la Montaña de Soma.— *Madrid, 1632 in 4°, fol. 76. (V. Quinones). (C. A.).*

DESVERGES M. — Sur l'éruption du Vésuve en janvier 1839. — *Nouv. Ann. de Voyage 1839. Jahrb. für Min. Geog. Geol. und Petrefaktenk. Heidelberg, 1839.*

DEVILLE (Sainte-Claire) C. J. — Quatre lettres à M. Elie de Beaumont sur l'éruption du Vésuve du 1.ᵉ mai. — *Bull. d. l. Soc. Géol. de France. Vol. XII, Paris 1854-55. Compt. Rend. d. l. Acad. des Sc. Vol. XL-XLI. Paris, 1855. Zeitsch. der Deuts. Geol. Gesell. Bd. VII, Berlin, 1855.*

DEVILLE (Sainte-Claire) C. J. — Sur la nature et la distribution des fumeroles dans l'éruption du Vésuve du 1 mai 1855. — *Bull. d. l. Soc. Géol. de France, Vol. XIII. Paris, 1855-56 in 8°, pp. 55.*

DEVILLE (Sainte Claire) C. J. — Recherches sur les produits des Volcans de l'Italie Méridionale. — *Compt. Rend. d. l'Acad. T. LXIII, Juin 16, 1856, pp. 5.* (C. A.)

DEVILLE (Sainte Claire) C. J.—Lettres à M. Elie de Beaumont sur les phénomènes éruptifs de l'Italie Méridionale. — *Compt. Rend. 5me, T. 43, pp. 204-214, july 28, 1855—6me, T. 43, 9me T. 43, pp. 431-435, aug. 25, 1855—7me, T. 43, pp. 533-538, sept. 8th 1855—8me, T. 43, pp. 606-610, Sept. 22nd, 1856 — pp. 681-686 — 10me, T. 43, oct. 20, 1856, pp. 745-751 — 11me, T. 54, jan. 13, 1862, pp. 99-109 — 12me, T. 54, feb. 10, 1862, pp. 241-252 — 13me, T. 54. feb. 17, 1862, pp. 328-339—14me, T. 54, march 5, 1862, pp. 473-483. (C.A.),*

DEVILLE (Sainte-Claire) C. J.—Sur l'éruption du Vésuve du 1858.—- *Bull. d. l. Soc. Géol. d. France. Vol. XV. Paris, 1858.*

DEVILLE (Sainte-Claire) C. J. and A. Scacchi — Sur la Cotunnite du Vésuve). —*Bull. de la Soc. Géol., 2me Sér., T. XV, pp. 376-377, 1858, also Compt. Rend. d. l'Acad. d. Sc., Vol. 46, pp. 496-497, Paris, 1858.* (C. A).

DEVILLE (Sainte-Claire) C. J. — Eruption du Vésuve. — *Compt. Rend. d. l'Acad. des Sc. Vol. LIII. Paris, 1861.*

DEVILLE (Sainte-Claire), C. J. Le Blanc F, and Fouqué.— Sur les émanations à gaz combustibles qui se sont échappées des fissures de la lave de 1794, à Torre del Greco, lors de la dernière éruption du Vésuve. — *Compt. Rend. l'Acad. d. Sc., Vol. LV, pp. 75-76. Paris, 1862.* (C. A.)

DEVILLE (Sainte-Claire) C. J. — De la succession des phénomènes éruptifs dans le cratère supérieur du Vésuve après 1861. — *Compt. Rend. d. l'Acad. d. Sc., Vol., LXIII, pp. 7. Paris, 1866.*

DEVILLE (Sainte-Claire) C. J. — Observations relatives à une communication de M. Palmieri, intitulée : Faits pour servir à l'histoire éruptive du Vésuve. — *Compt. Rend. de l' Acad. des Sc. Vol. LXVI. Paris, 1868.*

DEVILLE (Sainte-Claire) C. J. — Réflexions au sujet des deux communications de M. Diego Franco sur l'éruption actuelle du Vésuve. — *Compt. Rend. d. l' Acad. des Sc. Vol. LXVII. Paris, 1868.*

DEVILLE (Sainte-Claire) C. J. — Observations sur la prochaine phase d'activité probable du Vésuve. — *Compt. Rend. d. l'Acad. des Sc. Vol. LXXVI. Paris, 1873.*

DICKERT. — Relief du Vésuve & du Somma avec dessins géognostiques. — *Bonn, 1849.*

DIGENSTED E. — Olivin vom Vesuv analyrsirt. — *Min. Mittheil. Bd. III., Wien, 1873.*

DION. CASSIUS. — Vesaevi montis conflagratio, Giorgio Merula interprete.—*Milan, 1503, in 4°, fol. LXXVI, (very rare).* (C.A.)

DOGLIONI N. — Anfiteatro d'Europa etc. — *Venetia, 1632, in 4°. pp. 72 + 1377. At p. 694*

DOLOMIEU D. de. — Sur l'éruption du Vésuve de l'an 2. — *Journ. de Phys. etc. Vol. LIII, p. 1.* (C. A.).

DOMENICHI J. —Montis Vesuvii alluvio; ad Lillam. 4 Epigrams in "Castaliae Stillulae".—*Florentiae, 1667, in 8°, pp. 187-190.* (C. A.)

DONATI E. — Phenomena observed at the last eruption of Mount Vesuvius in 1828. — *Journ. of the R. Inst. of Great Britain Vol. I. London, 1831, Bibliot. Univ. d. Sc. Lett. et Arts. Part. d. Sc. 1 Sér. Vol. II. Genève, 1831.*

DONATO S. da. (Attributed to). — Discorso filosofico et Astrologico, nel quale si mostra quanto sia corroso il Monte Vesuvio dal suo primo incendio sino al presente, e quanto habbi da durare detto Incendio. — *Napoli, 1632 in 4°, fol IV.* (C. A).

DORMAN H. — Ein Vesuvausbruch erimering an den April 1872. — *Naples, 1882, in 8°, pp. 14.* (C. A.)

DUCHANOY.—Détails sur la dernière éruption du Vésuve 1780.—(?).

DUCHANOY. — Esatta descrizione dell'ultima eruzione del Vesuvio.— *Osservazione appartenente alla fisica, alla storia naturale ed alle arti. 1780, in 8°, pp. 36.* (C. A.)

DUFRENOY A. — Terrains vulcaniques des environs de Naples. — *Paris, 1838, in 8°. fol. 4, pp. 420, pl. 9, figs.* (O. V.). *See also Comptes. Rend. Acad. Sc. Paris.*

DULAC A.—Compte rendu de l'histoire du Vésuve par della Torre.— *Mélanges d'Hist. Nat. Vol. IV., Lyon 1735. pp. 375-401. pl. I.*

DUPERRON de CASTERA. — Histoire du Mont Vésuve.—*Paris, 1741.* (O. V.).

DURER G. — Saggio di cataloghi per ordine di materie della libreria antica e moderna. 1° Vulcani e tremuoti.—*Napoli, 1866 in 12, pp. 104.*

DURIER E. — Le Vésuve en septembre 1878 —. (?).

ELISEO N. A. — Rationalis methodus curandi febres , flagrante Vaesevo subortas. — *(Pars prima), pp. 160, (Pars secunda) fol. 2, pp. 160. Napoli, 1634, in 8°,* (C. A.). *Another edition 1645 , in 8°, fol. 2, pp. 160 , with a fresh frontisp. and dedication.*

EMANUEL MONACUS — Vita S. Januarii E. M., in greco c latino (At the end a description of the eruption of A. D. 472, followed by another of the eruption of 685. — *Ex typi Montis Cassini, 1875, in large 4°, pp. 32.* (C. A.).

EMILIO L. d'. — La conflagrazione Vesuviana del 27 aprile 1872.— *Napoli, 1872, in 8.°, fol. IV, pp. 27.* (C. A.).

ESQUILACE Principe di.—Sonetto. (1631).—*V. de Quinones.* (C. A.).

ESTATICO (pseudon.) — Dissertazione intorno all'eruzione del Vesuvio del 1751. — *? In 8.°, pp. 27.* (O. V.).

EUGENII F. (DE) – Il maraviglioso e tremendo Incendio del Vesuvio detto in Napoli la montagna di Somma nel 1631. — *Napoli, 1631, in 4.° pp. 20.* (C. A.).

EYLES Sir Francis Haskins. — An account of an Eruption of Mount Vesuvius, in a letter to Phillip C. Webb, from Sir Francis Haskins Eyles. — *Phil. Trans, Vol. LII, 1762, pp. 39-41. Another account of an eruption of Mount Vesuveus.— Ibid. pp. 41-43.* (C. A.).

F. A. A. — Dialoghi sul Vesuvio in occasione dell' eruzione della sera de' 15 giugno 1794. Parlono Aletoscopo e Didascofilo. — *Napoli, 1794, in 8.°. pp. 52.* (C. A.).

FALCO B. DE. — Antiquitates Neapolis, atquae amoenissimi ejusdem Agri, ex emendatione, post multas alias editionae latinae, ex italicis factae cura Sigiberti Havercampi. — *Lug. Bal. in fol. pp. 48.*

FALCONE DELLI B. A. — Gli terrori del titubante Vesuvio. — *Napoli, 1632, in 8.°, pp. 24.* (C. A.).

FALCONE N. C. — L'intera istoria della famiglia, vita, miracoli, translazione e culto del glorioso martire S. Gennaro. — *Napoli, 1713, in 4.°, fol. VIII, pp. 526 with figures.* (C. A.).

FALCONE S. — Discorso naturale delle cause ed effetti causati nell' incendio del Monte Vesuvio, con Relatione del tutto.—*Napoli, 1632, in 4.°, fol. XXII.*

FARIA L. — Relacion cierta y verdàdera de el incendio de la montaña de Somma, ecc. — *Napoles, 1631, in small. fol., pp. 8.*

FARIA L. — Relacion de l'incendio del Vesuvio. — *Napoles, 1632.*

FARRAR A. S. — On the late eruption of Vesuvius.—*Rep. of the British. Assoc. for the Adv. of Sc. London, 1855.*

FARRAR A. S. — The earthquake at Melfi in 1851, and recent eruption of Vesuvius in 1855.—*Abst. of Proceed. Ashmodean Soc. N. 34, Oxford, 1856.*

FAUGAS DE SAINT FOND. — Recherches sur les volcans éteint du Vivarais et du Valay avec un discours sur les volcans brulants. — *Grenoble, 1778, in fol. Max. fol. 2, pp. 20+460, pl. 20, maps. 2.*

FAUJAS DE SAINT-FOND B. — Sur l'éruption du Vésuve de l'année
dernière. — *1780*.

FAVELLA G. — Abbozzo delle ruine fatte dal Monte di Somma con
il seguito incendio insino ad hoggi 23 di Gennaro 1632.—*Napoli, 1632, in 4.°, pp. 16*. (C. A.).

FELBER. — 1831. — *V. Nesteman*.

FENICI G. —· Lo struppio della Montagna de Somma, in rima napoletana. — *Napoli, 1632: in 8.°, fol. IV*. (C. A.).

FERBER. — Lettres à Mr. le Chev. de Born sur la mineralogie, et
sur divers objets d'histoire naturelle de l' Italie, traduit de
l'Allemand, enrichi de notes et d'observations faites sur les
lieux par M. de Dietrich.—*Strasburg.1776, in 8°, pp. 16+508*.

FER C. G. N. (DE) — Description du mont Vésuve tel que l'auteur
l'a vu en 1667. — (?) *fol. I, pl. I*.

FERRARA M. — Lettera sull'analisi della cenere del Vesuvio cruttata dal 16 al 18 giugno 1794. — *Napoli, 1794. pp. 14*.

FERREIRA-VILLARINO G. — Vera relatione di un spaventoso prodigio seguito nell'isola di S. Michele alli 2 di settembre di
questo presente anno 1630, tradotta dal portoghese in italiano. — *1.ʳᵗ Edition, Roma, 1630, in 8°; 2.ᵈ Edit., Napoli, on
the occasion of the eruption of 1631.; 1632, in 12° (fol. 4)*.

FIGUIER L. — Revue scientifique (Eruption du Vésuve). — *Feuilleton de la Presse du samedi, 18 Janvier, 1862*. (C. A.).

FILERT J. C. — De montibus ignovomis. — *Willeb. 1661, in 4°,
fol. 11*. (O. V.).

FILLICIDIO (IL). —· del Vesuvio, anacreontica etc.—*1, in 4°, pp. 8*.

FIORDELISI N. — Lettera all'arcidiacono Cagnazzi sulla elettricità
della cenere del Vesuvio. — *Giornale Enciclopedico, Napoli
1806, pp. 7 in 8.°* (C. A.).

FLORENZANO G. — Accanto al Vesuvio. Salmo. — *Napoli, 1872,
in 8.°, pp. 12*. (C. A.).

FLORUS L. A. — L'Heracleade ou Herculaneum enseveli sous la
lave du Vesuve. Poeme traduit en vers français avec des
notes par I. F. S. Maizony de Laureal. — *Paris, 1837, in
8.°, fol. II., pp. XXI+458, I. map*. (C. A.).

F. M. D. C. A. T. — Dettaglio su l'antico stato ed eruzioni del
Vesuvio colla ragionata relazione della grande eruzione accaduta ai 15 giugno 1794. — *Loc? in 8.°, pp. 16*. (C. A.).

FONSECA F. (DE)—Observations géognostiques sur la Sarcolite et
la Melilite du Mont Somma.—*Bull. d. l. Soc. géol. d. France.
Vol. IV. Paris, 1846-47*.

FONTANELLA GIROLAMO.—L'incendio rinovato del Vesuvio. Ode.—
Napoli, 1632, in 12.°, pp. 24. (C. A.).

FORBES J. D. — Remarks on mount Vesuvius. — *The Edinburgh Journ. of Sc.*, *Vol. VII. Edinburgh, 1827. Not. aus dem Gebiele der Natur. und Heilkunde. Bd. XVIII, Erfurt, und Weimar, 1827.*

FORBES J. D. — Physical notices of the Bay of Naples. N° I. On Mount Vesuvius. — *The Edinburgh Journ. of Sc. Vol. IX. Edinburgh, 1828. (Publ. anonymously).*

FORBES J. D. — Physical notices of the Bay of Naples: N.° 2. On the buried cities of Herculanum, Pompeii and Stabiae ; With note on Mount Vesuvius. — *The Edinb. Journ. of Sc., Vol. X, Edinburgh, 1829.*

FORBES W. A. — A visit to Vesuvius. — *Rep. of the Winchester Coll. of the Nat. Hist. Soc. 1875.*

FOREST J. — Le Vésuve ancien et moderne. — *Lyon, 1858, in 8.°, pp. 22.*

FORLEO G. — Meteorico discorso sopra i segni e cause dei terremoti et incendii di diverse parti della terra a causa dell'incendio della montagna di Somma. — *Napoli, 1632. in 4° fol. 6.*

FOUGEROUX DE BOUDAROY. — Observation sur le Vésuve près de la Ville de Naples. — *Compt. rend. d. l' Ac. d. Sciences. Paris, 1765.*

FOUGEROUX DE BOUDAROY. — Recherches sur les ruines d' Herculaneum et sur les lumières qui peuvent en résulter relativement à l' état présent des Sciences et des Arts. — *Paris, 1770. pp. XVI+232, pl. III. (C. A.).*

FOUQUÉ F. LE BLANC ET SAINTE-CLAIRE-DEVILLE. — Sur les émanations à gaz combustibles, qui se sont échappées des fissures de la lave de 1794 à Torre del Greco, lors de la dernière éruption du Vésuve. — *Compt. rend. hebdom. d. Séanc. de l'Acad. des Sc., Vol. LV. 1862, Vol. LVI. Paris, 1863.*

FOUQUÉ F. — Étude microscopique et analyse médiate d'une ponce du Vésuve. — *Compt. rend. hebdom. d. Séanc. d. l'Acad. d. Sc. Vol. LXXIX. N. 18, Paris, 1874.*

FRANCESCO II. — Lettera a S.ª Em. il Cardinale Arc.° di Napoli pe' danneggiati di Torre del Greco. Eruzione del 1861. — *Napoli, 1862, pp. 14. (C. A.).*

FRANCO D. — Excursion au cratère du Vésuve le 21 février 1868. — *Comp. rend. hebdom. d. Séances d. l'Acad. d. Sc., Vol. LXVI. Paris, 1868. pp. 9.*

FRANCO D. — Excursion faite, le 17 mars 1868, à la nouvelle bouche qui s' est ouverte à la base orientale du Vésuve. — *Compt. rend. hebdom. d. Séanc. d. l' Acad. d. Sc., Vol. LXVII. Paris, 1868.*

FRANCO D. — Faits pour servir à l'histoire éruptive du Vésuve. — *Compt. rend. hebdom. d. Séanc. d. l' Acad. d. Sc. Paris, 1868. Vol. LXVI, pp. 159-162.*

FRANCO D. — L' acido carbonico del Vesuvio. — *Napoli, 1872, in 4.°, pp. 31.*

FRANCO D. — Sur l' éruption d' avril 1872 au Vésuve. — *Compt. rend. hebdom. d. Séanc. d. l' Acad. d. Sc., Vol. LXXV. Paris, 1872.*

FRANCO P. — Memorie per servire alla Carta Geologica del Monte Somma. Memoria Prima. — *Rend. R. Accad. Sc. fis. mat. Napoli, fasc. 4°, 1883, pp. 13.*

FRANCO P. — Il Vesuvio ai tempi di Spartaco e di Strabone. — *(Atti Accad. Pontaniana, Vol. XVII. Napoli, 1887).*

FRANCO P. — Ricerche micropetrografiche intorno ad una pirossenandesite trovata nella regione vesuviana. — *Rend. Acc. Sc. fis. e mat. S. II, Vol. 2.°, fasc. 11.ª Napoli, 1888.*

FRANCO P. — I massi rigettati dal Monte di Somma detti lava a breccia. — *Napoli, 1889, in 4°, pp. 16, pl. 1.*

FRANCO P. — Quale fu la causa che demolì la parte meridionale del cratere di Somma. — *Atti Soc. It. Sc. nat., Vol. XXXI, 1889, pp. 81.*

FREDA G. — Sulla presenza del molibdeno nella sodalite Vesuviana. — *Rend. d. R. Acc. d. Sc. fis. e mat. Fasc. VII. An. 17, Napoli, 1878, pp. 3.*

FREDA G. — Sulla presenza dell' acido antimonioso in un prodotto Vesuviano. — *Rend, d. R. Acc. d. Sc. Fis. e Mat., Fasc. I, An. 18., Napoli, 1879.*

FREDA G. — Millerite del Vesuvio. — *Rend. d. R. Acc. d. Sc. fis. e mat. Napoli, maggio e giugno, 1880.*

FREEMAN. — An Entract of a Letter dated May 2., 1750, relating to the ruins of Herculaneum.. — *Phil. Trans. 1751. Vol. XLVII, pp. 131-142.* (C. A.).

FROJO G. — Osservazioni geologiche su di un ramo della lava del Vesuvio della eruzione del 1 maggio 1855, — *Ann. Sc. d. V. Janni e N. Buondonno. Vol. III. 1856, in 8° pp. 5.*

FUCCI P.—La crudelissima guerra, danni, e minacce del superbo campione Vesuvio, etc. — *Napoli, 1632, in 4°, pp. 8. Frontisp. with view of Vesuvius.* (C. A.).

FUCHS C. W. C. — Die Laven des Vesuv. Untersuchung der vulcanischen Erupt. Producte des Vesuv in ihrer chronologischen Folge, von 11 jahrhundert an bis zur gegenwart. — *Neues Jahrb. f. Miner. Geol. u Palœont. Stuttgart, 1866.*

FUCHS C. W. C. — Untersuchungen der Vesuv-Laven. — *Jahrb.*

f. Min. Geol. und Pal. v. Leonh. und Geintz. Stuttgart,
1866-68-69.

FUCHS C. W. C. — Die laven des Vesuv. — *Neus Jahrb f. Min,*
Geol. und Pal. 1868-1869 — Quart. Journ. Geol. Soc. Vol.
XXV, 1869, - Verhand. d. Naturh. Med. ver. z. Heidelberg
Bd. V. 1871.

FUCHS C. W. C. — Bericht über die vulkanischen Ereiguisse des
Jahres 1872 (Vesuv).—*Tscherm. Min. Mitheil. Wien, 1873.*

FUCHS C. W. C. — Bericht über die vulkanischen Ereignisse des
Jahres 1874. — (*Vesuv, Aetna etc.*) *Tscherm. Min, Mitth.*
Wien, 1875.

FUCHS C. W. C. — Bericht über die vulkanischen Ereignisse des
Jahres 1875. — *(Aetna, Vesuv, etc.) und Erdbeben. Tscherm.*
min. Mitth. Bd. II, Wien 1876.

FUCHS C. W. C. — Die vulkanischen Ereignisse des Jahres 1879,
(Vesuv und Aetna, — *Miner. und Petrograph. Mitt. Bd. III,*
Wien, 1880.

FURCHHEIM F. — Bibliotheca pompejana. Catalogo ragionato di
opere sopra Ercolano e Pompei pubblicate in Italia ed all'e-
stero, dalla scoperta delle due città fino ai tempi più recenti
con appendice : Opera sul Vesuvio. — *Napoli, 1879, in 8°,*
pp. 37. (C. A.).

F . . . V . . . —Notizie storiche delle eruzioni del Vesuvio. —
Annali Civ. d. due Sicilie. Napoli, 1833-47, Vol. VII, pp.
31-38.

GALANTE G. — Breve descrizione della Città di Napoli e del suo
contorno, etc. — *Napoli, 1792, in 8°, pp, XVI+ 1 + 34.*
With an appendix, Napoli, 1803, pp. 28.

GALEOTA O. (Ab. Galiani). — Spaventosissima descrizione dello
spaventoso spavento che ci spaventò tutti coll' eruzione del
Vesuvio la sera degli 8 d'agosto 1770, ma (per grazia di Dio)
durò poco. — *Napoli, 1825, in 8°, pp. 20.* (C. A.).

GALEOTA O. — Spaventossisima descrizione dello spaventoso spavento, che spaventò a tutti quanti la seconda volta colla
spaventevole eruzione del Vesuvio alli 15 giugno dell' anno
1794 a due ore scarse di notte, pure come era sortito l'anno
1770., che se ne fece la prima descrizione, che questa è la
seconda.—*Napoli, 1794, in 4°, with portrait, pp. 18. (very*
rare). (C. A.). *Another edit. Napoli, 1825, in 8°, pp. 20* (O.V.).

GALEOTA D. O. — Opera estemporanea all' impronto. — *Napoli,*
1795, in 4°, pp. VIII+ 23. (O. V.).

GALIANI F. — Catalogo delle materie appartenenti al Vesuvio contenute nel museo con alcune brevi osservazioni. — *Londra,*

1772, in 12°, pp. VIII+ 184. 2nd edit. Napoli, 1780, in 4°,
pp. 18. 3rd edit. Napoli, 1825, in 8°, pp, 20.

GALIANI F. — Osservazioni sopra il Vesuvio e delle materie appartenenti a questo vulcano ed altri contenuti in questo Museo. — *In the work " Dei vulcani e monti ignivomi". (V. Anonymous) — Livorno, 1779.*

GALIANI F. — Spaventosissima descrizione della eruzione del Vesuvio della sera del dì 8 agosto 1779. — *Napoli, 1779.*

GALLO M. — Cenno storico sulla fondazione di Ercolano, e sua distruzione, corredato di utili riflessioni sulla natura del Monte Vesuvio, applicabili ancora agli altri Vulcani.—*1.st part (published only) Napoli, 1829, in 8°, pp. XV + 81.* (C. A.).

GALLO M. — Saggio storico su la fondazione e distruzione di Ercolano e Pompei.—*Napoli, 1835, in 8°, fol. II, pp. 79,* (C. A.).

GAMA G. — Descrizione del tremuoto di Napoli del 15 giugno 1794 e successivo scoppiamento flammifero del Vesuvio, etc. Canto. — *Napoli, 1794, in 4°, pp, 36.* (O. V.).

GAMBA B. — Lettere descrittive di celebri Italiani.—*2nd Edit. Venezia, 1819, in 8°, pp. 8 + 262.*

GARSIA G. A. — I funesti avvenimenti del Vesuvio principiati martedì 16 dicembre 1631. — *Napoli, 1632, in 4°, pp. 12.*

GARUCCIN G. — La catastrofe di Pompei sotto l'incendio vulcanico del 79 ed il Vesuvio colla produzione dei suoi fuochi.— *Napoli, 1872, in 8°, pp. 30.* (A. C.).

GAUDIOSI T. — Sonetto per l'incendio del Vesuvio del 1660. — *Arpa Poetica. Napoli, 1671, in 8°. al p. 63.* (C. A.).

GAUDRY A. — Sur les coquilles fossiles de la Somma. — *Bull. d. l. Soc. Géol, d. France, Vol. X. Paris, 1852-53.*

GAUDRY A. — Lettre sur l'état actuel du Vésuve. — *Compt. rend. hebdom. d. Séanc. d. l'Acad. d. Sc. Vol. XLI, pp. 486-87. Paris, 1855.*

GAVAZZA G. — Sonetto. — *V. Giorgi Urbano.*

GEMMELLARO G. — Eruzione del Vesuvio. — *Folio 29 dell'Album di Roma, Oct. 1834, pp, 2.* (O. V.).

GENNARO ANT. DI (Duca di Belforte). — Sonetto sul Vesuvio, diretto al P. Ant. Piaggio. — *Loose sheet.* (C. A.).

GENNARO ANT. DI. — Poesie. Il Vesuvio, poema in 3 canti. — *Napoli, 1795, in 12°, fol. IV, pp. 298 + 2.* (C. A.).

GENNARO ANT. DI, — Poesie from p. LIII to LIX letter to Amaduzzi on the eruption of 1779. — *Napoli, 1797, in 8° (large) IV Vols.* (C. A.).

GENNARO (DE) B. — Historica narratio incendii vesuviani, anno 1631. — *Napoli, 1631, in 8° (Mentioned by Soria).*

GENNARO (DI) B. A. — Lettera sopra l'ultima eruzione del Vesuvio dell' anno 1779. — *Vulcani e Monti Ignivomi. Vol. II., (V. Anonimo). Livorno, 1779. Autol. Rom. N. 10, 1779.*

GENTILI G. — Dei vulcani o monti ignivomi più noti, e distintamente del Vesuvio. — *Livorno. II Vols. in 4°, Vol I, pp. LXX + 149, Vol, II. pp. VIII + 228, I pl.* (C. A.).

GERARDI A. — Relatione dell'horribile caso et incendio occorso per l'esalatione del monte di Somma detto Vesuvio vicino la Città di Napoli. — *Roma, 1631, in 4°, fol. IV.*

GERARDI ANT. — Warhaffte Relation von dem erschröcklichen Erdbidem und Fewrsgelwalt so auss dem Berg zu Somma, Vesuvio genant, nit weit von Neaples entsprungen, im Monat dicember 1631. — *Augsberg, 1632, in 4° fol. 4.* (C. A.).

GERI F. — Osservazioni sul Vesuvio. — (?).

GERNING. — Nachricht von den letzten Ausbruch des Vesuvs. — *Mag. für Physick von Voigt. Bd. X. Gotha, 1795.*

GERONIMO (DI) B. — Ragguaglio del Vesuvio. — *Benevento, 1737, in 8°, pp. 24.*

GERVASI E SPANO. — Raccolta di tutte le Vedute che esistevano nel Gabinetto del Duca della Torre, rappresentanti l'eruzioni del Monte Vesuvio fin oggi accadute con le rispettive descrizioni, ora per la prima volta ricavate dalla storia etc. — *Napoli, 1805, pp. 21, pl. XXVII.* (C. A.).

GEUNS W. J. VAN. — De Vesuvius en zijne geschiedenis. — (?) *in 8°, pp. 267-287.* (C. A.).

GIACHETTI J. — Apuliae terraemotus defloratio. — *Romae, 1632, in 4° (p. 7.).*

GIANETTI G. — La vera relatione del prodigio novamente successo nel Vesuvio, etc. — *Napoli, 1631, in 4°, pp. 8.*

GIANETTI G. — Rime dell'Incendio del Vesuvio. — *Napoli, 1632, in 12°, fol. VIII.* (C. A.).

GIANNELLI B. — Sulle ceneri vesuviane dell'anno 1779. — (?).

GIANNETTASII P. N. — Ver Herculaneum. — *Neapoli, 1704, in 8°, fol. IV, pp. 256 engr. frontisp.* (C. A.).

GIANNONE P. — Lettera scritta ad un amico che lo richiedeva onde avvenisse che nelle due cime del Vesuvio, in quella che butta fiamme ed è la più bassa, la neve lungamente si conservi e nell'altra che è più alta ed intera non vi duri che per pochi giorni. — *Napoli, 1718.*

GIGLI G. — Discorso sulla zona vulcanica mediterranea. — *Napoli, 1857: in 8°, pp. 146.* (C. A.).

GIMMA G. — Storia naturale delle Gemme, delle Pietre e di tutti i minerali, ovvero fisica sotterranea. — *Nopoli, 1703, in 4°, Vol. I, pp. 46 + 551, Vol. II, pp. + 603.*

GIOENI G. Saggio di litologia Vesuviana. — *Napoli, 1791, in 8°, fol. 6, pp. 272* (O. V.) *In German, Wien, 1793, in 8°, pp. 392.* (O. V.).

GIOENI G. — Versuch einer Lithologie des Vesuvs. — *Vienna, 1793, in 8°, pp. 392. Napoli, 1790, in 8°, pp. XCII + 208 in Ital. also 1791 pp. 272.* (C. A.).

GIORDANO G.—Sur la dernière éruption du Vésuve, déc. 8, 1861.— *Moniteur, 31 janv. 1862, fol. III.*

GIORDANO G. — Succinta relazione dell' avvenuto durante l' eruzione del Vesuvio del di 8 dicembre 1861. — *Atti d. Ist. d. Incor. a Sc. Nat. ed Econ. Vol. X ,Napoli,1864,pp. 507-516.* ·

GIORGI URBANO. — Scelta di poesie nell'incendio del Vesuvio. — *Roma, 1632, in 4°, pp. 94, engrav. frontisp.* (C. A.).

GIOVANELLI D. — Sopra la non antica apertura, o manifestazione dei Lagone di Monte Cerboli nell' agro Volterrano. Note relativa al contenuto nella presente lettera rapporto ai sassi piovuti in Siena. — *Giornale Letterario di Napoli, 1793-1798 Vol. LXI, pp. 3-21.*

GIOVO N. — Del Vesuvio. Song. — *Napoli, 1737 in fol., fol. II, pp. XXVI.* (C. A.).

GIRARD A. — Académie des Sciences (Eruption du Vésuve). — *Feuilleton du Journal des Débats du 28 mars 1862.* (C. A.).

GIROLAMO MARIA DI S. ANNA F. — Aggiunta all'Istoria della vita, virtù e miracoli di S. Gennaro vescovo e martire principal padrone della fedelissima Città , e Regno di Napoli. Nella quale si rapportano varie erudizioni, e molte ~curiose notizie. — *Napoli, 1710, in 4°, pp. 70.* (C. A.). *Other editions.*

GIROS S. — Veridica relazione circa l'ultima eruzione del Vesuvio accaduta ai 15 giugno per tutto luglio del 1794. — *Napoli, 1794, in 8°, pp. XXXV.*

GIUDICE (DEL) F. — Brevi considerazioni intorno ad alcuni più costanti fenomeni vesuviani.—*Atti d. R. Ist. d'Incor. a Sc. Nat. ed Econ. Napoli, 1855, in 4°, pp. 67, pl. VII.*

GIUDICE R.—Lettera relativa all' eruzione del Vesuvio dell' anno 1804. — *Magazzino di Lett. Sc. Arti, Firenze , feb. 1805, in 8°, pp. 7.* (O. V.).

GIULIANI G. B. — Trattato del Vesuvio e de'suoi incendii. — *Napoli, 1632, in 4°, pp. X + 224, pl. II. Engrav. frontisp.*

GIUSTINIANI L. — La biblioteca storica e topografica del Regno di Napoli (From pp. 215 to 228 refers to Vesuvius).—*Napoli 1793, in 4°, pp. XV + 241.* (C. A.).

GIUSTO L. — Biverbio del Sebeto col Vesuvio su gl'insetti microscopici del Colera. — ?, in 8°, pp. 16.

GIUSTO P. — Progetto di Associazione per compensamento dei danni che il Vesuvio può recare ai paesi messi sul suo pendio ed alla sua base. — Napoli, 1862 and 1872, in 4.°, pp. 24. (C. A.).

GLIELMO A.—L' incendio del Monte Vesuvio, etc. del 1631. — Napoli, 1632 and 1634, in 12.°, pp, 185. Two editions.

GMELIN L. — Observationes oryctognosticae et chimicae de Haüyna. — Heidelbergae, 1814, in 8°, pp. 6 + 58, pl. 1, figs. (O. V.).

GMELIN L. — Chemische Untersuchung eines blauen Fossils vom Vesuv und des Lasursteins. — (?)

GORCEIX H. — Etat du Vésuve et des dégagements gazeux des Champs Phlégréens au mois de juin 1860. — Compt. Rend. hebdom. des Séanc. d. l'Acad. d. Sc. Vol. LXXIV. Paris, 1872.

GORI A. F.—Notizie del memorabile scoprimento dell'antica Città Ercolano, etc.—Firenze, 1748. in 8°, pp. XX + 106, pl. 2.

GRANDE DE LORENZANA F.—Brebe compendio del lamentabile ynzendio del Monte di Soma.—En Napoles, 1632, in 8° (p. 16).

GRANVILLE A. AND BORZI.—A report on a Memoire of Sig. Monticelli entitled : Descrizione dell' Eruzione del Vesuvio avvenuta nei giorni 25, 26 Dic. 1813. — (?)

GRAVINA C. — Poesie. — Catania, 1834, in 12°, al p. 10. (B. N.).

GRAYDON G. — On the Dykes of Monte Somma in Italy. — (?)

GRIFONI H. — Vue du cratère du Vésuve après l' éruption d' Octobre 1822. Dessin par Griffoni, écrit par Marco di Pietro.—(?) Plate and description. (C. A.).

GROSSI G. B. G. —Ragionamento per i Comuni Vesuviani, Isola del Cratere, ecc. contro il Commune di Sarno, ed altri. Nel Consiglio d' Intendenza di Salerno. — Napoli , 1817 , in 4.°, (B. N.).

GUARINI G. — Analisi chimica d' un prodotto vesuviano. — Resoconto d. Acc. d. Sc. d. Napoli, tornata 5 settembre 1833, in 4.°, fol. 2. (C. A.).

GUARINI G. — Analisi chimica della sabbia caduta in Napoli la sera de' 26 agosto 1834. — Atti R. Ac. d. Sc. Napoli 2 decembre 1834,Vol. V, pt. 2ª, pp. 233-237. Atti R. Ist. Incorag. Sc. Nat. Napoli. Vol. V, pt. 2, pp. 233-237.

GUARINI G. — Saggi analitici sopra taluni prodotti vesuviani (1832)— Atti d. R. Acc. Sc. fis. e mat. Vol. V. Parte 2.ª Napoli, 1843.

GUARINI G. — Analisi chimica di un prodotto Vesuviano (1833)—

5

Atti d. Acc. d. Sc. Fis. e Mat. Vol. V. Parte 2.ª Napoli, 1843.

GUARINI G. — Saggi analitici su talune sostanze vesuviane (1834)— *Atti d. Accad. d. Sc. fis. e mat. Vol. V. Parte 2.ª Napoli. 1843.*

GUARINI G., SEMENTINI L. — Saggi analitici su talune sostanze vesuviane. — *Atti d. R. Acc. d. Sc. d. Nap. 1834, in 4.°, fol. 2.* (C. A.). *Atti R. Ist. Incorag. Sc. Nat. Napoli. Vol. V. pt. 2, pp. 169-177.*

GUARINI G., PALMIERI L. and SCACCHI A.—Memoria sullo incendio vesuviano del mese di maggio 1855, preceduta dalla Relazione dell'altro incendio del 1850.—*In 4.°, pp. 208, with 7 plates, Napoli, 1855.*

GUARINI R. — Poemata varia.—*Neapoli, 1821, in 12°, pp. 5-192.*

GUICCIARDINI C. — Mercurius Campanus praecipua Campaniae Felicis loca indicans, et perlustrans.—*Neapoli, 1657, in 12°, fol. 6, pp. 273.* (O. V.).

GUIDICCIONI L.—De Vesevo Monte,epigramma.—*See Giorgi Urbano.*

GUILLAUMANCHES-DUBOSCAGE G. P. I. DE. — Relation de l'éruption du Vésuve en 1822, suivie 1.° de l'observation d'un phénomène qui constate les moyens que la nature employe pour alimenter les volcans; 2.° de la comparaison de l'éruption de 1822 avec celle ou Herculaneum et Pompeii furent engloutis; à la suite est un aperçu sur les anciens volcans. — *Aix, 1823, in 8.° pp. 54, fol. 1.* (C. A.).

GUIRAUD DR. — L'éruption du Vésuve en avril 1872. — *Recueil d. Soc. Sc. Bel. Lett. et Arts de Tarn-et-Garonne. Montauban, 1872, tir. à part, in 8.°, pp. 32.* ('C. A.).

GUISCARDI G. — Del solfato potassico trovato nel cratere del Vesuvio nel Nov. e Dic. 1848. — *Napoli, 1849, in 8°, pp. 11, pl. 1, figs.*

GUISCARDI G. — Lettera all'egregio prof. A. Scacchi sullo stato del Vesuvio. — *Napoli, 1855, in 8°, fol. 2, pl. 1, pp. 407-412, fig. 1.*

GUISCARDI G. — Sopra un minerale del Monte Somma (Guarinite)— *Atti R. Accad. d. Sc. fis. e mat., Vol. II. Napoli, 1855-57. figs.*

GUISCARDI G. — Fauna fossile vesuviana. — *Napoli, 1856, in 8°, pp. 16.*

GUISCARDI G. — Notizie del Vesuvio. — *" Giambattista Vico " Vol. I, pp. 132-134, Vol. II, pp. 137-139, Vol. III, pp. 457-461, Vol. IV, pp. 136-137, 314-315, Napoli, 1857, separate extracts in 8°, pp. 14.* (C. A.).

GUISCARDI G. — Notizie vesuviane.—" *Giambattista Vico* " *1857, in 8°, pp. 14.*

GUISCARDI G. — Sublimazioni verdiccie sulle scorie d'una fumarola apparsa nel Vesuvio. — *Ann. d. R. Oss. Vesuviano. Napoli, 1859.*

GUISCARDI G. — Analisi chimica della Wollastonite del Monte Somma—? *1861.*

GUISCARDI G. — Sechs Briefe ueber den Vesuv an Herrn Roth, Neapel, 16 Juni 1861.—*Zetts. d. d. geol. Gesells. t. IX, pp. 383-386. M. S. copy.* (C. A.).

GUISCARDI G. — Sur l'éruption du Vésuve, lettre à M. Deville.— *Compt. Rend. Ac. Sc., Vol. LIII, pp. 1233-1236, Paris, 1861.* (C. A.).

GUISCARDI G. — Notizie vesuviane. — *Rendiconto R. Acc. Sc. Napoli. Luglio 1862, in 4.° pp. 2.* (C. A.).

GUISCARDI G. — Lettres sur la dernière éruption du Vésuve. — *Compt. Rend. hebdom. d. l' Acad. d. Sc. Vol. LXXV. Paris, 1872.*

GUISCARDI G.—Sulla genesi della Tenorite nelle fumarole del Vesuvio. — *Rendiconto R. Acc. di Sc. d. Napoli. Fasc. 4.° 1873, in 4.° pp. 4.*

GUISCARDI G. — Sulla Guarinite. — *Rend. R. Accad. Sc. Napoli, 1.° fasc. 1876, in 4.° pp. 4.* (C. A.).

GUISCARDI G. — Ueber den Guarinite, eine neue Mineralspecies vom Monte Somma.—*(Abdruck a. d. Zeitschr. d. deutschen geolog. Gesells. 1858) fol. 2......* Risposta del Prof. Guiscardi. —*Rend. R. Acc. Sc. Napoli. Fasc. 1°, 1876, fol. 2.* (O.A.).

GUISCARDI G. UND ROTH H.—Ueber Erscheinungen am Vesuv.— *Neapel den 8 Februar 1880.*

GUISCARDI G. — Descrizione dello stato attuale del cratere del Vesuvio. — *Annali Scientifici, Giorn. Sc. Fis. Nat. Agric. etc. Vol. II, pp. 249-251.*

GUISCARDI G. — Sulla presenza di combinazioni del titanio e del boro in alcune sublimazioni vesuviane.—*Rendicont. d. Acc. d. Sc. d. Napoli, in 4.° pp. 3.* (C. A.).

GUISCARDI G. — Ueber die neuesten Kraterveränderungen und Ausbrüche des Vesuvs. — (?)

GUTTENBERG. — Eruption du Mont Vésuve du 14 mai 1771. — *Naples. A plate.* (C. A.).

GUTTENBERG. — Vue de la sommité et du cratère du Vésuve au moment de l'Eruption du 8 Août 1779. — ? (C. A.).

H... — Sur le Vésuve. — *Two articles in the "Journal d. l'Empire, 6 and 10 nov. 1807.* (C. A.).

HAIDINGER W. — On the sodalite of Vesuvius. - *The Edinburgh Philos. Journ. Vol. XIII. Edinburgh, 1825.*

HALL E. — 1872. — See *Haughton.*

HAMILTON W. — Observations on Vesuvius, Aetna and other volcanoes. —*London, 1773, in 8°, fol. 2, pp. 180, pl. 6, figs. Also same, London, 1774, and 1783.*

HAMILTON W. — Beobachtungen über den Vesuv, den Aetna und andere Vulcane in einer Reihe von Briefen. — *Berlin, 1773, in 8°, pp. 196, fol. 2, pl. 6, figs.*

HAMILTON W. — Observations on mount Vesuvius, mount Etna and other Volcanoes in a series of letters to the Royal Society— *Philosoph. Transactions of the R. Soc. of London. Vol. LVII-LXI. London. 1774, in 8°, pp. IV + 79, pl. 5, map. 1. (Separate edition in same year).*

HAMILTON W. — Campi Phlegraei. — *Napoli, 1776, Vol. I, pp. 90, map. I, pl. 2, frontisp., Vol. II, fol. 53, pl, 53, frontisp. Vol. III, pp. 29, fol. 6, pl. 5, frontisp.*

HAMILTON W. — Supplement to the Campi Phlegraei being an account of the great eruption of mount Vesuvius in August 1779 (in English and French). — *Naples, 1779, in fol. pp. 29, col. pl. 5.*

HAMILTON W. — Nachrichten von den neuesten Entdeckungen in der J. C. 79 am 24 August durch den Ausbruch des Vesuv verschuetteten Stadt Pompeji mit einigen Zusaetzen begleitet, von Cristoph Gottgeb. von Murr. — *Nuernberg, 1780, in 4° fol. 2, pp. 26, pl. 3.*

HAMILTON W. — Oeuvres complètes, traduites et commentées par l'abbé Giraud-Soulavie.—*Paris, 1781, in 8°, pp. XX + 506, map. 1. (C. A.).*

HAMILTON W. — Warnee mingen over de vuurbergen in Italie, Sicilie en omstreeks den Rhyn. — *Amsterdam, 1784, in 8°, pp. 16 + 55, fol. 4, pl. 2, (O. V.).*

HAMILTON W. — Some particulars of the present state of mount Vesuvius; with the account of a journey into the Province of Abruzzo, and a voyage to the island of Ponza. — *Philos. Transact. of the R. Soc. of London. Vol. LXXVI. London, 1786, pp. 19, map. 1. Dresden, 1787. (In German).*

HAMILTON W. —Bericht vom gegenwaertigen Zustande des Vesuv und Beschreibung einer Reise in die Provinz Abruzzo und nach der Insel Ponza. — *Dresden, 1787.*

HAMILTON W. — An account of the late eruption of mount Vesuvius. — *Philosph. Transact. of the R. Soc. of London. London, 1795, pp, 73-116, pl. 7.*

HAMILTON W. — Campi Phlegraei. — *Parts, l'an septième* (1799) *in fol. Atl. fol. 60, pl. 60.* (O. V.).

HANSEL V. — Mikroscopische Untersuchung der Vesuvlava vom Jahre 1878. — *Mineral. und Petrograph. Mittheil., Band II, Wien, 1879.*

HAUGHTON S. and HALL E. — Report on the chemical, mineralogical and microscopical characters of the lavas of Vesuvius from 1631 to 1868. — *Transact. of the R. Irish Ac., Vol. XXVI, Dublin, 1875.*

HEIM A. — Der Ausbruch des Vesuv im April 1872. — *Basel, 1873, in 8°, pp. XV+52, pl. 4.*

HEIM A. — Der Vesuv im April 1873. — *Zeitsch. d. Deut. Geol. Gesell. Bd. XXV, Berlin, 1873.*

HELBIG. — Untersuchungen über die campanische Wandmalerei. — *Leipzig, 1883, p. 105.*

HELLWALD (von) F. — Historische Nachrichten ueber den Vesuv.-(?).

HERBINII J. — Dissertationes de admirandis mundi Cataractis supra et subterraneis etc. — *Amstelodami, 1678 , in 4. pp. 14 + 267 + 17, figs.*

HESSENBERG F. — Magnesiaglimmer (Biotit) vom Vesuv. Mineral. Notiz von Hessen. — *Frankfurt, 1861.*

HESSENBERG F. — Titanit vom Vesuv. Mineral. Notiz von Hessen. — *Frankfurt, 1861.*

HIMMEL. — Nachricht von dem Ausbruche des Vesuvs am 15 Junius. — (?).

HOCHSTETTER (von) F. — Die Phlegräischen Felder und der Vesuv.-(?)

HOFFMAN F. — Mémoire sur les terrains volcaniques de Naples, de la Sicile, et des isles de Lipari. — *Bull. d. l. Soc. géol de France, Vol. III. 1833, pp. 170-180.*

HOMBRES-FIRMAS L. A. (D') — Souvenirs de voyage aux environs du Vésuve. — *Bull. d. l. Soc. Géogr. Tom. XVII. Paris, 1842. pp. 205-213.*

HON H. (LE) — Histoire complète de la grande éruption du Vésuve de 1631. — *Bull. d. l'Acad. R. d. Sc. Lett. et Beaux-Arts de Belgique. Vol. XX, Bruxelles, 1865-1866*

HOSPITAL MARCHESE DE L'. — Memoria sopra la Città sotterranea scoperta a' piedi del Monte Vesuvio. — *Raccolta di Opuscoli Sc. e Filos. Venezia, 1727-38-57, Vols. LI, in 12°, figs. See V. XLI, pp. 1-61.*

HOWARE J. — Observations on the heat of the ground of Vesuvius. — *Philosoph. Transact. of the R. Soc. of London , Vol LXI, London , 1771. (Also in French) Journ. d. Phys., T. XIII, pp. 224, 1779.*

HUERTA A. (DE) — Sonetto. — *See De Quinones.*

HURTADO DE MENDOÇA A. — Dezimas.—*See De Quinones.*

IL FILLICIDIO. — Vesuvio. Anacreontica.—? *in 4°, pp. VIII.* (C.A.).

IMBERT DE VILLEFOSSE. — Vue du Mont Vésuve et de son éruption arrivée le 25 octobre 1751 à 10 heures du soir.—*Fol. I.* (C. A.).

INCARNATO C. — Prodigium Vesevi Montis, etc. — *Neapoli, 1632, in 4°, pp. 7. Another edition in 8.°*

INCREDULO ACCADEMICO INCANTO. — Incendio del Vesuvio. Ode. — *Napoli, 1632, in 12°.* (B. N.).

INCREDULO ACCADEMICO INCANTO.—Le Querele di Bacco per l'incendio del Vesuvio. Ode. — *Napoli, 1632, in 4°, pp. 16,* (B. N.).

INOSTRANZEFF (von) A. — Historische Skizze der Thätigkeit des Vesuvs vom Jahre 1857 bis jetzt. — *St. Petersburg, 1872.*

INOSTRANZEFF (von) A.—Ueber die Mikrostructur der Vesuv-Lava vom Septemb. 1871, März und April 1872. — *St. Petersburg, 1872.*

INSENSATO ACCADEMICO FURIOSO. — L'afflitta Partenope per l' incendio del Vesuvio al suo glorioso Protettore S. Gennaro.— *Napoli, 1632, in 12°, fol. 8.* (C. A.).

ISABEY I.—Cratère du Vésuve après l'éruption de 1822.—(C. A.).

ISE A. — Fussreisse vom Brocken auf den Vesuv und Rueckkehr in die Heimat.—*Leipzig, 1820, in 8°, pp. XII+234, pl. 1.* (C. A.).

ITTIGIUS G. — De Montum incendiis. — *Lipsiae, 1671, in 8°, fol. 8, pp. 342, fol. V,* (O. V.).

IZZO 'S. — Altra Relazione del Monte Vesuvio. — *Gazzetta (Supplemento) Napolitana Civica Commerc. N.° 76, 1804 ?.* (B. N.).

JADELOT L'AB. — Mechanisme de la nature, ou Système du Monde fondé sur les forces du feu, etc. — *Londres, 1787, in 8°, pp. XVI+259. See pp. 209-259.*

JAMES C. — Voyage scientifique à Naples. — *Paris, 1844, in 8°, pp. 103* (O. V.).

JAMINEAU J. — An extract of the substance of three letters concerning the late eruption of Mount Vesuvius. — *Philosoph. Tansact. of the R. Soc. of London. Vol. XLIX, London, 1755, pp. 24-28.*

JAMINEAU J. — Eruption of Vesuvius in December 1754. — *Philosoph. Transact. of the R. Soc. of London, Vol. XLIX. London, 1755.*

JANNACE V. — La storia d'havere timore, e gran spavento dello foco dello inferno, lo quale si è scoperto per causa de li no-

stri peccati nella montagna di Somma la quale si è aperta,
e buttato lingue di foco, e cenere, e pietre che ha consu-
mato tridece tra terre e casali intorno di se, li quali segni
ci ha mostrato Iddio per nostro beneficio. E questo, e successo di martidi matino alli 16 di decembre 1631. — *Napoli,
1632 in 12°, fol. 6.* (C. A.).

JANUARIO F. (DE) — Felicis Campaniae Hilaritas tumolata. — *Neapoli, 1632, fol.* (C. A.).

JATTA G. — Discorso sulla ripartizione Civile, e Chiesastica del-
l'antico agro Cumano, Misenate, etc. — *Napoli, 1843, in 8°,
pp. VIII-+242.*

JAUCOURT DE. — Vésuve. — *Article of " Encyclopédie." Genève,
in 4°, pp. 330-334.* (C. A.).

JERVIS G. — Tesori sotterranei dell'Italia. — *4, Vol. in 8°, Torino, 1874-1888, numerous plates.*

JOHNS C. A. — Vesuvius previous to, and during the eruption of
1872 — (?)

JOHNSTON-LAVIS H. J. — A visit to Vesuvius during an eruption—
"Science Gossip." N.° 181, January 1880, pp. 9-10.

JOHNSTON-LAVIS H. J. — Note on the comparative specific gravi-
ties of molten and solidifed Vesuvian lavas. — *Quart. Journ.
Geol. Soc. Lond. Vol. XXXVIII, 1880, p. 240-241.*

JOHNSTON-LAVIS H. J.—Volcanic cones, their structure and mode
of formation. — *"Science Gossip." N.° 190, Oct. 1880, pp.
220-223, fig. 1.*

JOHNSTON-LAVIS H. J. — On the origin and structure of Volcanic
cones. — *"Science Gossip." N.° 193, Jan. 1881, pp. 12-14,
fig. 4.*

JOHNSTON-LAVIS H. J.—Diary of Vesuvius from Jan.1st to July 16th
1882. — *"Nature" Vol. XXIV, 1882, pp. 455-456, fig. 2.*

JOHNSTON-LAVIS H. J. — The late eruption of Vesuvius. — *"Nature," Vol. XXIX, 1884, p. 291.*

JOHNSTON-LAVIS H. J.—The Geology of Monte Somma and Vesuvius,
being a study in Vulcanology. — *Quart. Journ. Geol. Soc.
Lond., Vol. XL. 1884. pp. 35-112, with 2 woodcuts and 1
cromolithographic plate.*

JOHNSTON-LAVIS H. J. — First Report of the committee for the
investigation of the Volcanic phenomena of Vesuvius and
its neighbourhood.— *Brit. Assoc. Reports, 1885, pp. 2.*

JOHNSTON-LAVIS H. J.—The new outburst of lava from Vesuvius.—
"Nature." Vol. XXXII, pp. 55-108.

JOHNSTON-LAVIS H. J. — The physical conditions involved in the injection, extrusion and cooling of igneous matter. — *Quart. Journ. Geol. Soc. Lond. Vol, XLI, 1885, pp. 103-106.*

JOHNSTON-LAVIS H. J. — Notes on Vesuvius from February 4th to August 7th 1886. — *Nature, Vol. XXXIII, 1886, p. 557.*

JOHNSTON-LAVIS H. J. — On the fragmentary ejectamenta of Volcanoes. — *Proceed. Geol. Assoc. Vol. IX. 1886, pp. 421-432, fig. 3.*

JOHNSTON-LAVIS H. J. — Second Report of the committee for the investigation of the Volcanic phenomena of Vesuvius and its neighbourhood. — *Brit. Assoc. Reports. 1886, pp. 3, also "Nature" Vol. XXXIV, p. 481.*

JOHNSTON-LAVIS H. J. — Sounding a crater, fusion points, pyrometers, and seismometers. — *" Nature" Vol. XXXV, p. 197.*

JOHNSTON-LAVIS H. J. — The relationship of the activity of Vesuvius to certain meteorological and astronomical phenomena. — *Proceed. Royal. Soc. Lond. 1886, N.° 253, p. 1.*

JOHNSTON-LAVIS. H. J. — The relationship of the structure of igneous rocks to the conditions of their formation. — *Scientific, Proceed. R. Dublin Soc. Vol. V., N. S., pp. 112-156.*

JOHNSTON-LAVIS H. J. — Vesuvian eruption of February 4th 1886.— *Nature, Vol. XXXIII, 1886, p. 367.*

JOHNSTON-LAVIS H. J. — Diario dei fenomeni avvenuti al Vesuvio da luglio 1882 ad agosto 1886. — *"Lo Spettatore del Vesuvio e dei Campi Flegrei." Nuova serie pubblicata a cura e a spese della Sezione Napoletana del Club Alpino Ital. Furchheim Napoli, 1887, in 4°, pp. 81-103, with 13 photo-engravings.*

JOHNSTON-LAVIS H. J. — L'eruzione del Vesuvio nel 2 Maggio 1885. — *Ann. d. Accad. O. Costa d. Aspiranti Naturalisti. Era 3. Vol. I, 1887, Naples, pp. 8 with 1 photo-engraving and 1 cromolithograph.*

JOHNSTON-LAVIS H. J. — Third Report of the committee appointed for the investigation of the Volcanic phenomena of Vesuvius and its neighbourhood. — *Brit. Assoc. Reports, 1887, pp. 3.*

JOHNSTON-LAVIS H. J. — Fourth report of the committee for the investigation of the Volcanic phenomena of Vesuvius and its neighbourhood. — *Brit. Assoc. Reports, 1888, pp. 7.*

JOHNSTON-LAVIS H. J. — Further observations on the form of Vesuvius and Monte Somma. — *Geol. Magaz. dec. III, Vol. V, London, 1888, pp. 445-451, fig. 1.*

JOHNSTON-LAVIS H. J. — Note on a Mass containing metallic iron found on Vesuvius. — *Brit. Assoc. Report. 1888, pp. 2.*

JOHNSTON-LAVIS, H. J. — The Conservation of Heat in Volcanic Chimneys. — *Brit. Assoc. Reports, 1888, pp. 2.*

JOHNSTON-LAVIS H. J. — The ejected blocks of Monte Somma. Part I. Stratified limestones, — *Quart. Journ. Geol. Soc., Vol. XLIV, 1888, pp. 94-97.*

JOHNSTON-LAVIS H. J. — Fifth report of the committee appointed for the investigation of the Volcanic phenomena of Vesuvius and its neighbourhood. — *Brit. Ass. Reports, 1889, pp. 12 with 5 woodcuts.*

JOHNSTON-LAVIS H. J.—Il pozzo Artesiano di Ponticelli.—*Rend d. R. Accad. d. Sc. Fis. e Mat. d. Napoli, giugno 1889, pp. 7.*

JOHNSTON-LAVIS H. J. — L'état actuel du Vésuve. — *Bull. Soc. Belge de Géologie, Hydrologie et Paléontol., Vol. III, 1889, pp. 1-11, fig. 3.*

JOHNSTON-LAVIS H. J. — The new eruption of Vesuvius. — *"Nature" Vol. XL. 1889, p. 34.*

JOHNSTON-LAVIS H. J. — The recent activity of Vesuvius. — *Nature, Vol. XXXIX, 1889, pp. 184.*

JOHNSTON-LAVIS H. J.—The state of Vesuvius.—*Ibid, pp. 302-303.*

JOHNSTON-LAVIS H. J. — Viaggio scientifico alle regioni vulcaniche italiane nella ricorrenza del centenario del "Viaggio alle due Sicilie" di Lazzaro Spallanzani. — *(This is the programme of the excursion of the English geologists that visited the south Italian volcanoes under the direction of the author. It is here included as it contains various new and unpublished observations) Napoli, 1889, in 8°, pp. 1-10.*

JOHNSTON-LAVIS H. J. — Sixth report of the committee appointed for the investigation of Vesuvius and its neighbourhood. — *Brit. Assoc. Reports. Leeds meeting. London, 1890, pp. 14, fig. 3.*

JOHNSTON-LAVIS H. J. — Geological Map of Vesuvius and Monte Somma with a short explanation. Scale 1:10.000. Constructed intirely by the author during the years 1880-1888. — *Philip. and Son. 32 Fleet St. London , 1891. (Also Italian Edition, 1891).*

JUNGSTE DE CLASSENS E. A. — Souvenir d'une promenade au Mont Vésuve.— *Naples, 1841, in 8.°, pp. 61.*

JORDANUS FABIUS. — De Vesuvio monte (1631). — *M. S. in the Bibl. Brancacciana, in 8°, pp. 60, Copy. (C. A.)*

JORIO A. DE. — Notizie su gli scavi di Ercolano. — *Napoli, 1827 in 8°, pp. 122, pl. 5. (C. A.).*

JUDD W. J. — Contributions to the study of Volcanoes. — *Geol. Mag. Vol. II, London, 1876.*

KENNGOTT A. — Remerkungen über die Zusammensetzung einer Vesuvlava.—*Zeitsch. f. d. ges. Naturw. Bd. XV. Berlin, 1860.*

KENNGOTT A.—Pyrit, Calcit, Anorthit vom Vesuv.—*Vierteljahrsch.
der Naturf. Gesell. in Zuerich, Bd. XIV. Zuerich, 1869.*

KENNGOTT A. — Salmiak vom Vesuv.—*Vierteljahrssch. der Naturf. Gesell. in Zuerich. Bd. XV, Zuerich, 1870.*

KERNOT F. — L' Acqua Filangieri minerale acidolo-alcalina con
l'analisi quantitativa del Prof. R. Monteferrante. — *Napoli,
1873, in 8°, pl. 1.*

KIRCHERII A. —Diatribae de prodigiosis crucibus quae tam supra
vestes hominum, quam res alias non pridem post ultimum
incendium Vesuvii Montis, Neapolis comparuerunt. — *Rome,
1661, in 8°, fol. 4, pp. 103, pl. 1. (C. A.).*

KIRCHERII A. — Mundus subterraneus. — *Amstelodami, 1665 and
again in 1678, Vol. I, pp. 18 + 266 + 6, Vol. II, pp. 8 +
507 + 9. Numerous figs.*

KLAPROTH. — Risultato dell' analisi di alcune sostanze minerali.
Referring to incrustations, etc. of sulphates in an opening in
the Vesuvian conc. — *Giorn. Lett. di Napo'i, 1793-1798,
Vol, XC. pp. 81-104.*

KLUGE. — Verzeichniss der Erdbeben und vulkanischen Eruptionen
und der dieselben begleitenden Erscheinungen in den Jahren
1855 und 1856. — *Allg. Deut. Nat. Zeit. Vol. XVIII, pp.
361-416. (C. A.).*

KOBELL. (von) F. — Analyse eines sinterartigen Minerals vom
Vesuv. — *Gelehrte Anzeigen; herausg. v. Mitgl. d. Kön.
Bayr. Akad. d. Wissensch. Bd. XXI, München, 1845.*

KOESTLIN C. H. —Examen mineralogico-chemicum materiei, quae
Herculaneum et Pompejos anno 79 aerae Christ sepelivit.—
*Fasciculus animadversionum phisiologici atque mineralo-
gico-chimici argumenti. Stuttgardiae, 1780, in 4°, (C. A.).*

KOKSCHAROW (von) N. — Ueber den zweiaxigen Glimmer vom
Vesuv (1854). — *Ann. der Physik und Chemie. Bd. XCIV,
Leipzig 1855. Bull d. l. Classe Phys. Mathém. d. l' Acad.
Impér. des Sc. de St. Pétersbourg. Vol. XIII. St. Péters-
bourg, 1855.*

KOKSCHAROW (von) N. — Messungen eines besonders vollkommen
ausgebildeten Anorthitcrystalls vom Vesuv.—*Bull. d. l. Classe
Phys. Mathém. de l'Acad. Impér. des Sc. de St. Pétersb.
Vol. VII, N. S. St. Pétersbourg, 1864.*

KOKSCHAROW (von) N.—Ueber den Glimmer vom Vesuv.—*Mate-
rialien zur Miner. Russlands. B. VII. St. Petersburg, 1875.*

KREUTZ F. — Mikroskopische Untersuchungen der Vesuv-Laven
vom Jahre 1868. — *Anzeiger der Akad. der Wiss. Bd. IV.
Berlin, 1869.*

KURR. — Ueber den letzten Ausbruch des Vesuvs im December 1861. — *Jahresh. d. Ver. f. vaterl. Naturk. in Würtemberg Bd. XIX, Stuttgart, 1863.*

LA CAVA P. — Sulla efflorescenza della soda clorurata che trovasi in taluni fumaiuoli attivi del Vesuvio; memoria.—*Rendicont. R. Accad. Sci. Napoli, 1840, in 8°, pp. 6.* (C. A.).

LA CAVA P. — Rapporto sui cambiamenti avvenuti al Vesuvio dal 27 decembre al 19 marzo 1843. — *?, in 8°, pp. 12* (O. V.).

LALANDE (de).—Relation de la dernière éruption du Vésuve, Août 1779. — *Journ. d. Savants. Paris, janvier 1781, in 12°, pp. 103-114.*

LALANDE (de). — Du Mont Vésuve et de la nature des laves. — (*dans le "Voyage en Italie"). Paris, 1779, T. VII, pp. 153-206. 2nd Edit. 1786. 3me Edit. Genève, 1790, Vol. VII. in 8°, with an atlas in 4°, pl. 35.*

LANCELLOTTI J. — Epistolae tres: I, De Incendio Vesuvii; II, De Stabiis; III, De petitione Magistratura.—*Neapolis, 1784, in 8°, pp. 30.*

LANDGREBE G. — Mineralogie der Vulcane. — *Cassel und Leipzig, 1870, in 8°.*

LANFELFI. — Incendio del Vesuvio. — *Napoli, 1632. in 4°, fol. 8, figs.*

LANZA. — *See Liberatore.*

LASAULX (von) A. — Dünnschliffe der Vesuv-Lava der Eruption vom April dieses Jahres.—*Sitzungsb. der niederrhein. Gesell. in Bonn. B. XXIX. Bonn, 1872.*

LASAULX (von) A. — Microscopische Untersuchung der neuesten Lava vom Vesuv. — *Neues Jahrb. f. Mineral. Geol. u. Pal. B. XL. Stuttgart, 1872.*

LASENA P. — Dell'antico Ginnasio Napoletano. — *Roma, 1641, in 4°, pp, 3 + 292, At pp. 77-84.*

LATINA. — Lo scalco alla moderna, overo l' arte di ben disporre li conviti. — *Napoli, 1692-94, 2 Vol. in 4°, In Vol. II al pp. 234-238, the eruption of Vesuvius is described of April 12th 1694.*

LAUGEL A.— Sur l'éruption du Vésuve du 8 déc. 1861.—*Moniteur de la Côte d'Or. jan. 1862, fol. 3.*

LAUGIER A. — Examen chimique d'un fragment d'une masse saline considérable rejetée par le Vésuve dans l' éruption de 1822.—*Mém. du Museum d'Hist. Nat. Vol. X. Paris, 1823—Ann. de Chimie, N. S. Vol. XXVI. Paris 1824.—Quarterly Journ. of Sc. Liter. and Arts. Vol. XVIII, London. 1825.*

LAURENTIIS M. DE — Universae Campaniae Felicis Antiquitates.—
*Neapoli, 1826, in 4°, Vol. I, pp. 7 + 288, pl., Vol. II, pp.
303, pl. 1.*

LAVINI G.—Rime filosofiche colle sue annotazioni alle medesime.—
Milano, 1750, in 4°, pp. XXXII+ 232.

LAVINI G. — Analyse de la cendre du Vésuve de l'éruption 1822
(1828). — *Mém. d. R. Acc. d. Sc. d. Torino, Vol. XXXIII.
Torino, 1829.*

LEBERT II. — Le Golfe de Naples et ses volcans, et les volcans
en général. — *Lausanne, 1876, in 8°, pp. 120, pl. 1.*

LECOUTOURIER.—Phénomènes observés au Vésuve (par M. Palmieri). — *Musée des Sc. Mai 1856 (?)*

LE HON H. — Histoire complète de la grande éruption du Vésuve
de 1631.—*Bruxelles, par M. Hayez, 1865, in 8°. pp. 64, Map
of lavas since 1631. Sc. 1:25000.*

LEMMO G. — Pietosa istoria del danno accaduto nel paese detto
Somma, non già del foco ; ma di acqua, pietre, arena, e
saette, che ha spianato detto paese, con Ottajano. (Verses). —
(Napoli, 1794), in 8°. (B. N.).

LEMMO G. — Prodigioso miracolo del nostro gran Santone, e Difensore S. Gennaro, d'averci liberato dall'incendio del Vesuvio,
e dal terremoto nell'anno 1794. (Verses).—?, *in 8°.* (B. N.).

LEO M. DI.—Il Vesuvio nell' ultima eruzione dell' 8 agosto 1779.
Canto. — *Napoli, 1779, in 8°, pp. 26. Another edition. pp.
24,* (O. V.).

LEONE AMBROGIO DETTO NOLANO. — La storia di Nola. (This book
contains the oldest figure of Vesuvius. It is on this authority
that writers attribute an erupt. of Vesuvius in 1500). — *Venezia, 1514, in fol.* (O. V.).

LEONIS A.—Antiquatum nec non Historiarum Urbis ac Agri Nolae,
ut et de Montibus Vesuvio, et Abella descriptionis. — *Venetiis, 1514, and Lug. Bat. In fol. pp. 4 + 92 + 10, pl. IV.*

LE PERE G. — Deuxième recueil de divers memoires sur les Pouz-
zolanes naturelles et artificielles. — *Paris, 1807, in 4°, pp.
8 + 62, pl. 1.*

LE RICHE M. J.—Antiquités des Environs de Naples et dissérta-
tions qui y sont relatives par M. J. L. R. — *Naples, 1820,
in 8°, pp. 392 +5.*

LIBERATORE R. — Delle nuove ed antiche terme di Torre Annunziata e parere di V. Lanza sulle facoltà salutifere dell'acqua
termo-minerale Vesuviana Nunziante — *Annali Civile del Regno delle Due Sicilie, Vol. VI, pp, 95-109. Napoli, 1835, in
8°, pp. 56, pl. 1.* (C. A.).

Licopoli G. — Storia naturale delle piante che nascono sulle lave Vesuviane. — *Napoli, 1871, in 8°, fol. 2, pp. 58, pl. 3.* (C. A.).

Lidiaco T. — Stanza a Crinatea, (L'eruzione del Vesuvio dell'anno 70, e la morte di Plinio). — ? *in 4°, p. 12.*

Liguori F. S. — Cenni storico-critici della Città di Gragnano e luoghi convicini. — *Napoli, 1863. in 8°. See Chap. XI, pp. 36-37.* (B. N.).

Lippi C. — Dell'utilità della parte vulcanica. — 1807, *in 4°, p. 24.*

Lippi C. — Qualche cosa intorno ai vulcani all'occasione dell'eruzione del Vesuvio del 1° gennaio 1812. — *Napoli, 1813, in 8°, pp. 167, (O. V.)*

Lippi C. — Esposizione dei fatti che da novembre 1810 a febbraio 1815 han avuto luogo nell'accademia di Sci. di Napoli relativamente alla sua scoperta geologica-istorica dalla quale risulta che le due città Pompei ed Ercolano non furono distrutte e sotterrate dal Vesuvio, etc. Also Circolare Esaglotta. — *Napoli, 1816, in 4°, pp. 384+18, pl 1.* (C. A.).

Lippi C. — Fu il fuoco o l'acqua che sotterrò Pompei ed Ercolano?. Scoperta geologico-istorica. — *Napoli, 1816, in 8°, fol. 2, pp. 384, fol. 2, pl. 1.* (C. A.).

Lippi C. — Apologia sulla pretesa Zurite. — *Napoli, 1819, in 8°, pp. 15,* (C. A.),

Lobley L. J. — Mount Vesuvius; a description. Historical and Geological account of the volcano with a notice of the recent eruption, and an appendix containing letters by Pliny the younger, a table of dates of eruptions, and a list of Vesuvian minerals, — *Published by the Geologist's Association of London 1868, in 8°, pp. VI + 55, pl. 3.*

Lobley L. J. — Mount Vesuvius, a description. Historical and Geological account of the Volcano and its surroundings. — *London, 1889, in 8°, pp. 385, pl. 20.*

Longobardi P. — Musarum primi flosculi. See pp. 46, 90, 129 132. — *Napoli, 1714, in 4°, fol. 8, pp. 132.* (O. V.).

Longo G. B. — Il lagrimoso lamento del disaggio che à fatto il Monte di Somma, con tutte le cose occorse fino al presente giorno. — *Napoli, 1632, in 12°, fol. 6.* (C. A.).

Lope Felix de Vega Carpio. — Canzone. — *See De Quinones.*

Lopez de Zarate Francisco. — Sonetto, aludendo que en la tierra . del Vesuvio fue el leventamiento de los Titanos por su mucha abundancia. — *See De Quinones.*

Lopez Valderas Fernando. — Sonetto. — *See De Quinones.*

Lorenzano (de) F. G. — Breve compendio del lamentable incendio del Monte de Somma. — *Napoli, 1632, in 8°, pp. 16.*

LOTTI GIOVANNI (accad. errante). — L'incendio del Vesuvio in ottava rima, — *Napoli, 1632, in 12°, fol. 12.*

LUC (DE) J. A. — Formation des Montagnes Volcaniques. Observations au Vésuve et à l'Etna.—*La Haye et Paris, 1780, in 8° pp. 19.*

LUC (DE) J. A. — Remarks on the geological theory supported by M. Smithson in his paper on a Saline substance from mount Vesuvius. — *The Philosoph. Magaz. by Alex. Tilloch. Vol. XLIII. London 1814.—Journ. de Phys. et Chim et de l'Hist. Nat. par. J. C. de Lametherie et Ducrotay de Blainville. Vol. LXXVIII. Paris, 1814.*

LUCA (DE) P. — Memoria sull'eruzione del 1832. — *Nuova Antologia di Firenze, 1833, Bull. d. l. Soc. géol. de France. Paris, 1833.*

LUCA (DE) S. — Ricerche chimiche sopra talune efflorescenze vesuviane. — *Napoli, 1871.*

LUCA (DE) S. — Sopra talune materie raccolte in una fumarola del cratere vesuviano—*Rend. d. R. Acc. d. Sc. Napoli, 1876.*

LUCA (DE) S. — Ricerche chimiche sopra una cenere trovata negli scavi di Pompei. — *Rend. d. R. Acc. d. Sc. fis. e mat. Ann. XVII, Napoli, 1878.*

LUCA (DE) S. — Ricerche chimiche sopra una particolare argilla trovata negli scavi di Pompei. — *Rend. R. Accad. d. Sc. Fis. Mat. Napoli, 1878, Ann. XVII.*

LUDOVICI D. — Carmina et inscriptiones. Opus posthumum. Two parts.—*Part relating to earthquakes in Vol. I., pp. 42-46; and relating to Vesuvius pp. 46-47, 63-7, 143-145. Naples, 1746 (C. A.).*

LYTTON B. — The Last Days of Pompeii — *In three Volumes London 1834; many subsequent editions.*

MACKINLAY R. — Letter dated at Rome the 9th January 1761 concerning the late eruption of Mount Vesuvius etc. — *Philosoph. Transact. of the R. Soc. of London. 1761. Vol. LII. pp. 44,*

MACRINI J. — Vindimialium ad Campaniae usum libri duo, — *Neapoli, 1716, in 4°, pp. 12 + 36.*

MACRINO. G.—De Vesuvio, item ejusdem opuscula poetica.—*Neapoli , 1693 , in 8°, fol. 16, pp. 156.*

MADEMOISELLE * * * — Lettre sur l'éruption du Vésuve en Août 1756, — *In " Journal Etranger." mars 1757, pp. 159-168 (C. A.).*

M. A. D. O. — Ausführlicher Bericht von dem lezten Ausbruche des Vesuvs, am 15ten Juni 1794, die Geschichte aller vorher-

gegaugenen Ausbrüche und Betrachtungen über die Ursachen
der Erdbeben; von Herrn M. A. D. O. Professor der Arzney-
gelahrheit zu Neapel — Nebst einem Schreiben des Einsiedlers
am Vesuv und zwey Briefen des Duca della Torre über den
nämlichen Gegenstand. Als ein Anhang zu des Ritters Hamilton
Bericht vom Vesuv. Aus dem Italienischen überstzt. Mit
einem nach der Natur gezeichneten Kupfer. — *Dresden, 1795*
(O. V.).

M. A. D. O. (ONOFRIO (D') MICHELANGELO C. A. — Relazione ragionata della eruzione del nostro Vesuvio accaduta a' 15 giugno 1794. In seguito della storia completa di tutte le eruzioni
memorabili fino ad oggi, con una breve notizia della cagione
dei terremoti. — *Napoli, in 8°, pp.* (C. A.).

M. A. D. O. — Relazione ragionata della eruzione del nostro Vesuvio nel di 15 Giugno 1794. Breve fatica del professore di
medicina M. A. D. O. (Michele Arcangelo d' Onofrio). — ?,
in 4°. (B. N.) *See Anonymous.*

MAFFEI G. C. — Scala naturale, overo fantasia dolcissima intorno
alle cose occulte, e desiderate nella filosofia. — *Vinegia, 1573,
in 8°, fol. 140.*

MAFFEI S. — Tre lettere (The 2nd treats of the new discovery of
Herculaneum) a P. Bernardo de Rubeis. — *Verona, 1748, in
4°,* (C. A.).

MAGALOTTI L. — Salita sul Vesuvio. — *See « Dei Vulcani o Monti
Ignivomi. »* Vol. II. (*See Anonymous.*) *Livorno,* 1779.

MAIZONY DE LAUREOL. — L'Héracleade, ou Herculaneum enseveli
sous la lave du Vésuve, poème de L'A. Florus, trad. en vers
français avec des notes. — *Paris, 1837, in 8°, pp. XXIII+
458* (C. A.).

MAJONE D. — Breve descrizione della R. città di Somma Vesuviana.—*Napoli, 1793, in 4°, fol. 10, pp. 56, pl. 1, figs.*

MAJONE G. — Della esistenza del Sebeto nella pendice settentrionale del Monte di Somma. — *Napoli, 1865, in 4°, fol. 1, pp.
34, pl. 1, map. 1* (C. A.).

MAJO. — Trattato delle acque acidule che sono nella città di Castellammare di Stabia. — *Napoli, 1754.*

MALLET R.—Determination of Volcanic Temperatures.—*London,
Sept. 29th 1862, in folio, pp. 2.*

MALLET R. — The Great Neapolitan Earthquake of 1857. — *London, 1862, II Vols, in 8°. Numerous figures and plates.*

MALLET R. — Preliminary Report on the Experimental Determination of the Temperatures of Volcanic Foci, and of the
Temperature, state of Saturation, and Velocity of the issuing

gases and vapours. — *Reports. Brit. Assoc. 1863, in 8°, p. 1 (with an autograph letter of Mallet.* (C. A.).

MALLET R. — The eruption of Vesuvius in 1872, by Prof. Luigi Palmieri. Notes and an introductory sketch of the present state of knowledge of terrestrial vulcanicity, the cosmical nature and relations of Volcanoes and Earthquakes.—*London, 1873, pp. 148, in 8°, pl. 8.* (C. A.)

MALLET R. — On some of the conditions influencing the projection of discrete solid materials from Volcanoes and on the mode in which Pompeii was overwhelmed.—*Journ. of the R. Geol. Soc. of Ireland. Vol. IV. Part III. Dublin, 1876, in 8°, pp. 144-169.*

MALLET R. — On the mechanism of production of volcanic dykes, and on those of monte Somma. — *The Philos. Magaz. Vol. XII. London, 1876. Quarterly Journ. of the Geol. Soc. Vol. XXXII. London, 1876.*

MALPICA C. — La notte del 3 Gennaio in cima del Vesuvio. — *Poltorama Pittoresco. N. 23, pp. 181-183; 19, Gennaio, 1839, with figs.*

MANFREDI A. — *1632.* — See BARONIO.

MANNI P. — Saggio fisico-chimico della cagione de' baleni e della pioggia che osservasi nelle grandi eruzioni vulcaniche. In occasione dell' eruzione del Vesuvio del Giugno 1794. — *Napoli, 1795, in 8° pp. 16. See Santoli V. Narrazione de' fenomeni, etc.*

MANTOVANI P. — Un' escursione al Vesuvio durante l'eruzione del Gennaio 1871. — *Boll. Naut. Geogr. Vol. V. Roma, 1871.*

MANTOVANI P. — La pioggia di cenere caduta a Napoli e la lava del Vesuvio dell' Aprile 1872. — *Boll. Naut. Geogr. Vol. VI. Roma, 1872.*

MANZO G. B. (Marchese di Villa). — Lettera in materia del Vesuvio (Erupt. 1631) scritta da Napoli al sig. Antonio Bruni a Roma. — *M. S. in library of the Faculty of Medecine of Montpellier. See also Riccio L.* (C. A.).

MARANA. — Des Montagnes de Sicile et de Naples, qui jettent des feux continuels : de la nature de leurs effets. — *Lettre XLIII, de l' Espion Turc. t. I. pp. 153-157* (C. A.).

MARAVIGNA C. — Esame di alcune opinioni del sig. N. Boubée contenute nelle sue opere intitolate « Géologie populaire et Tableau de l'état du globe à ses différens âges » — *4th édit. 1834, in 4°, pp. 48.*

MARENA THOM. ANTONIUS. — Brevissimum terraemotuum examen, etc. — *Neapoli, 1632. in 4°. fol. 10,* (C. A.).

MARI C. — Il Vesuvio. — Cauto — *Napoli, in 16°, pp. 16.* (C. A.).

MARIGNAC (DE) C. — Notices minéralogiques. (Epidote, Humite ou Chondrodite du Vésuve, Pinite, Gigantolite). — *Supplément à la Biblioth. Univ. et Revue Suisse. 2.ᵉ Sér. Vol. IV. Genève, 18-17. Journ. de Pharm. et des Sc. accessoires. 2.ᵉ Sér. Vol. XII. Paris, 1847.*

MARTINIO (DE) C. — Osservazioni giornaliere del successo del Vesuvio dalli 16 Dic. 1631 fino alli 10 Aprile 1632. — *Napoli, 1632, in 4°, pp. 32.* (B. N.).

MARTINO DI CARLES FLAMINIO. — Ottave sopra l'incendio del Monte Vesuvio, — *Napoli, 1632, in 12.° fol. 12.* (B. N.).

MARTINO (DE) L. M. — Eruzione del Vesuvio 29 (79) dell'Era cristiana. (Signed M. P.). — *Melphis excidium. pp. 209-213.*

MARTINOZZI V. — Sonetto. — *See Giorgi Urbani,*

MARTORELLI J. — De Regia Theca Calamaria. — *Neapoli, 1756. Vols. 2, in 4°, fig. Vol. I, pp. 8, fol. 290 ; Vol. II, pp. 8 and from 291-738 (Vesuvius pp. 417 and foll, and 566).*

MASCULI I. B. — De incendio Vesuvii excitato XVII Kal. Januar. An. 1631. Libri X, cum chronologia superiorum incendiorum et ephemeride ultimi. — *Neapoli, 1633 , in 4°, fol. 4, pp. 312 + 37, fol. 5, pl. 2.*

MASINO D. M. A. — Distinta relatione dell'incendio del sevo Vesuvio alli 16 Dic. 1631 successo etc. — *Napoli, 1632, in 4°, pp. 36.* (C. A.).

MASSARII J. P. — Sirenis lachrymae effusae in Montis Vesevi incendio. — *Neapoli, 1632, in 4°, pp. 28.* (C. A.).

MASTRIANI. — L'eruzione del Vesuvio del 26 Aprile 1872. — *Napoli, 1872, in 8°, pp. 102, col. pl. 1, map. 1.*

MASTROJANNI D. G. — L'incendio del Vesuvio di Maggio e l'accensione dell'aria di decembre, del caduto anno, etc. — *Napoli, 1738, in 4°.*

MAUGET A. — Lettres à M. S.-C. Deville sur l'éruption du Vésuve du 27 Mai 1858. — *Bull. d. l. Soc. Géol. de France. Vol. XV. Paris, 1858. Compl. Rend. Hebdom. des Séanc. d. l'Acad. d. Sc. Paris, Vol. XLVI. p. 1098, Mai et Juin 1858.*

MAUGET. A. — Sur les phénomènes consécutifs de la dernière éruption du Vésuve. — *Compt. Rend. hebdom. des Séanc. d. l'Acad. d. Sc. Paris, Vol. LVI. pp. 926-928. 1862.*

MAUGET A. — Sur les phénomènes consécutifs de l'éruption de décembre 1861 au Vésuve. — *Compt. Rend. Acad. Sc. Paris. t. LXIII, pp. 7-8, 1866.*

MAUGET A. — Faits pour servir à l'histoire éruptive du Vésuve. Récit d'une excursion au sommet du Vésuve le 11 Juin 1867. —

Compl. Rend. d. l' Acad. des Sc. Vol. LXVI, pp. 163-166. Paris, 1868.

MAURI A. — Memoria sulla eruzione vesuviana de' 21 Ottobre 1822.—*Napoli, 1823, in 8°, fol. 6. pp. 22. pl. 1, figs. (C. A.).*

MAURINI G. — De Vesuvio — *Neapoli, 1693, in 8°, pp. 156.*

MAYORICA. — L'incendio di Vesuvio successo nell'anno del Signor 1631. a 16 xbre. — *M. S. in the S. Martino Library, Naples. Copy pp. 99 (C. A.).*

MAZZEI D. — Sonetti due. — *See G. Urbano.*

MAZELLA S. — Descrizione di Napoli. — *Napoli, 1586.*

MAZOCHI A. S. — In vetus marmoreum Sanctae Neapolitanae ecclesiae Kalendarium Commentarius—*Napoli, 1744. Diatriba V. De Vesuviani incendii celerarumque vulcaniarum flammarum origine, ex antiquorum christianorum sententia ex chronographo Gerasimo Monacho, pp. 392-402—Vol. III., pp. XL+1096, fol. I, pl. 3. (C. A.).*

MECATTI G. M. — Racconto storico filosofico del Vesuvio e particolarmente dell' eruzione principiata a' 25 Ottobre 1751 e cessata a' 25 Febbraio 1752. — *Napoli, 1752, in 4°, fol. 4, pp, 460+244 pl. 10. (Most copies differ and are incomplete).* (C. A.).

MECATTI G. M. F. — Esame o sia confronto di ragioni adottate dall'autore delle novelle letterarie di Firenze Dr. Gio. Lami di Santacroce, e dall'Ab. Giuseppe M. Mecatti sopra la pretesa Città di Pompei, e di Ercolano: sopra la Rettina, o sia Resina di cui parla Plinio: e sopra le scavazioni, che presentemente si fanno alla Real Villa di Portici di S. M. Siciliana, estratte tutte da alcune lettere de' medesimi, — *Napoli, 1752. in 4° p. 7-88.*

MECATTI G. M. — Descrizione della lava scorsa nel mese di Luglio dell' anno 1754 nel cratere ossia piattaforma del Vesuvio ed eruttata dalla cima di una montagnola. —*Napoli, 1754.*

MECATTI G. M.—Discorsi storici filosofici sopra il Vesuvio.—*Napoli, 1754, in 4°, pp. 290.*

MECATTI G. M. — Osservazioni che si son fatte sopra il Vesuvio dal Marzo 1752 al Luglio 1754, etc. — *Napoli, 1754, in 4°, pp. 7+298+1, pl. 1. (At the end of a copy in (O. V.). there is a view of erupt. of Vesuvius of 1767). Several other editions.*

MECATTI G. M. — Continuazione delle osservazioni sopra diverse eruzioni del Vesuvio. — *Napoli, 1761, in 4°, fol. 4°, pp. 298, pl. 2. (C. A.).*

MECATTI G. M. — Osservazioni che si sono fatte nel Vesuvio dal

mese di Agosto 1752 sino alla narrazione istorica di quel che
è occorso al Vesuvio nella eruzione cominciata la notte del
dì fra i dieci e gli undici d' aprile dell' anno 1766. — *Napoli?,
fol. 4, pp. 298, pl. 6.* (C. A.).

MECATTI G. M. — Narrazioni storiche di quel che occorse alla
rottura del Vesuvio dal dì 3 Dic. 1754 fino a quanto è poste-
riormente avvenuto. — *Napoli, 1776.* (*)

MEISTERS. — Beobachtungen über den Vesuv. — *Mag. der Wis-
senschaft. und der Litt. Bd. II, Göttingen, 1781, in 12°,
pp. 25, pl. 1.*

MELCHIORRE D. — 3.ᵃ lettera. Raccolta di monumenti sopra l' e-
ruzione del Vesuvio seguita nell'agosto 1779. — *Giornale delle
Arti e del Commercio, Vol. I, Macerata, 1780, in 8°. at
pp. 141 and following.* (O. V.).

MELE F. — De conflagratione Vessevi, Poema. — *Neapoli, 1632,
in 12°, fol. 10.* (C. A.).

MELLONI ET PIRIA. — Recherches sur les fumaroles. Lettre de M.
Melloni à M. Arago.—*Compt. Rend., Acad. Sc. t. XI, Paris,
1840, pp. 352-356,* (C. A.).

MENARD DE LA GROYE F. J. B.—Observations avec réflexions sur
l'état et les phénomènes du Vésuve pendant une partie des
années 1813-14. — *Journ. de Phys. de Chim. et de l' Hist.
Nat. Vol. LXXX et LXXXI. Paris, 1815. Soc. Roy. Trans.
Le Mans. 1820 (?) « Courcier » , 1815, in 4°, fol. 2, pp.
98+4.* (C. A.).

MERULA G. — *See Dion Cassius.* (C. A.).

MÉRY M. — Les amans du Vésuve. — *Paris 1856, in 12, p. 95.*

MESCHINELLI L. — La flora dei tufi del Monte Somma. — *Rend.
R. Accad. d. l. Sc. Fis. Mat. Napoli, 1890. pp. 8.*

MESSINA N. M. — Relatione dell'incendio del Monte Vesuvio del-
l'anno 1682. — *Napoli, 1682, in 4°. fol. 2.* (C. A.).

METHERIE (DE LA) J. C. — Note sur quelques cristaux de Ceyla-
nite trouvés parmi les substances rejetées par le Vésuve. —
*Journ. de Phys. de Chim. et de l' Hist. Nat. Vol. LI. Paris,
1800.*

METHERIE (DE LA) J. C. — Observations sur les dernières éruptions du Vésuve, — *Journ. de Phys. de Chim. et de l' Hist.
Nat. Vol, LXI, Paris, 1805.*

(*) All these works of Mecatti are much the same, with additional
accounts of new eruptions, and rarely two copies are alike.

MEUNIER S. — Fer natif trouvé au Vésuve.—*Le Naturaliste, 10ᵐᵉ ann. Paris, 1888. pp. 89-91, fig, 1,.*

MIERISCH BR.—Die Auswurfsblöcke des Monte Somma.—*(Tschermak, Mineral. und petrogr. Mittheilungen. B. VIII). Wien. 1887, pp. 78, figs.*

MILANO N. P. — Vera relazione del crudele, misero e lagrimoso prodigio successo nel Monte Vesuvio, etc. — *Napoli , 1632, in 4°, pp. 8.* (C. A.).

MILENSIO F. — Vesevus. Carmen. — *Napôli, 1595, in 4°, fol. 6. N. B. Is in appendix to authors work entitled " Dell' impresa dell' Elefante "* (C. A.)

MILESIO G. — Vera relazione del miserabile e memorando caso successo nella falda della nominatissima montagna di Somma. — *Napoli, 1631, in 4°, pp. 8.*

MILESIO G. — La seconda parte degli avvisi di tutto quanto è successo in tutta la seconda settimana (Vesuvio).—*Napoli, 1632, in 4°, pp. 8.*

MILESIO DA PONTA GIACOMO. — Warhaffte Relation erbaermlichen und erschroecklichen Zustands, so sich in der Scyten desz weitberumbten Bergs Vesuuii. — *München, 1632, in 4°, pp. 19. See pp. 10-19.*

MILESIUS LE R. P. J. — Récit véritable du misérable et mémorable accident arrivé en la descente de la très-renommée Montagne de Somma, autrement le Vésuve, environ trois lieues loing de la ville de Naples. Depuis le lundy 15 Décembre 1631, sur les 9 heures du soir, jusques au Mardy suivant 23 du mesme ; décrit jour par jour et heure. — *Lyon, 1632, in 8°, pp. 13.* (C. A.).

MILO (DE) D. A. — All' Ill.ᵐᵃ Sig.ᵃ Maria Selvaggia Borghini, ragguagliandola del Monte Vesuvio e dei suoi incendi. — *Bulifon. Letter. Memor. Vol. III, (?)*

MINERVINO S. C. — Lettera sopra la ultima eruzione del Vesuvio dell'anno 1779. Dans l' ouvrage . « Dei vulcani.o monti ignivomi. » — *(See Anonymous).*

MINERVINO D. C. S. — Altra lettera sopra l'eruzione dell'anno 1779. — *Ibid.*

MINERVINO S. C. — Due lettere sull'eruzione del 1794. — *Giôrn. Lett. di Napoli. Vol II. 1794, in 8°, pp. 86-97.*

MINTO (EARL OF). — Notice of the barometrical measurements of Vesuvius, and the new cone which was formed in the eruption of February 1822.—*The Edinb. Journ, of Sc. Vol. VIII. Edinburgh; 1827.*

MIOLA A. — Ricordi Vesuviani, Carmen pel centenario di Pompei. — *Napoli, 1879, in 4°, pp. 7.* (C. A.).

MIRANDA D. DE AND PACI G. M. — Osservazioni di Meteorologia elettrica sulle vulcaniche esalazioni. — *Napoli, 1845. in 4°, pp, 14.* (C. A.).

MISCELLANEA POETICA (M. S.)—"Parthenope terraemotu vexata".— *fol. 6. " Vesuvius morum magister " fol. 3 " De Vesuvio semper ignum ejactante " fol. 1. " Fontis descriptio : ubi Vulcani statica diaculantis. fol I.* (C. A.).

MISSON M. — Voyage d'Italie. Edition augmentée de remarques nouvelles et interessantes.—*Amsterdam, 1743, 4 Vols. in 12°, Vol. I. fol. XLVIII, pp. 352, pl. 10., Vol. II. pp. 366, pl. 37., Vol. III, pp. 290, pl 9., Vol IV, pp. 295, pl. 2* (C. A.).

MITROWSKY (VON) I. G. — Physikalische Briefe ueber den Vesuv und die Gegend von Neapel.—*Leipzig, 1785, in 8° pp. 142.*

MITSCHERLICH E. — *1851-1852.* — *Rose G.*

MITSCHERLICH R. — Ueber eine Vesuvianschlacke. — *Zeitschr. d. Deut. Geol. Gesell. Bd. XV. Berlin, 1863.*

MOCCIA P. — Ad Andream Fontanam de Vesuviano Incendio anno 1706 — *Epistola.* (?)

MODESTO P. — All'Eccellentissimo D. Francesco Conte Esterhazy (Concerning the controversy about the presence of metallic iron in Vesuvian Sand between D. Tata f. F. Viscardi) — *Napoli, 1795, in 4°, pp. 26.* (O. V.).

MOLES F. — Relacion tragica del Vesuvio. — *Napoles, 1632, in 8°, pp. 68.*

MONACO V. DI.—Lettera analitica sull'acqua della Torre del Greco, comunemente creduta prodigiosa al Sig. Ant. Sementini, etc. (Dated Aug. 4 th 1789. — *Napoli, ? in 12°.* (B. N.).

MONACO. — Eruption du Vésuve en 1754. — *A plate.* (C. A.).

MONFORTI F. A.—Ad divum Januarium, elogium. — *See G. Urbano.*

MONGES G. — Sulla terribile eruzione del Vesuvio accaduta ai 17 Giugno 1794. — Lettera responsiva a N. N. Dated from — *Salerno, 1794, pp. 24.*

MONITIO C. — La Talia dove si contiene la Fiasca con le lagrime del Vesbo furioso. — *Napoli, 1647, in 8°, pp. 200.* (O. V.). *Another copy of the " Fiasca" only, at the end of which is a leaf not numbered in the work with a view of Vesurius and some lines of verse. pp. 74, fol. 1.* (O. V.).

M. J. L. R. — See Le Riche M. J.

MONITIO C. — La Talia, dove si contiene la Fiasca sotto sensati scherzi di vario stile. Con le croiche lagrime del Vesbo furioso, e un assaggio del volume maggiore intitolato Crumena Sapientis. — *Napoli, in 8°, 1645, pp. 208. fol. 1,* (C. A.).

MONNIER M. — Le Vésuve et les tremblements de terre. — *L' I̅-lustration. Paris, Janvier, 1858.*

MONNIER M. — Promenade aux environs de Naples (Eruption du Vésuve, destruction de Torre del Greco). — *" Le Tour du Monde " N. 124, 3me année, 1864. pp. 305-319, with illust.* (C. A.).

MONTEFERRANTE R. — See Kernot.

MONTEIRO I. A. — Mémoire sur la chaux fluatée du Vésuve. — *Annales du Muséum d'Hist. Nat. Vol. XIX. Paris, 1812, pp. 171-188. — Journ. des Mines. Vol. XXXII. Paris, 1812.*

MONTEMONT A. — Des volcans en général et plus spécialement du Vésuve et de l'Etna.—*Bull. d. l. Soc. de Géog. Paris, Sept. 1841, pp. 137-158.*

MONTICELLI T.—Descrizione delle eruzioni del Vesuvio nel 1813.— *Giornale Enciclopedico, in 8°, pp. 47.*

MONTICELLI T. — Descrizione dell'eruzione del Vesuvio avvenuta nei giorni 25 e 26 Dicembre 1813. — *Napoli , 1815 , in 4°, pp. 47. 2nd edit. ?* (Napoli 1842), *in 4° pp. 40.* (O. V.).

MONTICELLI T. — Lettre sur la découverte de la Wollastonite dans le Vésuve par M. le Prof. Gismondi. — *Bibl. Univ. de Genève. Vol. II. Genève, 1817.*

MONTICELLI T. — Rapporto nell'eruzione di dicembre 1817. — *Giornale Enciclopedico, Napoli, Marzo 1818, in 8°, pp. 7.*

MONTICELLI T. — Report on the eruption of Vesuvius in Decemb. 1817. — *The J. of Sc. and the Arts. Vol. V. London, 1818.*

MONTICELLI T. — Altre escursioni fatte nel Vesuvio dal 1817 al 1820. — ?.

MONTICELLI T. — Notizie di un'escursione al Vesuvio e dell'avvenimento che vi ebbe luogo nel 16 Genn. 1821 in cui il Francese Coutrel si precipitò in una di quelle nuove bocche.—*Atti d. Acc. d. Sc. e Lett, Napoli, 1821, pp. 9.* (O. V.).

MONTICELLI T. — Collections des substances volcaniques. — ? , *1825, in 8°, pp. 16.* (O. V.).

MONTICELLI T. — Memoria sulle sostanze vulcaniche rinvenute nella lava di Pollena discoperta dalle ultime alluvioni del Vesuvio.—*Atti d. R. Acc. di Sc. e Lett. Vol. II, pt. 1. Napoli, 1825, pp. 77-86.*

MONTICELLI T. — Memorie sulle vicende del Vesuvio nell'anno 1827. — *Atti d. R. Sc. e Lett. Napoli, 1828, pp. 90-125.*

MONTICELLI T. — Memoria sull'origine delle acque del Sebeto di Napoli antica, di Pozzuoli etc. — *Atti d. Acc. d. Sc. di Napoli, 1828, pp. 56, pl. 2.* (C. A.).

MONTICELLI T. — Osservazioni dello stato del Vesuvio dal 1823 al 1829. — « *Opere* » *Vol. II, pp. 106-112.*

MONTICELLI T. — Lettera sullo stato del Vesuvio nel prossimo passato mese di marzo. — *Ateneo, Giornale di Scienze, Letteratura, Arti ed Industria, Vol. I, Napoli, 1831, in 8°, p. 84-87.* (C. A.).

MONTICELLI T. — Memoria sopra talune nuove sostanze Vesuviane. — *Atti d. Acc. d. Sc. di Napoli, 1832, Vol. V, pt. 2, pp. 157-159.* (C. A.).

MONTICELLI T. — Saggi analitici sopra alcuni prodotti Vesuviani. — *Atti d. Acc. di Sc. di Napoli, 1832 pp. 6.* (C. A.).

MONTICELLI T. — Analisi chimica di un prodotto vesuviano. Rapporto di G. Guarini. — *Rendic. d. Acc. d. Sc. di Napoli, 1833, pp, 161-163.* (C. A.).

MONTICELLI T. — Sulle origini delle acque del Sebeto di Napoli etc. (1828). — *Atti d. R. Istit. d'Incor. alle Sc. Nat. ed Econ. di Napoli, Vol. V. 1834.*

MONTICELLI T. — Ausbrüche des Vesuv's seit April 1835. — *Neues Jahrb. für Miner. Geogn. Geol. etc. Bd. III, Stuttgart, 1835.*

MONTICELLI T. — Memoria sopra altre vicende del Vesuvio del 1835. — *Atti d. Acc. d. Sc. di Napoli, 1835. Vol. V, pt. 2, pp. 183-186.*

MONTICELLI T. — Memoria sopra i danni che il fumo del Vesuvio reca ai vegetabili — *Atti. d. Acc. d. Sc. di Napoli, 1835, — Vol. V, pt. 2ª, pp. 186 — 189.* (C. A.).

MONTICELLI T. — Memorie sul Vesuvio. — *Atti d. R. Acc. di Sc. e Lett. Vol. IV. Napoli, 1839.*

MONTICELLI T. — Rapporto del Segretario perpetuo della R. Accad. d. Sc. sulla eruzione del Vesuvio del dì 22 a 26 Dic 1817 letto nella tornata de' 9 marzo 1818. — *Napoli, 1841, in 4°, pp, 15.* (C. A.).

MONTICELLI T. — Monografia del ferro di Cancarone. — *Atti d. R. Accad. d. Sc. di Napoli, 1841, Vol. V. pt. 2ª, pp. 217 227, pl. 3.* (C. A.).

MONTICELLI T. — Opere. — *Napoli, 1841-43, in 4°, 3 Vols. Vol. I, fol. 4, pp. 295, pl. 2, Vol, II, pp. 335, pl. 7, Vol. III, pp. 432, pl. 19, map. 1.*

MONTICELLI T. — Genesi del ferro di Cancarone. — *Atti d. R. Accad. d. Sc. di Napoli, 1842, Vol. V, pt. 2ª, pp. 229-232.* (C. A.).

MONTICELLI T. — Memoria sulla eruzione del 28 Luglio (1833). — *Atti d. R. Acc. di Sc. e Lett. Vol. V, part. 2.ª Napoli, 1843, pp. 169-177.*

MONTICELLI T. — Introduzione alla monografia delle Pelurie lapidee del Vesuvio. — *Atti d. R. Acc. d. Sc. Napoli, 1843, Vol. V. Parte 2ᵃ, pp. 191-194.*

MONTICELLI T. — Monografia delle Pelurie lapidee del Vesuvio (1837). — *Atti d. R. Acc. di Sc. e Lett. Vol. V. parte 2.ᵃ Napoli 1843, pp, 195-205.*

MONTICELLI T. — Muriato ammoniacale cruttato dal Vesuvio (1834). — *Atti d. R. Acc. d, Sc. e Lett. Vol. V. parte 2.° Napoli, 1843, 179-181.*

MONTICELLI T. — Sopra alcuni prodotti del Vesuvio (1832). — *Atti di R. Acc. di Sc. e Lett. Vol. V. pl. 2.ᵃ Napoli, 1843.*

MONTICELLI T. — Sopra alcune vicende del Vesuvio del 1805. — *Atti d. R. Acc. di Sc. e Lett. Vol. V. pl. 2.ᵃ Napoli, 18-13.*

MONTICELLI T. — Sopra talune sostanze vesuviane (1832). — *Atti d. R. Acc. di Sc. e Lett. Vol. V, pl. 2.ᵃ, Napoli, 1843.*

MONTICELLI T. — Sulle sublimazioni del Vesuvio (1832). — *Atti d. R. Acc. di Sc. e Lett. Vol. V, Pl. 2.ᵃ, Napoli, 1843, pp. 147-149, pl. 1.*

MONTICELLI T. — Storia e giacitura del ferro di Cancarone. — *Atti d. Acc. d. Sc. di Napoli, Vol. V, pl. 2, 1843, pp. 211-215,* (C. A.).

MONTICELLI T. — Memorie sopra alcuni prodotti del Vesuvio e alcune vicende di esso. — *Atti d. R. Acc. di Sc. e Lett. Vol. V, pl. 2°, Napoli, 1844, pp. 141-145.*

MONTICELLI T. — Continuazione alla monografia delle Pelurie lapidee del Vesuvio. — *Atti d. R. Acad. d. Sc. Napoli, 1843, Vol. V, Parte 2ᵃ, pp. 287-218, pl. I.*

MONTICELLI T. AND COVELLI N.—Osservazioni ed esperienze fatte al Vesuvio in una parte degli anni 1821 e 1822. — *Giorn. Arcad. di Sc. ecc. Vol. XVI. Roma, 1822.* — (In French). *Naples, 1822, in 8°, pp. 66.*

MONTICELLI T. E COVELLI N. — Storia dei fenomeni del Vesuvio avvenuti negli anni 1821-22 e parte del 23 con osservazioni e sperimenti. — *Giorn. Arcad. d. Sc. etc. Vol. XX. Roma, 1823. Napoli, 1823, in 8°, pp. XLX+208+7, pl. 4. In German trans. by Dr. Noggerath and Dr. Pauls. Elberfeld, 1824, in 8°, pp. 30+234, fol. 1, pl. 6.* (C. A.). 2ⁿᵈ *Ital. edit.. Napoli, 1842, in 4°, pp. 170, fol. 3, pl. 2.* (O. V.).

MONTICELLI T. AND COVELLI N. — Appendici al Prodromo della Mineralogia Vesuviana.—*Napoli, 1839, in 8°, pp. 28.* (C. A.).

MONTICELLI T. AND COVELLI N. — Atlante della Mineralogia vesuviana. — *Pl. 10, in 8°,* (C. A.).

MONTICELLI T. E COVELLI N. —Prodromo della mineralogia vesuviana. Orittognosia. — *Napoli. 1825, in 8°, pp. 34+483, pl. 19. 2nd edit. 1843.*

MONTICELLI T. E COVELLI N. — Descrizione dei prodotti minera-logici del Vesuvio. — (?)

MONTICELLI T. E COVELLI N. — Nuove specie minerali del Ve-suvio. — (?)

MONTICELLI T. E RICCIARDI FR. — Qual sia l'influenza del Vesu-vio, colle sue varie eruttazioni, sulle meteore, e sulla vege-tazione del Circondario. — *Programmi due per la Real Ac-cademie delle Scienze. Napoli, 1810, in 4°, pp. 4.*

MORGAN O. — On some phenomena of Vesuvius. — (?)

MORI F. — Ricordi di alcuni rimarchevoli oggetti di curiosità e di belle arti di Napoli (Erupt. 1822). — *Napoli, 1837, in 8°, fol. 31, pl. 23, figs.* (O. V.).

MORMILE G. — Nuovo discorso intorno all'Antichità di Napoli, e di Pozzuolo. — *Napoli, 1629, in 8°, p. 69. (pp. 31-32 Vesuvius).*

MORMILE G. (Napolitano). — L'incendio del Monte Vesuvio e delle straggi, e rovine, che hà fatto ne' tempi antichi, e moderni insino a' 3 di marzo 1632. — *Napoli, 1632, in 8°, pp. 48,* N. B. *At page 47 there is a " Nota di tutte le relazioni stampati fino ad hoggi del Vesuvio raccolte da Vincenzo Bove", containing 56 entries and is the earliest biblio-graphical list of Vesuvian litterature.* (C. A.).

MORMILE. G. — Descrizione della Città di Napoli e del suo ame-nissimo distretto, etc. - *Napoli, 1670, in 8°, fol. 4, pp. 264, figs,* (O. V.). *Other editions in 1617, and 1625.*

MUNTERUS M. T. L. — Parerga historico-philologica. — *Gottinga, 1749.* (O. V.).

MURATORI L. A. — Rerum Italicarum Scriptores, etc. — *Mediola-ni, 1723-51. Vols. 28, in fol. (Vesuvii descriptio, Vol. I, parte I, p. 278.).*

NAPOLI R. — Nota sopra alcuni prodotti minerali del Vesuvio. — (?)

NAPOLI R. — Sulla produzione del sale ammoniaco nelle fumarole vesuviane. — (?)

NAUDÉ G. —.Sur les divers incendies du mont Vésuve et parti-culièrement sur le dernier qui commença le 16 Décembre 1631. — *Paris, 1632, in 12°, pp. 37.*

NAUDÉ G. UND GIULIANI G. B. — Ueber den Vesuv und Aetna. — (?) *1632.*

NECKER L. A. — Ueber den Monte Somma. — *Elberfeld, 1825, in 8°, pp. 10+264, pl. 3, figs.* (O. V.). *Mem. Soc. Phys. et d'Hist. Nat. de Genéve. t. II. 1823, pp. 155-203, pl 2, also in French.* (C. A.).

NECKER — 1825. — *V*. *Raffles*.

NEGRONI O. — Sulle ceneri vesuviane del 1779. — (?)

NESTEMAN UND FELBER.—Notizen über den Vesuv im Mai 1830— *Archiv. für Mineral. Geogn. Bergbau, und Hüttenkunde.* *Bd. IV. Berlin, 1831.*

NETTI F. — Il Vesuvio. — *Article in the journal: L' Illustrazione Italiane. 19-26 dic. 1875, and 2, 9, 16, and 23 Gen. 1876. with figs.* (C. A.).

NICOLAI A. — De Vesevo Monte, epigramma. — *See G. Urbano.*

NICOLINI A. — Tavola Metrica-cronologica delle varie altezze tracciate dalla Superficie del mare fra la Costa di Amalfi ed il Promontorio di Gaeta nel corso di diciannove secoli.—*Napoli, 1839, in 4°, pp. 52.*

NICCOLINI A. — Descrizione della Gran-Terme Puteolana, volgarmente detta Tempio di Serapide.—*Napoli, 1846, in 4°, pp. 95, numerous col. and uncol. pl., maps, etc.*

NIGLIO M. — Saggio di Poesia (sul Vesuvio, pp. 70-74)—*Napoli, 1825, in 8.° pp. 120.* (C. A.).

N. N. — Lettera scritta dal Sig. N. N. al sig. N. N. in Calabria sulle cagioni delle tante mosse e minacce fatte dagli edificj di Napoli nella fine del prossimo scorso anno 1766, e nel principio del corrente.—*Napoli, 1767, in 4°, pp. 16.* (B. N.).

NOBILI (DE) G.—Analisi chimica ragionata del lapillo eruttato dal Vesuvio nel dì 22 Ottobre 1822, etc.—*Napoli, in 8° pp. 20.*

NOCERINO N. -- La Real Villa di Portici illustrata.—*Napoli, 1787, in 8° pp. 157+3.*

NOTO S. — Cenno storico della Cappella di S. Maria della Bruna in Torre del Greco. — *Napoli, 1851, in 12, pp. 36. (erupt. 1631).*

NUNZIANTE LE MARQUIS. — Eau vésuvienne —?. *in fol, pp. 2.* (C. A.).

NUNZIANTE (Marchese Vito) —Dimanda di privativa per la fabbricazione di Lastre e Cristalli, facendo uso per essa delle lave vulcaniche, ecc. — *Napoli, 1826, in 4°, pp. 8.*

NUZZO MAURO A. — Un Papiro, ossia i gladiatori nella Caverna del Vesuvio. — *Venezia, 1826, in 8°, p. 197.*

OESTERLAND C. & WAGNER P.—Analyses des Cendres du Vésuve— *Deutsche Chemische Gesell. 1873. — Bull. d. l. Soc. Chim. de Paris. Oct. 1873.*

OLEARIUS T. — Feuer flammen des Vesuvii. — *Hall, 1650.*

OLIVA N. M.—Lettera scritta all'abbate Flavio Ruffo nella quale si dà vera e minuta relazione degli segni, terremoti ed incendii del Monte Vesuvio, cominciando dal dì 16 del mese

Dicembre 1631, per in sino alli 5 Gennaio 1632. — *Napoli, 1632, in 4.° fol. 4.*

OLIVA N. M. — La ristampata lettera con aggiunta di molte cose notabili, nella quale dà vera e minuta relatione delli segni, etc. — *Napoli, 1632, in 4°, pp. 8.* (C. A.).

OLIVIERI G. M. — Breve descrizione istorico-fisica dell' eruzione del Vesuvio del 15 Giugno 1794. — *Napoli, 1794, in 4°, pp. 22.*

OLTMANS J. — Darstellung der Resultate welche sich aus den am Vesuv von A. von Humboldt und anderen Beobachtern angestellten Höhenmessungen ableiten lassen. — *Abhandl. d. könig. Akad. der Wiss. zu Berlin. Berlin, 1822-23.*

ONOFRII P. DEGLI. — Elogii storici di alcuni servi di Dio, che vissero in questi ultimi tempi e si adoperarono pel bene spirituale e temporale della città di Napoli. — *Napoli, 1803, in 8°, pp. XVI+172. (In the life of P. Gregoris M.ª Rocco from pp. 432-461 some Vesuvian eruptions are described, particularly those of 1794 and 1799.* (C. A.) and (O. V.).

ONOFRIO (D') M. A. — Nuove riflessioni sul Vesuvio, con un breve dettaglio de'paraterremoti, etc. — *Napoli, 1794, in 12°, pp. 20. 2 ª edit. Nap. 1794, in 8°, pp. 16, pl. 1, figs.* (O. V.).

ONOFRIO (D') M. A. — Relazione ragionata dell' eruzione del nostro Vesuvio nel dì 15 Giugno 1794. — *Napoli, 1794, in 4.° fol. 1, pp. 9. Dresden, 1795, in 8°, pp. 88, fol. 1, pl. 1, and several other editions.*

ONOFRIO (D') M. A. — Lettera ad un amico in Provincia sul tremuoto accaduto ai 26 di Luglio e seguito dalla eruzione Vesuviana dell'Agosto 1805, etc. — *Napoli, 1805, in 8°, pp. 44.* (O. V.).

OPITZ M. — Vesuvius, Gedichte (1631). — *Frankfurt am Mayn 1746, in 8°, pp. 19-44, Vol. I.* (C. A.).

ORBESAN (D') M. — Description du mont Vésuve, compita relazione di quanto e succeduto insino hoggi (24 Dic.) — (?)

ORIMINI P. — Nell'eruttazione della Montagna di Somma del 1767. (p. 158-163) from: Degli antichi signori del Gaudo. Poesie. — *Napoli, 1771, in 4°, pp. 174.*

ORIMINI P. — Poesie. — *Napoli, 1771, in 8°, fol.1, pp. 174.* (O. V.).

ORLANDI G. — Dell'incendio del monte di Somma. — *Napoli, 1631, in 4°, pp. 15.*

ORLANDI G. — La cinquantesima e bellisima relatione del Monte Vesuvio in stile accademico. — *Napoli, 1632, in 8°, pp. 12, fig. 2.* (C. A.).

ORLANDI G. — Nuova e compita relatione del spaventevole incendio del Monte di Somma, detto il Vesuvio. Dove s'intende

minutamente tutto quello che è successo fin' al presente
giorno. Con la nota di quante volte detto Monte si sia ab-
brugiato. Aggiuntovi un rimedio denotissimo contro il terremoto. — *Napoli, 1632, in 4°, figs, pp. 16.* (C. A.).

ORLANDI P. P. — Tra le Belle la Bellissima, esquisita, et entiera
e desiderata Relazione dell'incendio del monte Vesuuio detto
di Somma. — *Napoli, 1632, in 4°, fol. 4.*

ORLANDI S. — La tregua senza fede del Vesuvio. — *Napoli, 1632,
in 4°, fol. 4.* (C. A.).

ORME W. — View of the last eruption of Mount Vesuvious from
an original painted at Naples. Dedic. to Sir W. Hamilton.
(Problably erupt. 1794. — *London, ? Col. tranparency of a
steel engraving in 1 R. fol. (In the collection of Mr. L. Sambon, Naples.*

ORRIGONE C. G. — Pensieri poetici (erupt. 1631). — *Genova, 1636,
in 8°, pp. 108-119.* (C. A.).

OTTAVIANO C. — Alla Maestà di Ferdinando IV Re delle Due Sicilie per la terribile eruzione del Vesuvio. Sonetto. — *?, loose
sheet.* (B. N.).

PACICHELLI G. B. — Memorie de' viaggi per l'Europa Christiana
scritte a diversi in occasione de' suoi Ministeri. — *Napoli,
1685, vols. 5, in 12. Parto I. pp. 40+743+53. Parte II.
8+827+40 Parte III. 8+761+27. Parte IV. vol. I. 4+
541+20. Parte V, Vol. II. 4+438+18. (Parte IV. Vol. II.
pp. 255 and follow. Del Vesuvio).*

PACICHELLI G. B. — Lettere familiare, istoriche ed erudite, etc.—
Napoli, 1695, Vols. II, in 12°. See. Vol. II, pp. 343-353.
(B. N.).

PADAVINO M. A. (According to Castelli P.). — Lettera narratoria
a pieno la verità dei successi del Monte Vesuvio detto di
Somma, seguiti dalli 16 Dicembre fin alli 22 dell'istesso mese.
— *Roma, 1632, in 8°, pp. 14.* (C. A.).

PADERNI C. — An account of the late Discoveries of Antiquities
at Herculaneum. — *Phil. Trans. 1756, pp. 490-508.* (C. A.).

PADERNI C. — An account of the late Discoveries of Antiquities
at Herculaneum, and of an Earthquake there. — *Phil. Trans.
1758, Vol. L, pp. 620-623.* (C. A.).

PALATINO L. — Storia di Pozzuoli e contorni con breve trattato
historico di Ercolano e Pompei. — *Napoli, 1826, in 8°, pp.
336, pl. 1, maps 2.* (C. A.)

PALMERI P. — Sulla cenere lanciata dal Vesuvio a Portici e a
Resina la notte del 3 a 4 Aprile 1876. Ricerche chimiche. —
Rend. R. Acc. Sc. Fis. Mat. An. XVI, 1876. pp. 73-74,

87-93. (C. A.). *Also*: *Ann. d. R. Sc. Sup. Agric. Portici, Napoli, 1878*.

PALMIERI P. — Il pozzo artesiano dell'Arenaccia del 1880 confrontato con quello di Palazzo Reale di Napoli del 1847. — *Lo Spettatore del Vesuvio e dei Campi Flegrei. Nuova Serie, Vol. I, Napoli, 1887, col. pl. 1*.

PALMIERI L. — Studj Meteorologici fatti sul Real Osservatorio Vesuviano. — *Napoli, 1853, in 4°, pp. 22*.

PALMIERI L. — Disquisizioni accademiche sulle scoperte Vesuviane attinenti alla elettricità atmosferica. — *Napoli, 1854, in 4.° p. 33. 1 pl.*

PALMIERI L. — Eruzione del Vesuvio del 1° Maggio 1855 studiata dal R. Osservatorio Meteorologico Vesuviano. — *Il Nuovo Cimento, Giorn. d. Fis., Chim. e St. Nat. Vol. I, Pisa, 1855. Also, Giornale Ufficiale del Regno delle Due Sicilie, Napoli, 25 Mag. 1855, pp. 1, in fol.*

PALMIERI D. — Alcune osservazioni sulle temperature delle fumarole che si generano sulle lave del Vesuvio. — *Il Nuovo Cimento, Giorn. d. Fis. etc. Vol. V, Pisa, 1857*.

PALMIERI L. — Osservazioni di meteorologia e di fisica terrestre fatte durante l'eruzione del Vesuvio del Maggio 1855. — *Il Nuovo Cimento. Giorn. di Fis. etc. Vol. V. Pisa, 1857*.

PALMIERI L. — Sur l'éruption actuelle du Vésuve. — *Lettre à Deville—Compt. Rend. d. l'Acad. d. Sc. Vol. XLV. Paris, 1857, pp. 549-550*.

PALMIERI L. — Sur le Vésuve Lettre a Deville. — *Compt. Rend. Acad. Sc. Vol. LXVI, Paris, 1858, pp. 1219-1220*.

PALMIERI L. — Annali del Reale Osservatorio Meteorologico Vesuviano. Anno Primo 1859. — *Napoli 1859.—Biblioteca Vesuviana — Anno Secondo 1862. Napoli 1862. vol. 2, in 8° (Vol. I, pp. 80+XVIII+2. Vol. II, pp. VII+88+1)*.

PALMIERI L. — Sur l'éruption du Vésuve. Lettre à Deville. — *Compt. Rend. d. l'Acad. d. Sc. Vol. LIII. Paris, 1861, pp. 1231-1233*.

PALMIERI L. — Relazione intorno allo incendio del Vesuvio cominciato il dì 8 Dic. 1861. — *Il Nuovo Cimento. Vol. XV. Pisa, 1862. Also, Rend. Accad. Pontaniana, 1862, Napoli, in 8°, pp. 36, pl. 2*.

PALMIERI L. — Notizie sulle scosse di terremoto segnate dal sismografo elettro-magnetico dopo l'incendio del Vesuvio cominciato il dì 8 Dic. 1861. — *Napoli, 1862*.

PALMIERI L. — Sur les phénomènes électriques qui se sont produits dans la fumée du Vésuve pendant l'éruption du 8 décembre 1861.—*Comp. Rend. Acad. Sc., t. LIV, Paris, 1862, pp. 14.* (C. A.).

PALMIERI L. — Sur les secousses de tremblement de terre ressenties à l'Observatoire du Vésuve pendant les mois de Décembre 1861 et Janvier 1832. — *Compt. Rend. d. l'Acad. d. Sc. Vol. LIV. Paris, 1862, pp. 608-611.*

PALMIERI L. — Delle scosse di terremoto avvenute all'Osservatorio Meteorologico Vesuviano nell'anno 1863, quali furono registrate dal sismografo elettro-magnetico — *Rend. R. Accad. Sc. Fis. Mat. Napoli, Vol. III. 1864.*

PALMIERI L. — Il Vesuvio dal 10 Febbraio al 5 Marzo del 1865. — *Rend. d. R. Acc. d. Sc. Fis. e Mat. Vol. IV. Napoli, 1865.*

PAEMIERI L. — Il Vesuvio, il terremoto di Isernia e l'eruzione sottomarina di Santorino — *Rend. d. R. Acc. d. Sc. Fis. e Mat. An. V. Napoli, 1866, pp. 2.*

PALMIERI L. — Dell'incendio vesuviano cominciato il 13 Novembre 1867. — *Napoli, 1867.*

PALMIERI L. — Di alcuni prodotti trovati nelle fumarole del cratere del Vesuvio. — *Rend. d. R. Acc. d. Sc. An. VI. Napoli, 1867, pp. 2.*

PALMIERI L. — Nuove corrispondenze tra i terremoti del Vesuvio e l'eruzioni di Santorino. — *Rend. d. R. Acc. d. Sc. An. VI. Napoli, 1867, pp. 2.*

PALMIERI L. — Sur les produits ammoniacaux trouvés dans le cratère supérieur du Vésuve. — *Compt. Rend. d. l'Acad. d. Sc. Vol. LXIV. Paris, 1867, pp. 668-669.*

PALMIERI L. — Faits pour servir a l'histoire éruptive du Vésuve.— *Compt. Rend. d. l'Acad. d. Sc. Vol. LXV, Paris, 1868, pp. 897-898; Vol. LXVI, pp. 205-207, 756-757, 917-918.*

PALMIERI L. — Dell' incendio del Vesuvio cominciato il 13 novembre del 1867. Sunto di una relazione dell' Autore.—*Rend. R. Accad. Sc. Fis. Mat. An. VII, Napoli, 1868, pp. 76-77.*

PALMIERI L. — Relazione delle eruzioni del Vesuvio dal 13 Novembre 1867 fino al 30 Maggio 1868. — *Atti d. R. Acc. d. Sc. Vol. VI. Napoli, 1868.*

PALMIERI L. — Ueber den neuen Ausbruch des Vesuv. — *Verhandl. der K. K. Geologisch. Reichs-Anst. Wien, 1867. Und Fortsetzung, 1868.*

PALMIERI L. — Nuovi fatti di corrispondenza tra le piccole agitazioni del suolo al Vesuvio ed i terremoti lontani. — *Rend. d. R. Acc. Sc. An. VIII, Napoli, 1869, pp. 179.*

PALMIERI L. — Osservazioni sul terremoto del 26 Agosto 1869.—

Rend. d. R. Acc. d. Sc. di Napoli, An. VIII. Napoli, 1869, pp. 179.

PALMIERI L. — Il terremoto di Calabria ed il sismografo vesuviano. — *Rend. d. R. Acc. d. Sc. di Napoli. Vol. IX. 1870.*

PALMIERI L. — Indicazioni del sismografo dell'osservatorio Vesuviano del 1° Dic. 1869 al 31 Dic. 1870. — *A note inserted in « Memoria sopra i terremoti della Prov. di Cosenza nell'anno 1870 del Sig. Doll. Conti ».*

PALMIERI L. — Qualche osservazione spettroscopica sulle sublimazioni vesuviane. — *Rend. d. R. Acc. d. Sc. di Napoli, An. IX, Napoli, 1870, pp. 58-59.*

PALMIERI L. — Ultime fasi delle conflagrazioni vesuviane del 1868. *Rend. d. R. Acc. d. Sc. di Napoli. An. VIII, 1869, pp. 44-48. — Il Nuovo Cimento. Ser. 2ª, Vol. III. Pisa, 1870.*

PALMIERI L. — Indicazioni del Sismografo dell' Osservatorio Vesuviano del 1° dicembre del 1869 al 31 dicembre del 1870.— *Rend. R. Accad. Sc. Fis. Mat. An. X, Napoli, 1871, pp. 16-17.*

PALMIERI L. — Il Litio ed il Tallio nelle sublimazioni vesuviane.— *Rend. d. R. Acc. d. Sc. di Napoli, An. X, 1871, pp. 124.*

PALMIERI L. — Intorno ad un Lapillo filiforme eruttato dal Vesuvio. — *Rend.' d. R. Acc. d. Sc. di Napoli. An. X. 1871. pp. 51-52.*

PALMIERI L. — Le iave del Vesuvio guardate con lo spettroscopio.— *Rend. d. R. Acc. d. Sc. di Napoli An. X, 1871, pp. 33-34.*

PALMIERI L. — Osservazioni microscopiche sulle sabbie eruttate dal Vesuvio nei mesi di Gennaio e Febbraio del 1871. — *Rend. d. Acc. R. d. Sc. di Napol. An. X. 1871, pp. 34-35.*

PALMIERI L. — Sopra qualche legge generale cui obbediscono le sublimazioni del Vesuvio, delle fumarole delle lave del Vesuvio. — *Rend. d. R. Acc. d. Sc. di Napoli, An. X, 1871, pp. 90-93.*

PALMIERI L. — Trasformazione di alcuni cannelli di vetro rimasti per lungo tempo in nna fumarola. — *Rend. R. Accad. Sc. Fis. Mat. An. X, Napoli,' 1871, pp. 124.*

PALMIERI L. — Il solfato di zinco fra le sublimazioni vesuviane.— *Rend. R. Accad. Sc. Fis. Mat. An, X, Napoli, 1872, pp. 13.*

PALMIERI L. — L'incendio vesuviano del 26 Aprile 1872. Conferenza tenuta nel di 9 maggio coll'analisi chimica delle ceneri cadute il 28 Aprile, del Prof. C. Catalano. — *Napoli, 1872, in 12°, pp. 12. (C. A.).*

PALMIERI L. — Annali del R. Osservatorio meteorologico Vesuviano. 1° Ser. — *Napoli, 1869-72. 2ª Ser. Napoli, 1873.*

PALMIERI L. — Carbonati Alcalini trovati tra' prodotti vèsuviani.—
Rend. R. Acc. Sc. Fis. Mat. An. XII, Napoli, 1873, pp. 92.

PALMIERI L. — Dell'incendio vesuviano del 26 Aprile 1872.—*Atti
d. R. Acc. d. Sc. e Belle Lett. An. IX, Napoli, 1872, pp.
157-158. — In German: Leipzig, and Berlin, 1872, pl. 7. In
English, with introd. by R. Mallet. London, 1873, pl 8.*

PALMIERI L. — Del sale ammoniaco giallo e della Cotunnia
gialla. — *Rend. R. Accad. Sc. Fis. e Mat. An. XII, Napoli,
1873, pp. 92-94.*

PALMIERI L. — Indagini spettroscopiche sulle sublimazioni vesuviani. — *Rend. R. Accad. Sc. Fis. Mat. An. XII, Napoli,
1873, pp. 47-18.*

PALMIERI L. —La conflagrazione vesuviana del 26 Aprile del 1872,
riferita all'Acc. delle Scienze Fis. e Mat. — *Napoli, 1873, in
4°, pp. 64, pl. 5. (C. A.).*

PALMIERI L. — Recherches spectroscopiques sur les fumeroles de
l'éruption du Vésuve en Avril 1872 et état actuel de ce volcan.—
Compt. Rend. d. l'Acad. d. Sc. Vol. LXXVI. Paris, 1873.

PALMIERI L. — Sepolcri antichi scoperti sul Vesuvio. — *Rend. d.
R. Acc. d. Sc. Fis. ecc. An. XI, Napoli, 1872, pp. 2.*

PALMIERI L. -- Sopra alcuni fenomeni notati nell'ultimo incendio
vesuviano del 26 Aprile 1872.—*Rend. d. R. Acc. d. Sc. An. XI,
Napoli, 1872, pp. 108.*

PALMIERI L. — Sul ferro oligisto trovato entro le bombe dell'ultima eruzione del Vesuvio — *Rend. R. Accad. Sc. Fis. Mat.
An. XII, Napoli, 1873, pp. 48.*

PALMIERI L. — Sulle fumarole eruttive osservate nell' incendio vesuviano del 26 Aprile 1872. — *Rend. R. Acc. Sc. Fis. Mat.
An. XII, Napoli, 1873, pp. 143.*

PALMIERI L. — Cronaca del Vesuvio, Sommario della storia dei
principali accendimenti del Vesuvio dal 1840 fino al 1871,
seguito da estesa relazione dell'ultimo incendio del 1872. —
Napoli, 1874.

PALMIERI L. — Del peso specifico delle lave vesuviane nel più per-
fetto stato di fusione. — *Rend. d. R. Acc. d. Sc. Fis. e Mat.
An. XVI, Napoli, 1875, pp. 214-215. Riv. Scient. Napoli,.
1876.*

PALMIERI L. — Il cratere del Vesuvio nel di 8 Novembre 1875
(estratto di una lettera del giovane Alpinista, G. Chiarini.—
*Rend. R. Accad. Sc. Fis. Mat. An. XV, Napoli, 1876, pp,
9-10, fig. 1.*

PALMIERI L. — Il terremoto del 6 Dic. 1875. — *Rend. d. R. Acc.
d. Sc. Fis. etc. An. XIV, Napoli, 1875, pp. 215-216.*

Palmieri L. — Il Tallio nelle presenti sublimazioni vesuviane.— *Rend, d. l. R. Acc. d. Sc. Fis. etc. Napoli, An. XVI, 1877, pp. 179-180.*

Palmieri L. — Sulla cenere lanciata dal Vesuvio a Portici e Resina la notte del 3 al 4 Aprile 1876. — *Rend. d. R. Acc. d. Sc. Fis. etc. Fasc. IV. Napoli, 1876.* — *Ann. d. R. Sc. Sup. di Agr. in Portici. Napoli, 1878.*

Palmieri L. — Del Vesuvio dei tempi di Spartaca e di Strabone e del precipuo cangiamento avvenuto nell'anno 79 dell'Era volgare. — *See, Num. and Not. publ. by the Uff. Tecn. degli scavi: Pag. 91-94 pl. 1. Napoli, 1879. See also « Pompei e la Regione Sotterrata »* etc.

Palmieri L. — Specchio comparativo della quantità di pioggia caduta nell'anno meteorico 1880 nelle stazioni di Napoli (Università) e Vesuvio (O. V.). — *Rend. R. Accad. Sc. Fis. Mat. An. XIX, Napoli, 1880, pp. 179-180.*

Palmieri L. — Della Riga dell'Helium apparsa in una recente sublimazione vesuviana. — *Rend. R. Accad. Sc. Fis. Mat. An. XX, Nopoli, 1881, pp. 233.*

Palmieri L. — Nuove esperienze che rifermano le antecendi sull'origine dell'elettricità atmosferica. Appendice alla memoria inscrita nel tomo IV della Società Italiana delle Scienze. — *Atti. R. Acc. Sc. Fis, Mat. ?, Napoli, 1886, pp. 24.*

Palmieri L. — Il Vesuvio e la sua storia. — *Milano, 1880. See also: Lo Spettatore del Vesuvio e dei Campi Flegrei, Nuova serie, Vol. 1.° Napoli, 1887.*

Palmieri L. — L'elettricità negl'incendi vesuviani studiata dal 1855 fin'ora con appositi istrumenti. — *Lo Spettatore del Vesuvio e dei Campi Flegrei. Nuova serie, Napoli, 1887, in 4°, pp. 77-79.*

Palmieri L. — Azione de'terremoti dell'eruzioni vulcaniche e delle folgori sugli aghi calamitati. — *Rend. R. Accad. Sc. Fis. Mat. Napoli, 1888, pp. 3.*

Palmieri L. — Le correnti telluriche all'Osservatorio vesuviano osservate per un anno intero non meno di quattro volte al giorno. — *Rend. R. Accad. Sc. Fis. Mat. Napoli, 1890, pp, 6.*

Palmieri L. — Osservazioni simultanee sul dinamismo del Cratere vesuviano e della grande fumarola della Solfatara di Pozzuoli fatte negli anni 1888-89-90. — *Rend. R. Acc. Sc. Fis. Mat. Napoli, 1890, pp. 3.*

Palmieri L. — Elettricità atmosferica, continuazione degli studi meteorologici fatti sul Reale Osservatorio Vesuviano. — *Po-*

9

*liorama Pilloresco. N.° 23, 24, 25, Anno XV, in 4°, pp.
8-11. pl.*

PALMIERI L. e DEL GAIZO M. — Il Vesuvio nel 1887. — *(Annuario
Met. It. anno III). Torino, 1888.*

PALMIERI, SCACCHI E GUARINI. — Memoria sullo incendio vesuviano del mese di Maggio 1855. Preceduta dalla relazione del-
l'altro incendio del 1850 fatta da A. Scacchi—*Napoli, 1855,
in 4°, pp. VIII—207, con 7 lavole litografiche.*

PAOLI F. — Per l'andata al Vesuvio de Marchese di Palombara.
Sonetti due. — *S.e Giorgi Urbano. (C. A.).*

PAPACCIO G. S. (venditor d'oglio). — Relatione del fiero, et ira-
condo incendio del Monte Vesuviano flagello occorso a 16 di
Decembre 1631, nella montagna di Somma all'incontro sei miglia della fedelissima e famosissima Città. In ottava rima.—
Napoli, 1632, in 4°, fol. 4. (C. A.).

PARAGALLO G. — Istoria naturale del monte Vesuvio, divisa in 2
libri. — *Napoli, 1705, in 4°, fol. 10, pp. 430.*

PARKER J. — Part of a letter concerning the late Eruption of
Mount Vesuvius 1751. — *Philos. Trans. of the R. Soc. of
London. London, 1752,Vol. XLVII, pp. 474-475.*

PARRINO D. A. — Relazione dell'eruzione del Vesuvio nel 1694. —
Napoli, 1694.

PARRINO D. A. — Succinta relazione dell'incendio del Vesuvio nel
1696. — *Napoli, 1696.*

PARRINO D. A. — Moderna distintissima descrizione di Napoli Città
nobilissima, antica e fedelissima, e del suo Seno Cratere. —
Aggiunte, osservazioni, e correzioni a questo primo tomo
della nuova descrizione di Napoli. — *Napoli, 1703-1704, vol.
in 12.° (Vol I. pp. 20 + 438 + 51. 46+2. Vol. II. pp. 16+
292+23. pl. XXVIII, fig. See. Vol. II, pp. 205-235.).*

PARRINO D. A. — Nuova guida per l'antichità di Pozzuoli, e di
tutte le Città, luoghi e Isole, che sono alla veduta presso il
mare dalla parte destra della Città di Napoli.—*Napoli, 1751,
in 12. pp. 257+7+XXVII, pl. (See Parte II, pp. 182-220.).*

PARRINO D. A. — Nuova guida dei Forestieri per osservare e godere le curiosità più vaghe della fedelissima gran Napoli.—
Napoli, 1709, and 1725, in 12°, fol. 18, pp. 382, fol 6, pl. 40
(C. A.). *Another edition 1751, in 12°, fol. 2, pp. 269 + 18
pl. 38, maps 9. See pp. 181-218. (C. A.).*

PARTENIO (accademico). — La morte; Idillio fatto in occasione
dell'incendio del Monte Vesuvio, ed una canzonetta sopra la
stella apparsa nel medesimo tempo sopra detto monte. —
Roma, 1632, in 4°, fol. 4. (O. V.).

PASQUALE G. A. — Flora Vesuviana, o catalogo ragionato delle piante del Vesuvio confrontate con quella dell'isola di Capri e di altri luoghi circostanti. — *Atti d. R. Accad. d. Sc. di Napoli, 1868, pp. 142.* (C. A.).

PASSE C. (DE).—Uvaerachtige af-beeldinge van den schricklijcken Brandende Bergh Somma (anders genoemt Vesuvi,) gelegen vande wijtberoemde Stadt Neepolis een uyre gaens, die meteen onuytspreechenlijck Dyer en Water noch dagelijer der Stadt groolelijer beschadicht, als blijckt wt dit nac-volgende. As ghebeeldt ende overgeset uyt het Italiens nae de Roomsche Copye. — *Engraved plate of erupt. 1631, with 3 columns of explanation.* (C. A.).

PASSERI G. — Saggio di Poesie.—*Napoli, 1766, in 8°, pp. 7÷202 +1. Cantata sul Vesuvio (pp. 114-126.).*

PAYAN D. — Notice sur quelques volcans de l'Italie méridionale. — *Bull. Soc. Stat. Arts Utiles, Sc. Nat. du Départ. Drôme, t. III, in 8°, 1842, pp. 145-163.* (C. A.)

PELLEGRINI G. — Il Vesuvio, poemetto. — *Bassano, 1785, in 8°. pp. 112, frontisp.* (C. A.). *Bassano, 1798, in 8°, fol I, pp. 301.* (O. V.). *Palermo, 1814, in 8°, pp. 108, fol. 1.* (O. V.).

PELLEGRINO C.—Discorso istorico dell'incendii naturali del Monte Vesuvio ed altri luoghi di Terra di Lavoro detti anticamente Campania, raccolto in un manoscritto a di 16 dicembre 1631.— *M. S. (fol. 15) Copy of a M. S. belonging to Signor Adolfo Parascandolo.* (C. A.).

PELLICER (DE) TOVARJ. — Estancias al Vesuuio ed un epigramma. — *See de Quinones.* (C. A.).

PEPE A. — Il medico clinico o sia dessertazione fisico-medica (speaks in Chap. I, of the eruption of Vesuvius 1767). — *Napoli, 1768, in 4°. fol. 6, pp. 178, fol 1°.* (C. A.).

PERENTINO GIANO (Pietro Giannone). — Lettera ad un suo amico che lo richiedeva onde avvenisse che nelle due cime del Vesuvio, ecc. — *Napoli, 1718, in 4°, p. 3.* (O. V.).

PEREZ DE MONTALVAN J.—Sonetto (Erupt. 1631). — *See de Quinones.* (C. A.).

PERI D. — Sull'eruzione del Vesuvio del 15 giugno 1794.—*Anacreontica.* (O. V.).

PERILLO D. — Vero e distinto ragguaglio di ciò etc. — spaventevole fiumana di fuoco scoppiato dal Monte Vesuvio in caminavasi al di lei danno e sterminio.—*Napoli, 1755, in 4°, p. 76.* (O. V.).

PERROTTA F. — Relatione del nuovo incendio del Monte Vesuvio delli 3 luglio 1660 del medico fisico Francesco Perrotta di

Piedimonte d'Alife, medico della Torre del Greco. — *M. S. in* (O. V.).

PERROTTI A. — Discorso astronomico sopra li quattro Ecclissi del 1632 et uno del 1633.—*Napoli, 1632, in 4°, fol. 26.* (C. A.).

PESCE D. — Il povero lacrimante sopra alcune dimostrazioni di Fisica naturale del Gran Monte Vesuvio. — *M. S., 1767, in 4°, fol. 34.* (O. V.).

PETRIS (DE) FR. — De Vesuvij conflagratione. — *Distico, in Mormile.*

PETRIZZI ANT. DA. — Lettera a Sua Eccellenza il signor D. Francesco Ant. Marmigola Duchino di Petrizzi sulla lava eruttata dal fianco, o pendici del Vesuvio ad ore due, e minuti 10, di notte circa, del dì 15 Giugno, 1794.—*?, in 12°, and in 4°,* (B. N.).

PHILIPPI R. A.—Nachricht über die letzte Eruption des Vesuvs.— *Neues Jahrb. für Mineralog. Geognos. etc. Stuttgart, 1841, in 8°, pp. 59-69.*

PHILIPPI R. A. — Relief des Vesuvs und seiner Umgegend. — *Bericht über die Versamml. der Deutsch. Naturforsch. u. Aerzte. Bd. VII. 1842.*

PHILLIPS J.—Vesuvius.—*Oxford, 1869. London 1872, pp. XVII+ 355, pl. 10, map. 1, figs. 35.*

PIAGGIO A. — Furentis Anno MDCCLXXXIX Vesuvii prospectus.— *Plate engraved by Cataneo at the back of which is a sonnet by the Duca di Belforte.* (C. A.).

PIAZZAI S. — Sonetti due (1631). — *See G. Urbano.* (C. A.)

PICCININI D. — Per la eruzione del Vesuvio, accaduta nell'anno 1822. Verses in neapolitan dialect.—*Poesie italiane e in dialetto napolitano, Napoli, 1827, in 8', pp. 49-64.* (C. A.).

PIETRO FR. (DI). — I problemi accademici ove le più famose quistioni proposte nell'Ill. Accademia degli Otiosi di Napoli si spiegano.—*Napoli, 1642, in 4°, pp. 40+317+25.—V. Problema LXXX. Dell'incendio del Monte Vesuvio avvenuto ai 16 di Decembre 1631. (pp. 217-220.)*

PIETRO M. (DI). — 1822. — *See Grifoni E.*

PIETROSIMONE N.—Descrizione istorica-cronologica delle principali eruzioni del Vesuvio tolte dalle opere di Luigi Galante e riportato nell'istoria dei monumenti di Napoli da Camillo Napol e Sasso, con due sonetti sul Vesuvio del Pietrosimone.— *« L'Ateneo Popolare » Napoli, 1868, in 8°, pp. 80, pl. 62.* (C. A.).

PIGHIUS CAMPENSIS. — Hercules prodicius seu principis juventutus, etc. — (*Describes the Vesuvian crater previous to 1631)* Antuerpiae, 1587). (O. V.).

PIGNANT.—Sur une éruption du Vésuve le 11 Mars 1866.—*Compt. Rend. Acad. Sc. Paris, 1866, t. LXII, p. 749.* (C. A.).

PIGONATI A. — Descrizione delle ultime eruzioni del Vesuvio dai 25 Marzo 1766 fino a' 10 Dicembre dell'anno medesimo.—*Napoli, 1767, in 8°, fol. 4°, pp. 28, pl. 3.* (O. V.) *Varying editions.*

PIGONATI A. — Descrizione dell'ultima eruzione del Vesuvio dei 19 ottobre 1767, in seguito dell'altra del 1766—*Napoli, 1768, in 8°, pp. 23, pl. 4, fig. Varying editions, one in 4.°*

PIGONATI — Relazione della straordinaria eruzione del Monte Vesuvio nel dì 8 Aprile 1779.—*Opuscoli Scelti Sulle Scienze e Sulle Arti, Milano, 1778, t. II, parte IV, pp. 310-312, in 4°,* (C. A.).

PILLA L.—Sur l'éruption du Vésuve en Juillet et Août 1832 — *L'Osservatore del Vesuvio. N.° 3; Bibl. Univ. T. 52, Avril 1833, pp. 351-356.*

PILLA L. — Narrazione d'una gita al Vesuvio fatta nel dì 26 gennaio 1832. — *Il Progresso d. Sc. Lett. ed Arti. Vol. I. Napoli, 1832.*

PILLA L.—Bollettino geologico del Vesuvio e dei Campi Flegrei, destinato a far seguito allo Spettatore del Vesuvio.—*Ann. d. R. Oss. Vesuv. Napoli, 1833-34, in 8°, pp. 35, 30, 28, 31, 40, complete.* (O. V.).

PILLA L. — Esposizione dei fenomeni osservati nel cratere del Vesuvio durante l'eruzione del 1833. — *(?)*

PILLA L. — Ausbrüche des Vesuvs im Anfange Aprils 1835. — *Neues Jahrb. für Mineralog. etc. Stuttgart, 1835.*

PILLA L. — Bollettino geologico del Vesuvio e dei Campi Flegrei.— *Il Progresso d. Sc. etc. Vol. VIII, IX, X, e XVI. Napoli, 1834-1837.*

PILLA L. — Observations tendantes à prouver que le cône du Vésuve a été primitivement formé par soulèvement. — *Compt. Rend. d. l' Acad. des Sc. Vol. IV. Paris, 1837.*

PILLA L. — Parallelo fra i tre Vulcani ardenti dell'Italia. Napoli, 1835. — *Atti d' Acc. Gioenia d. Sc. Nat. di Catania. Vol. XII. Catania, 1837, p. 39. Jahrb. f. Min. Stuttgart, 1836, p. 347.*

PILLA L. — Sur des coquilles trouvées dans le Fossa Grande de la Somma. — *Bul. Soc. Géol. France. t. VIII, pp. 199-201, 3 Avril, 1837, and, pp. 217-224, 17 Avril, 1837. (C. A.).*

PILLA L. — Ventesimo viaggio al Vesuvio il 21 e 22 Agosto. — *Il Progresso d. Sc. etc. Vol XVI. Napoli, 1837.*

PILLA L. — Ventitresima gita al Vesuvio nel'a notte del 13 al 14 Settembre, 1834. — *Il Progresso d. Sc. etc. Napoli, 1838, fol. 7. Spettatore del Vesuvio. Fasc. XI. Napoli, 1838.*

PILLA L. — Ausbruch des Vesuvs Anfangs Januar 1839. — *Neues Jahrb. für Mineralog. etc. Stuttgart, 1839.*

PILLA L. — Relazione dei fenomeni avvenuti nel Vesuvio nei primi del corrente anno 1839. — *Il Progresso d. Sc. etc. Vol. XXII. Napoli, 1839, pp. 29-41.*

PILLA L. — Sur la dernière éruption du Vésuve. — *Compt. Rend. d. l'Acad. d. Sc. Vol. VIII, Paris, 1839.*

PILLA L. — Sur l'éruption du Vésuve en Janvier, 1839. Lettre à M. Elie de Beaumont. — *Compt. Rend. Acad. Sc. Paris, t. VIII, 1839, pp. 250-253. (C. A).*

PILLA L. — Observations relatives au Vésuve. — *Compt. Rend. d. l'Acad. d. Sc. Vol. XII. Paris, 1841.*

PILLA L. — Sur quelques minéraux recueillis au Vésuve et à la Rocca Monfina. — *Compt. Rend. l'Acad. d. Sc. Vol. XXI. Paris, 1845.*

PILLA L. — Catalogue de collections des minéraux et de laves du Vésuve à vendre. — *Without date or locality, in 8°, fol. 2. (C. A.).*

PILLA L. AND CASSOLA — Lo Spettatore del Vesuvio e dei Campi Flegrei. — *Napoli, 1832-33, in 8°.*

PILLA N. — Geologia Volcanica della Campania. — *Napoli, 1823. 2, Vols in 8°, (Parte I. pp. XIX+124+1) Parte II. pp. 159+1.*

PINA JUAN (DE).—Al Bolcan di Soma; Soneto.— *See de Quinones. (C. A.).*

PISANI F. — Rapport sur l'éruption du Vésuve du 24 au 30 Avril 1872. — *Bull. d. l. Soc. Géol. de France, 2.ª Sér. Vol. XXIX. Paris, 1872.*

PLACIDO F. — Dialogo sopra il miracolo del Gloriosissimo Protettore della Città e del Regno di Napoli S. Gennaro.—?(B. N.).

PLANGENETO U. — La lacrima del Monte Vesuvio volgarmente Lacryma Christi. — *Ditirambo. Napoli, 1811 in 12, p· 67.*

PISTOLESI E. — Real Museo Borbonico descritto ed illustrato. — *Roma, 1836; in 4°, See Vol. I, pp. 5-91.*

PITARO A. — Esposizione delle sostanze costituenti la cenere vulcanica caduta in questa ultima eruzione de' 16 di Giugno 1794. — *Napoli, 1794, in 8°, pp. 22.*

PLINY C. S. C. — Letters, with observations on each letter, and an essay on Pliny's Life, addressed to Charles Lord Boyle by John Earl of Orrery. — *London, 1751, 2 Vols, in 8°, with figs(Vol.I.pp. IV+LXX+397 +34.)(Vol. II.pp. 2+450+53.).*

PLINII CAECILII SECUNDI C. — Epistolae lib. IX. ejusdem et Trajani Imp. Epist. etc. adjectae sunt Isaaci Casauboni notae in Epist. variae lectiones ultra precedentes, in hac posteriori editione marginis accesserunt. — *1632, in 16°, p. 862.*

POLI G. AND LAURIA G. A. — Prosa Elegiaca per Giacinto Poli e fotografia morale del giovane Vitangelo Poli vittima della esiziale eruzione del Vesuvio nella notte del 25 Aprile 1872. — *Napoli, 1872. in 4°, p. 20. (C. A.).*

POLI G. S. — Saggio di Poesie. — *Palermo, 2 Vols, in 4°, divisi in quattro parti. (Vol. I, Parte I e II, pp. 10+337. Vol. II. Parte I. e II. pp. 366 + 6.) Il Vesuvio , Poemetto. Vol. I. part. I. pp. 1-| 21. Dissertazione intorno al Vesuvio, etc. Vol. II, pt. II, pp. 247+292.*

POLI G. S. — Dissertazione intorno al Vesuvio in cui si ragiona del suo stato si antico che recente—*(?) in 8°, pp. 46. Ext. works of author.*

POLLERA G. D. — Relatione dell'incendio del monte di Somma successa (sic) nell'anno 1631, nella quale si rendono le ragioni di molte cose le più desiderabili. — *Napoli, 1632, in 8°, fol. 8. (B. N.).*

POMPEI e la regione sotterrata dal Vesuvio nell'anno LXXIX. Memorie e notizie pubblicate dall'Ufficio Tecnico degli Scavi delle Provincie Meridionali. — *Napoli, 1879, in fol. pl. I, pp. 291, pl. 2., pl. II, p. 243. (C, A.).*

PONTANO.—La seconda parte delli avisi del Rev. Pad. Pontano, etc. Di tutto quello, ch' è successo in tutta la seconda settimana. Et così l'haverete d' ogni sette in sette giorni. — *Napoli, 1632, in 4°, (B. N.).*

PORRATA S. — Discorso sopra l'origine dei fuochi gettati dal monte Vesevo, ceneri, etc. — *Lecce, 1632, in 4°, fol. 3, pp. 55.*

PORTII L. A. — Opera omnia, etc. in unum collecta atque ad meliorem, commodioremque formam redacta cura restudio Francisci Portii. — *Vol. II, (O. V.).*

PORZIO L. A. — Lettere e discorsi Accademici.— *Napoli, 1711, in 4°, pp. 8+347. See. pp. 174-186).*

PREVOST C. — Etudes des phénomènes volcaniques du Vésuve et de l'Etna. — *Compt. Rend. d. l'Acad. d. Sc. Vol. XLI. Paris, 1855, pp. 794-797.*

PREVOST C. — Sur les coquilles marines trouvées à la Somma.— *Compt. rend. d. l'Acad. d. Sc. Vol. IV. Paris, 1837.*

PRINCIPE DI ESQUILACE. — Sonetto (1631). — *See de Quinones.* (C. A.),

PRINA L. G. — Ascensione al Vesuvio. — *Novara, 1874, in 8°, pp. 15.* (C. A.).

PRISCO C. — Componimento in versi latini sull'incendio del Vesuvio. — *Napoli, 1832, in 4°, pp. 31,* (O. V.).

PROCTOR R. A. — Le Vésuve et Ischia. — *Revue Mens. d'Astron. Pop. Paris, Sept. 1883, pp. 340-343.* (C. A.).

PROST. — Trépidations du sol à Nice pendant l'éruption du Vésuve. — *Compt. Rend. Acad. Sc. Paris t. XLIV, 1862, pp. 511-512.* (C. A.).

PULSINII V. — Sonetto (erupt. Vesuvius, Aug. 12th, 1805. — *See Cillunzio, Neante.* (B. N.).

P. R. — 2ª lettera. Raccolta di monumenti sopra l'eruzione del Vesuvio seguita nell'agosto, 1779. — *Giornale delle Arti e del Commercio. Vol. I, Macerata, 1780, in 8°, at pp. 141, and following.* (O. V.).

PUTIGNANI J. D. — De redivivo sanguine D. Januarii Episcopi ed Martyris. — *Napoli, 1723-26. Vols. III, in 4 parts, in 4°. See pt. I, pp. 155-188, and pl. IV, pp. 130-154.*

QUARANTA A. — Tre fugitivi, Dialogo, ove brevemente si dà ragguaglio dei principali successi, nell'Incendio di Vesuvio. — *Napoli, 1632, in 12°, pp. 35.* (C. A.).

QUEVEDO VILLEGAS F. DE. — Al Vesuvio que interpoladamente es jardin y Bolcan. Soñetto. — *See Quinones.* (C. A.).

QUINONES J. DE. — Epigramma. Acrostico — *See Quinones: Los incendios de la Montana de Soma.* (C. A.).

QUATTROMANI L. — Per Napoli salvata dal terremoto, e dalle lave del Vesuvio, ad intercessione di S. Gennaro. Sonetti. — ?, in *8°, pl.* (B. N.).

QUINONES J. DE. — El Monte Vesuvio aora la Montaña de Soma, dedicado a D. Felipe quarto el grande mestro Señor, ecc. — *Madrid, 1632, in 4°, fol. 16+56. Contains in first part. 1. El Principe de Esquillace — Soneto, 2. Lopez de Zarate Francisco — Soneto, 3. Solis — Messia Juvan — Soneto, 4. Villayçan Garces (de) Gironimo — El Bolcan que aborto la Montana di Soma — Soneto, 5. Ramirez de Arellano Luis — Soneto, 6. Cardoso doll. Fernando — Al Vesuvio — Soneto, 9. Coruna Conde (de) — Soneto, 10. Lope de Vega Felix Canzone, — 11. Valdivielso (de) Joseph — Silva, 12. Quevedo Villegas (de) Francisco — Al Vesuvio que interpoladamente es jardin y Bolcan — Soneto, 13. Perez de Montalban Juan — Soneto, 14. Bocangel y Vusuela — Epitafio*

*al Vesuvio, y sus incendios — Soneto, 15. Velez de Guevara
Luis — a la montana de Soma — Soneto, 16. Andosilla
Larramendi Juan — Al Vesuvio — Soneto, 17. Pina de Juan,
Al Bolcan de Soma — Soneto, 18. Huerta (de) Antonio —
Soneto, 19. Pellicer de Tovar — Estancias al Vesuvio, 20.
Pellicer de Tovar — Epigramma, 21, Ruiz de Alarcon y
Mendoça — Al Bolcan y incendios del Vesuvio — Soneto,
22. Hurtado de Mendoça Antonio — Dezimay, 23. Quino-
nes — Epigramma — acrostico , 24. Caruna (de) Conde —
Versi.* (C. A.).

RAFFLES, NECKER UND DAUBENY. — Ueber die Vulkane auf Java,
in den Auvergne und über den Monte Somma. — *Elberfeld,
1825.*

RAMIREZ DE ARELLANO LUIS — Soneto — *See Quinones.* (C. A.).

RAMMELSBERG G. — Ueber Humit und Olivins. — ? (1851), *in 8°,
fol. 8.* (O. V.).

RAMMELSBERG C. — Uber den magnoferrit vom Vesuv und die
Bildung des magneteisens und ähnlicher Verbindungen durch
sublimation. — *Monatsb. der K. Preuss. Akad. d. Wiss. zu
Berlin, 1859. — Journ. für prakt. Chemie, Bd. LXXVII.
Leipzig, 1859. — Ann. der Phys. und Chemie, Bd. CVII.
Leipzig, 1859.*

RAMMELSBERG C. — Ueber mineralogische Zusammensetzung der
Vesuvianen und das Vorkommen des Nephelins in denselben. —
Zeitsch. Deutsch Geol. Gesell. Bd. XI, Berlin, 1859, pp. 492-506.

RAMMELSBERG C.—Ueber die Chemische Zusammensetzung einiger
mineralien des Vesuvs. — *Ann. der Phis. und Chemie, Bd.
CIX, Leipzig, 1860.*

RAMMELSBERG C. — Uber den letzten Ausbruch des Vesuvs vom
8 December 1861. — *Zeitsch. Deutsch. Geol. Gesell. Bd. XIV.
Berlin, 1862, pp. 567-574.*

RAMMELSBERG C. — Ueber die chemische Natur des Vesuvasche
des Ausbruchs von 1872. — *Zeitschr. d. Deutsch. Geol. Gesell. Bd. XXIV. Berlin, 1872.*

RANIERI A. — Sale ammoniaco marziale raccolto sulla lava del
Monte Vesuvio. — *Ann. di Chim. Vol. XLIX, Milano, 1869.*

RATH (VOM) G. — Ueber die Zusammensetzung des Mizzonits vom
Vesuv. — *Zeitschr. d. Deutsch. Geol. Gesell. Bd. XV. Berlin, 1863.*

RATH (VOM) G. — Der Zustand des Vesuv am 3 April 1865. —
Sitz. Ber. d. Niederrhein. Naturforsch. Gesell. Bonn, 1865.

RATH (VOM) G. — Oligoklas vom Vesuv. — *Ann. der Phys. und
Chemie. Bd. CXXXVIII. Leipzig, 1869.*

10

RATH (VOM) G. — Crystallisirter Lasurstein vom Vesuv. — *Ann. der Phys. und Chemie, Bd. CXXXVIII. Leipzig, 1869.*

RATH (VOM) G. — Ueber den Wollastonit vom Vesuv. — *Ann. der Phys. und Chemie, Bd. CXXXVIII. Leipzig, 1869.*

RATH (VOM) G. — Orthit vom Vesuv. — *Ann. der Phys. und Chemie, Bd. CXXXVIII. Leipzig, 1869.*

RATH (VOM) G. — Ueber die Zwillingsbildungen des Anorthit vom Vesuv. — *Ann. der Phys. und Chemie, Bd. CXXXVIII. Leipzig, 1869.*

RATH (VOM) G. — Ueber die Zwillingsgesetze der Anorthits vom Vesuv. — *Sitz. Ber. d. Niederrhein. Naturf. Gesell. Bonn, 1869.*

RATH (VOM) G. — Ueber Humitcrystalle des zweiten Typus vom Vesuv. — *Ann. der Phys. und Chemie. Bd. CXXXVIII. Leipsig, 1869.*

RATH (VOM) G. — Orthit und Oligocklas in den alten Auswürflingen des Vesuvs. — *Sitz. Ber. der Niederrhein Naturf. Gesell. Bonn, 1870.*

RATH (VOM) G. — Der Vesuv am 6 und 17 April, 1871. — *Zeitsch;. d. Deut. Geol. Gesell. Bd. XXIII. Berlin, 1871.*

RATH (VOM) G. — Ein interessanter Wollastonit-Auswürfling vom Monte Somma. — *Sitz. Ber. der math. physic. Class. der K. Bayr. Akad. der Wissen. Bd. III. München, 1871, pp. 228-231.*

RATH (VOM) G. — Ueber die letzte Eruption des Vesuv und über Erdbeben von Cosenza. — *Verhandl. des Naturh. Vereins der Preuss. Rheinl. und Westph. Bd. XXVIII. Bonn, 1871.*

RATH (VOM) G. — Ueber den Zustand des Vesuv vor der letzten Eruption. — *Sitz. Ber. d. Niederrhein. Naturf. Gesell. Bd. XXIX. Bonn, 1872.*

RATH (VOM) G. — Ueber einige Leucit. Auswürflinge vom Vesuv. — *Ann. der Phys. und d. Chemie, Bd. CXLVII. Leipzig, 1872.*

RATH (VOM) G. — Ueber Vesuvische Auswürflinge der Eruption vom 26 April 1872. — *Sitz Ber. der Niederrhein. Naturf. Gesell. in Bonn, Bd. XXIX. Bonn, 1872.*

RATH (VOM) G. — Ueber einem merkwürdigen Lavablock des Vesuv. — *Sitz. der Niederrhein Gesell. in Bonn. Bd. XXIX. Bonn, 1872,* — *Report of the Brit. Assoc. for the Advanc. of Sc. London, 1872. Ann. Phys. und Chemie. Leipzig, pp. 562-568.*

RATH (VOM) G. — Der Aetna und der Vesuv. — *Bonn, 1872-73, 2 Vols, figs.*

RATH (VOM) G. — Der Vesuv. Eine geologische Skizze. — *Berlin, 1873, in 8°, pp. 53, pl. 2.* (C. A.).

RATH (VOM) G. — Geognostische mineralogische Fragmente aus Italien. XI. Ein Beitrag zur Kenntniss des Vesuv's. — Zeit-schr. d. Deut. Geol. Gesell. Bd. XXV. Berlin, 1873.

RATH (VOM) G. — On a Remarkable block of lava of the great Eruption of Vesuvius of April 1872. — Reports 42.nd Meet. of the Brit. Assoc. London, 1873.

RATH (VOM) G. — Ueber den angeblichen Epidot vom Vesuv. — Ann. der Phys. und Chemie, Bd. VI. Ergänz. Leipzig, 1873.

RATH (VOM) G.—Ueber die chemische Zusammensetzung der durch Sublimation in Vesuvischen Auswürflingen gebildeten Kry-stalle von Augit und Hornblende. — Ann. der Phys. und Chemie; Suppl. Bd. VI. Leipzig, 1873.

RATH (VOM) G.—Ueber die Glimmerkrystalle vom Vesuv.—Ann. der Phys. und Chemie. Bd. VI. Ergänz. Leipzig, 1873.

RATH (VOM) G. — Ueber die verschiedenen Formen der Vesuvi-schen Augite.—Ann. der Phys. und Chemie, Bd. VI. Ergänz. Leipzig, 1873.

RATH (VOM) G. —Ueber die chemische Zusammensetzung des gelb-ben Augits vom Vesuv.—Monatsb d. K. Ak. der Wissensch. Berlin, 25 Juli, 1875.

RATH (VOM) F.—Ueber die oktaëdrischeny Krystalle des Eisenglan-zes vom Vesuv, über die Verwachsungen von Biotit, Augit und Hornblende mit grösseren Augitkrystallen vom Vesuv und über Augit von Traversella. Brief an Prof. Leonhard. — Neue Jahrb. für Min. Geol. und Pal. Bd. IV. Stuttgart, 1876.

RATH (VOM) G. — Ueber die sogenannten oktaëdrischen Krystalle des Eisenglanzes vom Vesuv. — Verh. des Naturh. Ver. d. Preuss. Rheinl. und Westph. Bd. XXXIV. Bonn, 1877.

RATH (VOM) G.—Ueber einige durch vulkanische Dämpfe gebildete Mineralien des Vesuv. — Verh. d. Naturh. Ver. d. Preuss. Rheinl. und Westph. Bd. XXXIV. Bonn, 1877.

RATH (VOM) G.—Orthit von Auerbac, calcit von Lancashire, Dan-burit von Russel, St. Lawrence Co. N. Y. und cuspidin-ähn-liches Mineral vom Vesuv, — Sitz. Ber. d. Niederrhein Ges. f. Natur=und Heilkunde. 3 Jan. und 7 Febr. 1881.

RATH (VOM) G. — Mineralien von Monteponi und Montevecchio auf Sardinien-Vesuvische Mineralien. — Ueber den Zustand des Vesuvs im December 1886.—Ueber die Tuffbrüche von Noce-ra. — Sitz. Ber. d. Niederrhein. Ges. f. Natur und Heilkunde, 6 Juni. Bonn, 1887.

RAZZANTI F. — Récit véritable d'un misérable et mémorable ac-cident arrivé en la descente de la trés renommée Montagne de Somma autrement le Vésuve environ trois lieues loin de la ville de Naples depuis le Lundy 15 Décembre 1631 sur les neuf heure du soir jusque au Mardi suivant 23 du mesme, par un Observantin reformé du Couvent Royal de Naples. — *Ext. du Mercure François, 1631, 2me partie, pp. 67-73, in 8.°, 1632, pp. 478-480, fol. 11.* (C. A.).

RECUPITO G. C. — De Vesuviano incendio nuntius. — *Neapoli, 1632, in 4°, fol. 4,° pp. 120. Also , 1633, in 8°, pp. 124.* (C. A;) — *Mediolani, 1633, in 4°, fol. 4, pp, 114, fol. 3.* — *Pictavis, 1636, in 12°, fol. 3, pp. 195.* — *Lovanii,' 1639, in 8°, pp. 180, index.* — *Romae, 1644, in 4°, pp. 140, fol. 5 —* *Romae, 1670, in 8°, pp. 140.*

RECUPITO G. C. — Avviso dell' incendio del Vesuvio, tradotto dalla lingua latina in italiana.—*Napoli, 1635, in 8°, fol. pp. 264.*

REGNAULT H. — Ascension au Vésuve, le 10 Janvier 1868. — *Compt. Rend. d. l'Acad. d. Sc. Vol. LXVI. Paris, 1868, pp. 166-169.*

REQUIER. — Recueil général, historique, et critique de tout ce qui a été publié de plus rare sur la ville d'Hercolane. — *Paris, 1754, pp. 135, in 4°.* (C. A).

REYNAUD J. D. — On the ancient and present state of Vesuvius (1831). — *Proceed. of the Geol. Soc. of London. Vol. I. London, 1834.*

RICCI G. — Analisi dell' acqua ferrata con un appendice sopra un nuovo liquido vesuviano. — *Giornale Enciclopedico, 1820, in 8°, fol 9.* (O. V.).

RICCI G. — Analisi dell'acqua termo-minerale della Torre dell'Annunziata. — *Napoli, 1831, in 4°, pp. 32.* (C. A.).

RICCI G. — Raccolta di osservazioni sull' uso dell' acqua termo-minerale Vesuviana e Nunziante. — *Napoli, 1833-34, in 8°, Vol. I, pp. 76, maps 1. Vol. II, pp. XXXIX + X +145, maps 1, pl. 1.* (C. A.).

RICCI G. — Analisi dell' acqua termo-minerale Vesuviana-Nunziante. — *Napoli, 1834, in 8°, pp. 49.*

RICCIO L. — Prodigiosi portenti del Monte Vesuvio. Invettiva di Camillo Tuttini contro gli Spagnuoli, in occasione dell'incendio dell'anno 1649. E note riguardanti quella eruzione. — *Archiv. Storico per le Prov. Napoli, Anno II, Fasc. I, 1887, pp. 28.* (C. A.).

RICCIO L. — Un altro documento sulla eruzione del Vesuvio del 1649. — *(Lo Spettatore del Vesuvio e dei Campi Flegrei. Nuova serie, Vol. 1ª.). — Napoli, 1887, in 4°, pp. 61-64.*

RICCIO L. — Nuovi documenti sull'incendio vesuviano dell'anno 1631 e bibliografia di quella eruzione. — *Archiv. Stor. p. l.*

Province Napolitane, An. XIV., fasc. III. Napoli, 1889, in 8°, pp. 69.

Riso B. (DE). — Relazione della pioggia di cenere avvenuta in Calabria ulteriore nel dì 27 Marzo 1800. — *Atti Accad. Pontaniana, T. I., Napoli, pp. 163-165.* (O. V..)

Rissler J. B. — Neuer Ausbruch des Vesuv. Ein feuerspeihender Berg des Königreichs Neapel in der Nähe von dessen Hauptstadt —*Mülhaussen, gedruckt bei J. B. Rissler 1855, in 4°, pp. 8.* (C. A.).

Rivinus A. (BACHMANN). — Vesuvius, in promotione Batalorium VI *idus Martii MDCXXXII,* Lipsiae, declamatus,— *Lipsiae, 1632, in 4°, fol. 22.* (C. A.).

Rivinus A. — Tripus Delphichus de Monte Campaniae Somma ejusque fatitidico incendio. —*Lipsiae, 1635, in 4°, See Soria.* R.***, L.***. — *See Liberatore. L.*

Rocca Romana.—Cratere del Vesuvio.—*?, 1805, pl. in fol.* (O. V,).

Rocca F. — Osservazioni (sull'acqua Vesuviana-Nunziante). — *See Ricci G.*

Rocco A. — Lettera, nella quale si dà vera, e minuta relatione delle Gratie fatte dalla Gloriorissima Vergine e Madre di Dio dell'Arco Maggiore a beneficio della sua Casa e della Gente, che in essa si salvò in questi travagliati tempi del nuovo incendio del Monte Vesuvio nel 1631, e della carità usatali dai Padri dell'Arco. — *Napoli, 1632, in 8°, pp. 40.* (C. A.).

Rocco A. — Oratione devotissima alla Gloriosa Vergine Maria dell'Arco. (Descript. of erupt. 1631.) — *Napoli, 1632, in 8°. Frotisp. wanting.* (B. N.).

Rodwell G. F. — La récente éruption du Vésuve et son état actuel. — *La Nature. Paris, 8 Mars 1879. — In English: Nature. London, February, 1879. The Academy, Feb. 15th, 1879, N.° 354, pp. 149.*

Romanelli A. D. — Viaggio a Pompei, a Pesto, e di ritorno ad Ercolano ed a Pozzuoli. — *Napoli, 1817, in 12°. Vol. I, pp. 288, pl. 11, Vol. II, pp. 35+275, pl. 3. See pp. 53-63.* (C. A.).

Rose G. — Abhandlung des Herrn C. Rammelsberg über die chemische Zusammensetzung des Condrodits, Humits und Olivins, etc. — ?, (1851), *in 8°, fol. 3.* (O. V.).

Rosini C. — Dessertationis isagocicae ad Herculanensium voluminum explanationem. Pars prima (alone published)—*Neapoli, 1797, in fol., fol. 3, pp. 104, maps 2, pl. XX.* (C. A.).

Rossi M. S. — Intorno ai fenomeni concomitanti l'ultima eruzione vesuviana, avvenuti nella zona vulcanica dell'Italia. —

*Atti d. Acc. Pont. d. N. Lincei. An. XXV, Sess. VI, del
26 Maggio 1872, in 4°, pp. 7. (C. A.).*

ROSSI (DE) M. S. — Studii intorno al terremoto che devastò Pompei nell'anno 62 e ad un basso rilievo votivo pompeiano che lo rappresenta. — *Boll. d. Vulc. Ital. Fasc. VIII. XI. Ann. VI. Roma, 1879.*

ROTH J.—Analysen: I—Dolomitischer Kalkstein, sogenannter Auswürfling vom Rio della Quaglia von der Somma.—II—Dolomitischer Kalkstein von der Punta delle Coglione an der Somma.—III –Stängliger Braunspath aus Mexico.—IV—Kluftgestein aus dem Gypse des Schildsteins bei Lüneburg.—V—Stinkstein von Segeberg. — *Zeitschr. d. Deuts. Geol. Gesell. Bd. IV. Berlin, 1852. — Journ. für prakt. Ch. Bd. LVIII. Leipzig, 1853.*

ROTH J. — Der Vesuv und die Umgebung von Neapel. — *Berlin, 1857, in 8°, pp. XLIV+540, pl. IX, figs.*

ROTH J. — Litteratur über den Vesuv, besonders der Ausbrüche.— *Berlin, 1857, in 8°, pp. 98. (C. A.).*

ROTH J. — Ueber den Ausbruch des Vesuv vom Jahre 1861. — *Zeitschr. der Deut. Geol. Gesell. Bd. XV. Berlin, 1863.*

ROTH J. — Geschichte des Vesuvs. — *Berlin, 1869.*

ROTH J. — Ueber Vesuv und Actnalaven. — *Zeitschr. d. Deut. Geol. Gesell. Bd. XXV. Berlin, 1873.*

ROTH J. — Ueber eine neue Berechnung der Quantitäten der Gemengtheile in den Vesuvlaven. — *Zeitschr. der Deut. Geol. Gesell. Bd. XXVIII. Berlin, 1876.*

ROTH J. — Studien am Monte Somma. — *Abhandl. der Kön. Ak. der Wissensch. Berlin, 1877.'— Abstract in Italian: Boll. d. R. Com. Geol. d'Italia, N. 11-12. Roma, 1877.*

ROTH J. — Ueber die Gänge des Monte Somma. — *Monatsb. der K. preuss. Ak. der Wissensch. Berlin, 1877.*

ROTH H. — 1880. — *See Guiscardi,*

ROTH J. —Zur Geologie der Umgebung von Neapel. — *Gesammtsitz. Akad. Berlin, 10 Nov. 1881, pp. 990-1006. (C. A.).*

ROZEL. — Sur les volcans des environs de Naples. — *Bull. d. l. Soc. Géol. d. France, Vol. I. Paris, 1843-44.*

RUGGIERO M. — Sopra una massa di pomici trovata in Pompei, con una lettera del Prof. Scacchi.—*Napoli, 1877, in 4°, pl. 1.*

RUGGIERO M. — Della eruzione del Vesuvio nell'anno LXXIX.— *Napoli, 1879.*

RUIZ DE ALARCON Y MENDOÇA J. — Al Bolcan y incendios del Vesuvio. Sonetto — *See de Quinones.* (C. A.).

SACCO G. — Ragguaglio storico della calata nel Vesuvio e relazione del suo stato dei 16 Luglio 1794. — *Portici, 1794, in 8°, pp. 14.*

S. A. F. — Relazione dell' ultima eruzione del Vesuvio accaduta in Agosto di quest'anno, secondo le osservazioni del S. A. F.— *Giornale Enciclopedico d'Italia, Napoli, 1787, in 8°.* (B. N.).

SALIS-MESSIA (DE) J. — Sonetto — *See De Quinones.*

SALMON. — Descrizione del Monte Vesuvio. Included in the Storia del Regno di Napoli.—*Napoli, 1763, in 8°, pp, 65-79,* (B. N.).

SALVADORI G. B. — Notizie sopra il Vesuvio e l' eruzione dell'Ottobre 1822. — *Napoli, 1823.*

SALVADORI G. B. — Notizen über den Vesuv, und dessen Eruption. v. 22 Oct. 1822 verdeutch durch C. F. C. II.— *Neapel, 1823, in 4°, pp. 75, pl. 3.* (C. A.).

SAMBIASI O. — Sonetto, (1631). — *See G. Urbano.* (C. A.).

SANCHEZ G. — Il Monte Vesuvio deificato. — *Il Progresso delle Scienze, etc. Vol XII. Napoli, 1835, in 8°, pp. 145-149.* (C. A.).

SANCTIS (DE) A. — Il mostruoso parto del Monte Vesevo hora dal volgo detto, Monte Diavolo la cui mostruosità e crudeltà è qui descritta. — *Napoli, 1632, in 12°.*

SANCTIS (DE) A. — *See Genovesi Ab.*

SANDRANT J. — Wartsaffte Contrafactur des Bergs Vesuvii, und desselbigen Brandt sambt des umbligenden gelegenheit nach dem leben gezeichnet. — ?, *1631, a plate in 4°.* (C. A.).

SANDULLI P. — Gli Eroi del Virginiano celebrati con epistole, idilli, ed altre rime eroiche sagre. — *Napoli, 1708, in 8°, p. 134. From. pp. 60 lo 65 : Il Vesuvio a posteri.*

SANFELICE A. — Campania. — *Amstelaedami, 1656, in 12°, fol. 3, pp. 64, pl. 1, frontisp.* (C. A.).

SANFELICE A. — Campania notis illustrata. — *Neapoli, 1722, in 4°, pp. 13 +256, pl. 1.* (C. A.).

SANFELICE A. — La Campana recata in volgare italiano da Girolamo Aquino Capuano. — *Napoli, 1779, in 8°, pp. LXXI+ 117. pl. 1, portrait.* (C. A.). *Also 1796, with map.*

SAN MARTIN A. — Un viaje al Vesubio. Novità originale, historica. — *Madrid, 1880, in 8°, pp. 236.* (C. A.).

SANNICOLA G. — Biografia di Nicola Covelli. — *Palermo, 1845, in 8°, p. 16. Another edition with portrait. Napoli, 1846, in 8°, p. 19.*

SANTA MARIA (AGNELLO DI). — Trattato scientifico delle cause che concorsero al fuoco ed al Terremoto del Monte Vesuuio vicino Napoli. — *Napoli, 1632, in 8', pp. 100.*

SANTA MARIA ANDREA — Sonetti tre (1631). — *See G. Urbano.*

SANTARELLI A. — Discorsi della natura, accidenti e pronostici dell'incendio del Monte di Somma nell'anno 1631. — *Napoli, 1632, in 4°, fol. 2, pp. 58.* (C. A.).

SANTELET DE LAGRAVIERE M. — Étude sur les pierres précieuses suivie de l'Eruption du Vésuve en 1872. — *Avellino, 1876, in 8°, pp. 74, Maps, 2.* (C. A.).

SANTELIA A. — Contentio inter Coridonem Partenopeum et Mocridem ex Septemtrione. An Vesuvius Neapolitanis deliciis obstet, an vero sit emolumento. Egloga.—*M. S. in fol. (small), About 1681.* (C. A.).

SANTOLINI G. M. — Egloga in lode di S. Gennaro difensore contro l'incendio vesuviano. At p. 55 of a work entit. "Carmina Latina et Italica". — *Neapoli, 1784, in 4°, fol 8, pp. 112, portrait.* (O. V.).

SANTOLI V. M. — De Mephiti et vallibus Auxanti, libri 3. Cum observationibus super nonnullis urbibus Hirpinorum, quorum lapides et antiquitatum relliquiae illustr. — *Napoli, 1783, in fol., pl. 6.*

SANTOLI V. M. — Narrazione de'fenomeni osservati, sul suolo Irpino contemporanei all'ultimo incendio del Vesuvio accaduto a Giugno di questo anno 1794. — *Napoli, 1795, in 8°, pp. VII+160, pl. 1, figs.*

SANZMORENO F. — Ampla, copiosa y verdadera Relacion dell'incendio della montaña de Soma o Vesubio, etc.—*Napoles, 1632, in 4°. fol. 8, pp. 80.*

SARACINELLI M. — Guerra della Montagna o sia eruzione del Vesuvio del dì 24 Agosto 1834. — *Loose sheet in fol.* (C. A.).

SARNELLI P. — La vera guida de' forestieri curiosi di vedere, e d'intendere le cose più notabili della Real Città di Napoli, e del suo amenissimo distretto, etc.—*Napoli, 1752, in 12°, pp. 302. 2nd edit, 1788, in 12°, pp. VIII+396, pl. 13. See pp. 337-356, and pl. 13.*

SARNELLI P. — Nova guida de' forastieri e dell'istoria di Napoli. — *Napoli, 1791, in 12, fol. 1, pp. 394, pl. 10, figs.* (O. V.).

SASSO C. N. — Il Vesuvio, Ercolano e Pompei, con una pianta geometrica della città di Pompei e con l'indicazione di quanto ivi si è rinvenuto sino a tutto il 1855. — *Storia dei Monumenti di Napoli. Napoli, 1857, in 8°, Fasc. 16, 17, 18, pp. 60, pl. 1.* (C. A.).

SASSONE A. F. — Sonetto (1631). — *See G. Urbano.* (C. A.).

SAUSSURE (DE) H. — Sur l'éruption du Vésuve en Avril 1872. — *Compt. Rend. d. l'Acad. d. Sc. Vol. LXXIV. Paris, 1872.*

SAUSSURE (DE) H. — La dernière éruption du Vésuve en 1872.—
Act. d. l. Soc. Elvét. d. Sc. Nat. Vol. LV. Fribourg, 1873.

SAUSSURE (NECKER DE) L. A. — Description du cône du Vésuve
le 15 Avril 1820.—*Bibl. Univ. d. Sc. etc. Vol. XXIII. 1e Sér.
Genève, 1823.*

SAUSSURE (NECKER DE) L. A. — Mémoires sur le mont Somma,
(1822). — *Mém. d. l. Soc. d. Phys. et d'Hist. Nat. de Genève, Vol. II. Genève, 1823. — Elberfeld, 1825.*

SAUSSURE (NECKER DE). — 1825. — *V. Raffles.*

SAVARESE A. — Lettera seconda sui vulcani. — *Napoli ?, Nov.
1798, in 8°, fol. 16.* (O. V.).

SAVARESE A. — Lettera sui vulcani al Sig. Gugl. Thomson. —
Giornale Letterario di Napoli, April, 1798, in 8°, fol, 12.
(O. V.).

SAVARESE. — Troncs d'arbres trouvés à Pompei. — *Feuilleton de
la Presse, 19 mai 1860* (C. A.).

SAVASTANO G. — Acqua termo-minerale Vesuviana-Nunziante. —
See *Ricci.*

SCACCHI A. — Della periclasia, nuova specie di minerale del Monte
Somma. — *Mem. Mineral. Napoli, 1841, pp. 16. Ann. d.
Mines. 4me, Sér. Vol. III, Paris, 1843, pp. 369-384. In Ger-
man, München Gelehrte Anz. XVI, 1843, pp. 345-348.
Erdm. Journal für prakt. Chemie, XXVIII, 1843, pp. 486-
489. Atti R. Acc. Sc. Fis. Mat. Napoli, 1850.*

SCACCHI A. — Esame cristallografico del ferro oligisto e del ferro
ossidulato del Vesuvio. — *Napoli, 1842, in 8°, pp. 34, figs.*

SCACCHI A. — Notizie geologiche e conchiliologiche ricavate da
una lettera del Dr. R. A. Philippi ad A. Scacchi.—*Rend. R.
Accad. Fis. Mat. Vol. I, Napoli, 1842, pp. 86-88 (186-188.)*

SCACCHI A. — Sulle forme cristalline della Sommità.—*Rend. d. R.
Acc. d. Sc. Fis. e Mat. Vol. I. Napoli, 1842, pp. 129-131.*

SCACCHI A.—Lezioni di Geologia.—*Napoli, 1843, in 8°. pp. 153-
174* (C. A.).

SCACCHI A.—Osservazioni critiche sulla maniera come fu seppellita l' antica Pompei. — *Napoli, 1843, in 8°, Bull. Archeol.
Nap. An. I, N. 6, marzo 1843, pp. 41-45.* (C. A.).

SCACCHI A. — Notizie geologiche dei Vulcani della Campania
estratte dalle Lezioni di Geologia. — *Napoli, 1844. in 8.°*

SCACCHI A. — Campi ed Isole Flegree, Vesuvio. Specie orittogno-
stiche del Vesuvio e del Monte Somma. — See: *Napoli ed i
Luoghi Celebri delle sue Vicinanze, Vol. II, Napoli, 1845,
in 4.°, pp. 361-413.*

SCACCHI A. — Istoria delle eruzioni del Vesuvio accompagnata

11

dalla bibliografia delle opere scritte su questo Vulcano. — *Il
Pontano, Vol. I, Napoli, 1847, in 4°, pp. 16-21 and 106-131.*

SCACCHI A. — Notice sur le gisement et sur la cristallisation de
la sodalite des environs de Naples. — *Ann. des Mines, 4.me
Sér. T. XII, Paris, 1847, pp, 385-389, figs. 11-14 of pl. 3.*

SCACCHI A. — Notizie su l'ultima eruzione del Vesuvio, composizione della lave, delle cenere, de'lapilli, emanazioni gassose,
etc. — *Il Propagatore delle Scienze Naturali, etc. Napoli,
1847, in 4, pp. VIII + 416, pl. 4, See pp. 150-184.*

SCACCHI A. — Sopra una straordinaria eruzione di cristalli di Leucite. — *Racc. d. Lett. etc. intorno alla Fis. ed alla Mat.
Vol. I, Roma, 1845, and Ann. Civ. del Regno di Napoli,
Vol. XLVI, fasc. LXXXVII, 1847, pp. 62-66, (C. A.). Raccolta Scient. An. I, 1845, pp. 185-189, pl. 1.*

SCACCHI A. — Relazione dell'incendio accaduto nel Vesuvio nel
mese di febbraio del 1850, seguita dai giornalieri cambiamenti
osservati in questo vulcano dal 1840 sinora. — *Rend. d. R.
Acc. d. Sc. Fis. e Mat. Vol. IX, Napoli, 1850, pp. 13-48,
pl. 3, Ann. d. Mines, 4.me Sér. Vol. XVII, Paris, 1850, in
8°, pp. 323-380. pl. 4.*

SCACCHI A. — Della Humite e del Peridoto del Vesuvio (1850). —
*Atti d. R. Acc. d. Sc. e Lett. Vol. VI, Napoli 1851, pp.
241-273, pl. 1. Journ. für Prakt. Chemie, Bd. LIII, Leipzig,
1851. pp. 156-160.— The American Journ. of Sc. and Arts,
Vol. XIV. New-Haven, 1852, pp. 175-182, Annal. der Phys.
und d. Chemie, Ergänzungs, Bd. III, Leipzig, 1853, fol. 14,
pl. 1. Atti R. Accad. Sc. Napoli, Vol. VI, 1852.*

SCACCHI A. — Sopra le specie di Silicati del Monte di Somma e
del Vesuvio le quali in taluni casi sono state prodotte per
effetto di sublimazioni.—*Rend. d. R. Acc. d. Sc. Fis. e Mat.
N. S., An. I, Napoli, 1852, pp. 104-112.*

SCACCHI A. — Notiz über den Sommit, Mizzonit und Mejonit, —
*Poggend. Ann. Ergänz. Bd. III, 1853, pp. 478-479, fig. 16-
18 of pl. 2.*

SCACCHI A. — Uebersicht der Mineralien, welche unter den un-
bezweifelten Auswürflingen des Vesuvs und des Monte di
Somma bis jetzt mit Bestimmtheit erkannt worden sind. —
*Neues Jahrb. für Mineral. Geol. u. Pal. Stuttgart, 1853,
pp. 257-263.*

SCACCHI A. — Memoria sull'incendio vesuviano del mese di maggio 1855. — *V. Guarini e Palmieri.*

SCACCHI A. — Sur la dernière éruption du Vésuve. — *Bull. d. l.
Soc. Géol. de France. Vol. XV, Paris, 1857-58.*

SCACCHI A. — Dell' criocalco e del melanotallo nuove specie di minerali del Vesuvio.—*Rend. d. Acc. d. Sc. Fis. e Mat. An. IX, Napoli, 1870, pp. 86-89,.*

SCACCHI A. — Note mineralogiche , Memoria Prima. Leucite del Monte Somma metamorfizato, etc. Cristalli geminati di ortosia di Monte Somma, etc. — *Atti d. R. Acc. d. Sc. Napoli, Vol. V, 1870, pp. 40, pl. 1* (C. A.).

SCACCHI A. — Durch sublimationem entstandene Mineralien bei dem Vesuvausbruch im april 1872.—*Zeitschr. d. Deut. Geol. Gesell. Bd. XXIV, Berlin, 1872.*

SCACCHI A. — Notizie preliminari di alcune specie mineralogiche rinvenute nel Vesuvio dopo l'incendio di aprile 1872. — *Rend. d. R. Acc. d. Sc. fis. e mat. An. XI. Napoli, 1872 pp. 210-213 Zeitschr d. Deut. Geol. Gesell. Bd. XXIV, Berlin, 1872, pp. 505-506.*

SCACCHI A. — Sopra l'eruzione di ceneri vulcaniche avvenuta nell'aprile 1872.—*Rend. d. R. Accad. d. Sc. Fis. e Mat. Napoli, 1872.*

SCACCHI A. — Sulla origine della cenere vulcanica. — *Rend. R. Accad. Sc. Fis. Mat. Napoli, 1872. Zeitschr. d. Deut. Geol. Gesell. Bd. XXIV, Berlin, 1872.*

SCACCHI A. — Contribuzioni mineralogiche per servire alla storia dell'incendio Vesuviano del mese di aprile 1872. Napoli, 1872—*Atti d. R. Acc. d. Sc.' Vol. V, Napoli, 1872, pl. 1. Zeitsch. der deutsch geol. gesells. Bd. XXIV, 1872 , pp. 493-504. Idem Part. 2.ª Napoli, 1873. Atti d. R. Acc. d. Sc. Vol. VI, Napoli 1873, pl. 4.*

SCACCHI A. — 1ª e 2ª appendice alle contribuzioni mineralogiche sull'incendio Vesuviano del 1872. — *Rend. d. R. Acc. d. Sc. Fis. e Mat. An. XIII, Napoli, 1874, pp. 179-180, and An. XIV, pp. 77-79.*

SCACCHI A. — Della Cuspidina e del Neocrisolito, nuovi minerali vesuviani. — *Napoli , Rend. R. Accad. Sc. Fis. Mat. An. XIV, 1876, pp. 208-209. Zeitschr. f. Kryst. und Miner. P. Groth, Bd. 1, Leipzig, 1877, pp. 398-399, figs. 2.*

SCACCHI A. — Microsommite del monte Somma. — *Rend. d. R. Acc. d. Sc. Fis. e Mat. An. X, Napoli, 1876 pp. 3, 27-69, fig. 1.*

SCACCHI A. — Sulla regolare scambievole posizione dei cristalli di olivina congiunti a quelli di Humite e dei cristalli di oligisto congiunti a quelli di magnetite.—*Lettera al Prof. vom Rath (versione tedesca). Neues Jahrbuch für Mineralogie etc. 1876 pag. 637. fig.*

SCACCHI A. — Dell' Anglesite rinvenuta sulle lave vesuviane. — *Rend. d. R. Acc. d. Sc. Fis. e Mat. An. XVI, Napoli, 1877, pp. 226-230.*

SCACCHI A. — Sopra un masso di pomici saldate per fusione trovato a Pompei. — *Atti d. R. Accad. Archeol. Lettere e Bel. Arti, Vol. VIII, Napoli, 1877 in 4°, pp. 199-207, pl. 1, Map. 1. (C. A.).*

SCACCHI A. — 1877. — *See Ruggiero M.*

SCACCHI A. — Le case fulminate di Pompei. — *Pompei e la Regione sotterate dal Vesuvio nell' anno LXXIX. — Napoli, 1879, in fol. pp. 117-129, col. pl. 3. (C. A.).*

SCACCHI A. — Ricerche chimiche sulle incrostazioni gialle della lava vesuviana del 1631.—*Atti d. R. Acc. d. Sc. Fis. e Mat. Vol. VIII, Napoli, 1879, pp. 19.*

SCACCHI A. — Le incrostazioni gialle della lava vesuviana del 1631. Risposta di A. Scacchi ad una domanda rivoltagli dal Collega A. Costa.—*Rendiconto della R. Accad. delle Scienze Fis. e Mat. An. XIX, Napoli, Aprile, 1880, pp. 40-41.*

SCACCHI A. — Nuovi sublimati del cratere vesuviano trovati nel mese di ottobre del 1880. — *Atti d. R. Acc. d. Sc. Fis. Mat. Vol. IX. Napoli, 1881. p. 4, fig. 1.*

SCACCHI A. — Breve notizia dei vulcani fluoriferi della Campania.— *Rend. d. R. Acc. d. Sc. Fis. e Mat. di Napoli. Ottobre 1882:*

SCACCHI A. — Della silice rinvenuta nel cratere vesuviano nel mese di Aprile del 1882.—*Rend. d. R. Acc. d. Sc. Fis. Mat. An. XXI, Napoli, 1882, pp. 176-182.*

SCACCHI A. — Della lava Vesuviana dell'anno 1631. Memoria prima. — *Mem. Soc. It. d. Sc. (detta dei XL), Ser. 3ª, Vol. II, N.° 8, Napoli, 1883, in 4°, pl. 4.*

SCACCHI A. — Sopra un frammento di antica roccia vulcanica inviluppato nella lava vesuviana del 1872. — *Atti R. Acc. Sc. Fis. e Mat. di Napoli; serie 2ª, vol. I, Napoli, 1883, pl. 1.*

SCACCHI A. — Le eruzioni polverose e filamentose dei Vulcani.— *Atti R. Accad. Sc. Fis. Mat. Napoli, Ser. 2ª, Vol. II, 1886.*

SCACCHI A. — Catalogo dei minerali vesuviani con la notizia della loro composizione e del loro giacimento. — *Lo Spettatore del Vesurio e dei Campi Flegrei, N. serie Vol. 1°, Napoli, 1887, pp. 13.*

SCACCHI A. — Catalogo dei minerali e delle rocce vesuviane per servire alla storia del Vesuvio ed al commercio dei suoi prodotti.—*Atti R. Ist. d' Incoraggiamento di Napoli, 1888, 4ª Ser. Vol. I, pp. 57, pl. 4. Riv. di Mineral. e Cristall. Ital. Vol. V, Padova, 1889, pp. 84-87, pl. 4.*

SCACCHI A. — Katalog der vesuvischen Mineralien mit Angabe ihrer Zusammensetzung und ihres Vorkommens. — *Neues Jahrbuch für Mineralogie, etc.*, B. II, 2ª II. *Stuttgart, 1888, pp. 123-141. Riv. di Mineral. e Cristall. Ital. Vol. III, Padova, 1888, pp. 58-73.*

SCACCHI A. — Appendice alla prima memoria sulla lava vesuviana del 1631. — *Mem. Soc. Ital. d. Sc., 1889, Vol. VII, ser. 3.ª N. 7, pp. 26, pl. 1.*

SCACCHI A. — I projetti agglutinanti dell'incendio vesuviano del 1631. — *Rend. R. Acc. Sc. Fis. Mat. Ser. 2ª, An. XXVIII, fasc. 10°, Napoli, 1889, pp. 220-223.*

SCACCHI E. — Dei lapilli azzurri trovati nel cratere del Vesuvio nel mese di giugno del 1873. — *Rend. d. R. Acc. d. Sc. 1880, pp. 7.*

SCACCHI E. — Notizie cristallografiche sulla Humite del M. Somma. — *Rend. R. Acc. d. Sc. Fis. e Mat. Napoli, 1883, pp. 9.*

SCACCHI E. — Facellite, nuovo minerale del Monte Somma. — *Rend. R. Acc. Sc. Fis. Mat. Napoli, Ser. 2ª, Vol. II. N. 12.*

SCACCHI E. — Contribuzioni mineralogiche : Memoria quarta. — *Rend. R. Acc. Sc. Fis. e Mat. Ser. 2ª, Vol. II, n.° 12 Napoli, 1888.*

SCHAFHAEUTL D. — Aufsatz über den gegenwärtigen Zustand des Vesuvs und sein Verhältniss zu den phlegräischen Gefilden. — *Gel. Anz.; herausgeg. v. Mitgl. d. Königl Bayer. Akad. d. Wiss. Bd. XX München, 1845.* (C. A.).

SCHIAVONI F. — Osservazioni geodetiche sul Vesuvio. — *Atti d. Soc. Pontan. di Napoli. Napoli, 1855, in 8°. pp. 5, pl. 1. Annali Sci. Giorn. Sc. Fis. Mat. Agric. etc. Vol. II. Napoli, 1855, pp. 418-422. Naples, 1872, pl. 1, in 8°, pp. 6 with plates.*

SCHMIDT J. — Die Eruption des Vesuvs im Mai 1855 — Nebst Beiträgen zur Topographie des Vesuvs, der Phlegräischen Crater, Roccamonfinas und der alten Vulkane im Kirchenstaate. — *Wien, und Olmütz, 1856, in 8°, pp. 12+213, with figs.* (C. A.).

SCHMIDT J. — Neue Höhen — Bestimmungen am Vesuv, im den phlegräischen Feldern, zu Roccamonfina und in Albaner gebirge. — *Wien, 1856, in 4°, pp. 41.* (C. A.).

SCHMIDT J. F. J. — Die Eruption des Vesuv ihren Phaenomenen in Mai 1855, nebst Beiträgen zur Topographie des Vesuvs und anderer italiänischen Kratern. — *Olmütz, 1856, in fol. Atlas, pp. 24. in 4°, pl. 0, Atlas.* (O. V.). *Mittheil. auf dem Gesammt. der Geogr. von Dr. Petermann, Gotha, 1856.*

SCHMIDT J. F. — Vulkanstudien. Santorin 1866-1872, Vesuv, Bajae, Stromboli, Aetna 1870. — *Leipzig, 1874-78, in 8°, fol. 4, pp. 235. map. 1, col. pl. 7., fig. 13.* (C. A.).

SCHNETZER C. — Sur l'éruption du Vésuve du 22 Octobre 1822. (traduction). — *Bull. d. Sc. Nat. et d. Géol. Vol. I. Paris, 1824.*

SCHOOK M. — De Vesuvio ardente disputationes (1631). — *See* ✦ *Morhof.* (C. A.).

SCHOTT F. — Visite de Pighius (Etienne) au Vésuve vers 1575.— *Itin. Italiae, Vicentiae 1601, in 8°, pp. 222-225.*

SCORIGGIO L. — L'incendio del Monte Vesuvio, rappresentazione spirituale, composta da un devoto saccerdote e data in luce da Lazaro Scoriggio.—*Napoli, 1632, in 12°, pp. 185.* (C. A.).

SCOTTI E. — Ragionamento della eruzione del Vesuvio accaduta il dì 15 giugno 1794. — *Gazzetta (Supplemento) Napolitana Civica, Napoli, 1794, in 4°, pp. 48. See also his: Elementi di Fisica, 1831.*

SCOTTI E. — Lettera a Dom. Cotugno sulla eruzione del Vesuvio del 1804. — *Gazzetta (Supplemento) Napolitana Civica, N. 70, Napoli, 1804, in 4°, fol. 2.*

SCROPE (POULETT) G. — An account of the eruption of Vesuvius in October 1822. — *The Journ. of Sc. and the Arts. Vol. XV, London, 1823.*

SCROPE (POULETT) G. — On the volcanic district of Naples. — *?, 1827, in 4°, pp. 16, pl. 1, figs.* (O. V.).

SCROPE (POULETT) G. — Volcanoes. The character of their Phenomena, etc. — *London, 1862, Paris, 1864, Berlin, 1872.*

SCROPE (POULETT) G. — Vues du Vésuve et de l'Etna. — ?

SEMENTINI E GUARINI. — Saggi analitici su talune sostanze vesuviane. — *Rend. R. Acc. d. Sc. Fis. e Mat. Napoli, 1831, pp. 165-168* (C. A.).

SEMMOLA E. — Sulle emanazioni acriformi delle fumarole collocate a diversa distanza dall'attuale bocca d'eruzione del Vesuvio.—*Rend. d. R. Acc. d. Sc. Fis. e Mat. Napoli, agosto e settembre 1878.*

SEMMOLA E. — Sur l'état actuel du Vésuve. — *Compt. Rend. d. l'Acad. d. Sc. Vol. LXXXVIII, N. 17. Paris, 1879. Boll. d. R. Com. Geol. d'Italia, fasc. III-IV. Roma 1879., Riv. Scient. Ind. N. 10, Firenze, 1879. Rend. R. Accad. Sc. Fis. Mat. Napoli, 1879, pp. 2.*

SEMMOLA F. — Relazione ragionata dell'analisi chimica delle ceneri vesuviane eruttate nell'ultima deflagrazione de' 16, 17 e 18 giugno dell'anno 1794. — *? in 4°, pp. 15,* (C. A.).

SEMMOLA G. — Del rame ossidato nativo, nuova specie minerale del Vesuvio (Tenorite) Nelle opere minore. — *Napoli, 1845, in 8°, fol.2, pp. 42-491.*

SEMMOLA M. — Analisi chimica delle acque potabili dei dintorni del Vesuvio e del Somma. — *Napoli, 1857.*

SERAO D. F. — Vesuviani incendii anui 1737 mensis maje Historia; curavit Academia Scientiarum neapolitana. — *Neapolis 1738, in 4°, fol. 4, pp. 163, pl. 2, Latin and Italian. Two editions followed by one in 1740, in 8°, pp. 226, pl. 2, fol. 8,* (C. A.). *Fifth edition, Napoli, 1778, in 4°, fol. 4, pp. 244, pl. 2* (C. A.).

SERAO F. — Histoire du Mont Vésuve avec l'explication des phé- nomèues qui ont coûtume d'accompagner les embrasements de cette montagne. Traduite de l'italien de l' Academie des Sciences de Naples par M. Duperre de Castera.—*Paris, 1741, in 8°, pp. XX, fol, 2, pp. 362, map, 1, table 1, pl. 2, Another edition in 12°, pp. 361, pl. 1.* (C. A.).

SERIO. — Ottave sul Vesuvio. — *Napoli, 1775, in 4.°, pp. 24.* (C. A.).

SEVERINO N. — *See Ulloa.*

S. FR. S. - *See Silvestro F. S.*

SICA FRA GERONIMO DE GIFONI. — Morale discorso fatto tra l'effeti cagionati dalla voragine del Vesuvio , e li motivi visti nelli Cristiani. — *Napoli, 1632, in small 8°, fol. 8.* (C. A.).

SICURO F. — Prospetto della Villa del Principe di Aci (oggi Favorita) con veduta del Vesuvio. — *Pl. 2 , in largest fol.* (O. V.).

SIGISMONDO G. — Lettera ad un suo amico di Benevento con la quale gli dà notizia dell'ultima eruzione del Vesuvio seguita nella sera del 15 giugno 1794, e con un confronto della medesima con quella accaduta nel 1631. — *M. S. in* (O. V.).

SIDERNO D. DA. — Discorso filosofico ed astrologico, nel quale si mostra quanto sia corroso il monte Vesuvio dal suo primo Incendio fino al presente, e quanto habbi da durare detto Incendio.—*Napoli, 1632, in 4°, fol. 4.* (C. A.).

SIEMENS W. — Physikalisch-mechanische Betrachtungen, veranlasst durch einige Beobachtungen der Thätigkeit des Vesuvs im Mai 1878. — *Berlin, 1878.*

SILLIMAN B. —Miscellaneous notes from Europe. 1. Present condition of Vesuvius. 2. Grotta del Cane and Lake Aguano. 3 Sulphur Lake of Campagna near Tivoli. 4. Meteorological Observatory of mount Vesuvius, etc. — *Am. Journ. of Sc. 2nd Ser. Vol. XII, 1851, pp. 250-200.* (C. A.).

SILOS M. — Vesuvius crumpens. Ode. — *In the " Pinacotheca sive Romana pictura et scultura " Romae, 1673, in 8°, pp. 344-346.* (C. A.).

SILVESTRI O.—Ricerche chimiche sulla eruzione del Vesuvio. — *Atti d. Acc. Gioenia d. Sc. Nat. Ser. 3.ª , Vol. II, Catania, 1868*

SILVESTRI O. — Sur l' éruption actuelle du Vésuve. — *Compt. Rend. d. l'Acad. d. Sc. Paris, 1868.*

SILVESTRI O. — Sulla eruzione del Vesuvio incominciata il 12 Nov. 1867. — *Atti d. Acc. Gioenia d. Sc. Nat. Ser. 3.ª Vol. III, Catania, 1869*

SILVÉSTRO F. S. — All'inclito Martire S. Giorgio, singolar protettore del Villaggio S. Giorgio a Cremano. Ringraziamento per aver fermato la lava del Vesuvio nella notte del 12 maggio (1855) alle ore 12 p. m. dopo la processione fatta nello stesso giorno. Ode. — *Napoli, 1855, in 8°, pp. 4.* (C. A.).

SILVEYRA Dr. — Sonetto. — *See de Quinones.* (C. A.).

SINCERO (ACCAD. INSENSATO) –Il Vesuvio fiammeggiante Poema.— *Napoli, 1632, in 8°, fol. 8, pp. 155.* (C. A.).

SINISCALCO C. — Compendio delle principali eruzioni vesuviane dall' anno 79 E. V. infino alla descrizione delle recenti. — *Napoli, 1863, in 8°, pp. 28. In the text mention is made of plates which were never published* (C. A.).

SMITHSON J. — On a saline substance of mount Vesuvius. — *Philosph. Transact. of the R. Soc. of London, 1813.*

SOLIO MESSIA JUAN. — Sonetto.— *See Quinones* (C. A.).

SORIA F. A. — Scrittori vesuviani. Memorie storico-critiche degli storici napoletani. — *Napoli, 1781, in 4°, pp. 621-641.*

SORRENTINO I. — Istoria del monte Vesuvio divisa in due libri.— *Napoli, 1734, in 4°, fol. 8, pp. 224, fol. 2.*

SOTIS B. — Dissertazione fisico-chimica dell'ultima eruzione vesuviana dei 12 agosto 1804. — *Napoli, 1804, in 8°, pp. 55.*

SOYE L. R. — Ode cantada no felis dic natalicio d'Augusta Maria Carolina d'Austria Rainha das Duas Cecilias.—*Napoles, 1792, in 8°, pp. XX, 2, figs.*

SPALLANZANI L. — Viaggio alle due Sicilia. — *Pavia, 1792-1797, VI Vols, in 8°, pl. 11.*

SPALLANZANI L. — Lettera nella quale si tratta de' sassi caduti dall'aria nella Campagna Sanese il dì 16 giugno 1794.—*Giorn. Lett. di Napoli, 1793-1798, Vol. XXXI, pp. 81-102.*

SPALLANZANI L. — Travels in the Two Sicilies and some parts of the Apennines.— *Translated from the Original Italian. 4 vols. with 11 plates, London, 1798.*

SPALLANZANI L. — Lettera scritta al Wilseck sul fenomeno della pioggia di pietre avvenuta a Siena nel 1794. — *The original M. S. letter presented by Prof. G. Uzielli* (C. A.).

SPALLANZANI L. — Relation de l'éruption du Vésuve arrivée le 15 Juin 1794. — (?).

SPINOSA S. — Dichiarazione geneologica fisico chimica naturale apologetica, ed epidemica del Signor Vesuvio. — ?, *in 4°. See Crisippo Vesuvino.*

STAIBANO V. — Risolutiones forensis (Centuria II , Resolutio CXLIV). — *Napoli, 1654,in fol* (C. A.).

STAS. — Sur la découverte par le prof. Scacchi, de Naples, d'un corps simple nouveau dans la lave du Vésuve. — *Bull. d, l'Acad. R. d. Sc. de Belgique, 2ᵉ Sér. Vol. XLVI, XLVII, XLVIII, XLIX. N. 1-4, Bruxelles, 1878-80.*

STILES F.—Eruption of mount Vesuvius on 23 Dicembre, 1760.— *Philosph. Transactions of the R. Soc. of London. Vol. LII, London, 1761.*

STOPPA G. — Memorie istorico-fisiche sulle eruzioni vesuviane - etc. — *Napoli, 1806, in 4°, pp. 92, pl, 1.*

STOPPANI A. — Osservazioni sulla eruzione vesuviana del 24 aprile 1872. — *Atti d. R. Ist. Lom. d. Sc. e Lett. Ser. 2.ᵃ, part. II, Vol. V, Milano, 1872.*

STROZZI N. — Sonetto. — *See G. Urbano* (C. A.).

STRÜVER G. — Sodalite pseudomorfa di Nefelina del monte Somma. — *Atti d. R. Acc. d. Sc. di Torino , Vol. VII, disp. 3.ᵃ 1872.*

STÜBEL A. — Die Laven des Somma bei Neapel. — *Sitz. Ber. d. Naturwiss. Gesell, Isis. zu Dresden. Dresden, 1861.*

SUAREZ F. M. — De Monte Vesuvio. — *M. S. Bibl. Brancacciana in 8°, pp. 21. Copy* (C. A.).

SUPO (Padre Gesuita Matematico et Meteorista, nel collegio di Napoli). — Relation del nuovo incendio del Vesuvio ai 3 di Luglio 1660. — *M. S. in* (O. V.). (*This determines the authorship of the articles under " Anonymous " referring to this eruption*).

SUPPLE R. — An account of the Eruption of mount Vesuvius, from its first Begining to the 28 Oct. 1751. — *Philosph Transactions of the R. Soc. of London, Vol. XLVII, London 1751-52, pp. 315-317.*

SZEMBECH F.—Relazione composta di varie relazioni intorno all'ultimo incendio del Vesuvio. — (*In Polish) Cracovia, 1632, in 4°.*

TADINI CONTE F. — *See T. C. F.* (C. A.).

12

TARGIONI TOZZETTI G. — Saggio de' Monti ignovomi della Toscana e del Vesuvio. — *See Anonymous : Dei Vulcani o Monti Ignovomi, etc. 1779.*

TARI A, — Reliquie di lava sul lido di Resina. — *?, in 12°, fol. 3.* (O. V.)

TARINO G. A. — Continuatione de' successi del prossimo Incendio del Vesuvio, con gli effetti della cenere, e pietre da quello vomitate, e con la dichiaratione, e espressioni delle croci maravigliose apparse in varii luoghi dopo l'incendio. — *Napoli, 1661. in 4°, pl. 1.* (B. N.).

TATA D. — Descrizione del grande incendio del Vesuvio successo nel giorno 8 Agosto 1779. — *Napoli, 1779, in 8°, pp. 38.*

TATA D. — Relazione dell'ultima eruzione del Vesuvio accaduta in Agosto di quest'anno. — *Mem. d. R. Acc. d. Sc. Fis. e Mat. Vol. V. Napoli, 1787.* (O. V.)

TATA D. — Breve relazione dell' ultima eruttazione del Vesuvio. (Agosto e Settembre 1790) etc. — *Napoli, 1790, in 8°, pp. 24.*

TATA D. — Lettera al Sig. Barbieri sull'eruzione del Vesuvio. — *Napoli, 1794, in 8°, pp. 26.*

TATA D. — Lettera sulla figura, ecc. del Vesuvio. 21 Agosto 1794. Con breve risposta di F. Viscardi. — *?, 1794.*

TATA D. — Memoria sulla pioggia di pietre, avvenuta nella Campagna Sanese il dì 16 di Giugno di questo corrente anno (1794). — *Napoli, 1794, in 8°, pp. 74.*

TATA D. — Relazione dell' ultima eruzione del Vesuvio nel 15 Giugno 1794. — *Napoli, 1794, in 8°, pp. 42.*

TATA D. — Continuazione delle notizie riguardanti il Vesuvio. — *?, in 12°, pp. 24.* (C. A.).

TAYLOR (LE BARON.) — Lettre à M. Charles Nodier sur les villes de Pompéi et d' Herculanum. — *Nouv. Ann. des Voy. t. XXIV, pp. 424-425, Déc. 1824.* (C. A.).

T. C. F. (Tadini Conte Franc.). — L'Eruzione del Vesuvio della notte de' 15 giugno 1794, poeticamente descritta. — *? in 8°, pp. 30, pl. 1.* (C. A.).

TCHIHATCHEFF (von) P. — Lettre sur l'éruption du Vésuve du 1r Mai 1855. — *Compt. Rend. d. l'Ac. d. Sc. Paris 1855. T. XL, pp. 1229-1238.*

TCHIHATCHEEF (von) P. — Nouvelle éruption du Vésuve. — *Compt. Rend. d. l'Acad. d. Sc. Vol. LIII, Paris, 1861, pp. 1090-1092 — Zeitschr. d, Deut. Geol. Gesell. Bd. XIII. Berlin, 1861, pp. 453-458.*

TCHIHATCHEFF (VON) P. — Der Vesuv im Dezember 1861. — *Ver-*

handlungen d. K. K. Geolog. Reichsanstalt. Bd. XII. Wien. 1861-62.

TENORE M. — Relation de l'éruption du Vésuve aux premiers jours de 1839. — *Bull. d. Soc. Géol. de France, Vol. X. Paris, 1839.*

TENORE M. — Congetture sull' abbassamento altra volta avvenuto nel Vesuvio e l'innalzamento avuto luogo successivamente nelle posteriori eruzioni. — *Nota letta nella tornata d. Acc. d. Sc. Napoli, 16 Giugno, 1846. — Ann. Civil. d. Regno d. Due Sicilie. Fasc. LXXIII. Napoli, 1846.*

TENORE M. — Storia del Vesuvio intorno ad un passo del Cosmos concernente l'altezza del Vesuvio. — *" Il Lucifero " N. 36, Vol. IX. Napoli, 1847, in 8°, pp. 6.*

TENORE G. — Notizia di una gita al Vesuvio nel giorno 10 Febbraio, 1850.—*Rend. d. R. Acc. d. Sc. Fis. e Mat. Vol. VIII. pp. 379-380, Napoli, 1849. — Ann. d. Fis., Chim. etc. Vol. II. Torino, 1850*

THOMPSON G. — Breve catalogo di alcuni prodotti ritrovati nell' ultima eruzione del Vesuvio. — *Giorn. Lett. Vol. CII. Napoli, 1794.*

THOMPSON G. — Notizia sul marmo bianco del Vesuvio. — *Giorn. Lett. Vol LXXXIX. Napoli, 1781. In French, 1797, in 8° pp. 5.* (O. V.).

THOMPSON G. — Sur l'origine de l'oxigène nécessaire pour entretenir le feu souterrain du Vésuve. — *Giorn. Lett. Vol. CVI, Napoli, Septembre 1798, pp. 3-46.*

TOMASELLI (AB.). — Ricerche sulla natura e generi delle lave compatte, lettera al sig. Ab. Olivi, e risposta del Sig. Ab. Olivi al Sig. Ab. Tomaselli. — *Giornale Letterario di Napoli per servire di continuazione all' Analisi ragionata de' libri nuovi.—Napoli, 1793-98, vol. 112, in 8°, Vol. I, pp. 85-93.*

TOMMASI (DE) D. — Altro avviso al pubblico sulla nuova analisi delle ceneri eruttate dal Vesuvio ne' dì 16, 17, 18 del corrente mese di giugno 1794. — *Pag. 1, in 4°, (O. V.).*

TOMMASI (DE) D. — Esperienze et osservazioni del sale ammoniaco vesuviano. — *Napoli, 1794, in 8°, pp. 3-15.* (C. A.). *Also 1794, in 4°, pp. 16.* (O. V.).

TORCIA M. — Relation de la dernière éruption du Vésuve, en août 1770. — *Naples, 1779, in 12°, pp. 135, also in Italian, Napoli, 1779, in 8°, fol. 5, pp. 136, col. pl. 1, fig.* (O. V.).

TORCIA M. — Lettere al Sig. D. Biagio Michitelli Regio Assessore nella Piazza di Longone. — *Napoli, 1795, in 8°, pp. 16. Antologia, Num. XXX, 1796.* (B. N.).

TORRE DUCA (SENIORE) (DELLA)—Incendio trentesimo del Vesuvio accaduto gli 8 Agosto 1770. — *Napoli, 1779, in 4°, fol, 2 pp. 15.*

TORRE DUCA (SENIOR) (DELLA). — Estratto dalla prima lettera sulla eruzione del Vesuvio dei 15 giugno 1794. — *Napoli, 1794, in 8°, pp. 8.* (C. A.).

TORRE DUCA (DELLA). — Breve descrizione dei principali incendi del monte Vesuvio e di molte vedute di essi, ora per la prima volta ricavata dagli storici contemporanei, ed esistenti nel gabinetto del Duca della Torre. — *Napoli, 1795, in 8°, pp. 16, fol. 3.*

TORRE DUCA (DELLA). — Lettera prima e seconda sulla eruzione del Vesuvio del 15 Giugno 1794, — *Napoli, 1794, in 8°, pp. 8+25. Several editions — In German, Dresden, 1795.*

TORRE DUCA (SENIORE) (DELLA) — Il Gabinetto Vesuviano. — *1ª Ediz. Napoli (?) — 2ª ediz. 1796, in 8°. fol. 1, pp. 108, fol. 4, pp. 22 — 3ª ediz. 1797, in 8°, fol. 2, pp. 86, fol. 1, pl. 22.*

TORRE DUCA (DELLA). — Descrizione dei principali incendi del Monte Vesuvio e di molte vedute di essi. —*?, in 4°, p, 86-2, From p. 55 to 63: Catalogo delle pietre vesuviane. From p. 67 to 86: Biblioteca vesuviana esistente nel Gabinetto.*

TORRE DUCA (JUNIORE) (DELLA)—Eruption du Vésuve.—*Ann. du Museum d'Hist. Nat. Vol. V, Paris, 1804.*

TORRE DUCA (JUNIORE) (DELLA). — Relazione prima dell'eruzione del Vesuvio dagli 11 Agosto fino ai 18 Sett. 1804. — *Napoli. 1804, in 8°, pp. 61, fol. 1.*

TORRE DUCA (JUNIORE) (DELLA). —Veduta di una apertura formatasi all'orlo del Vesuvio nell'eruzione del 22 Novembre 1804.— *Napoli, 1804, a plate in 4°, with descript.* (O. V.).

TORRE DUCA (JUNIORE) (DELLA) — Observations sur les dernières éruptions du Vésuve. —*Journ. d, Phys d. Chimie. etc. Vol LI, Paris, 1805.*

TORRE DUCA (JUNIORE) (DELLA). — Pianta topografica dell'interno del cratere del Vesuvio formata nel mese di Giugno 1805. — *Napoli, 1805, pl. in 4°, with description.* (O. V.).

TORRE DUCA (JUNIORE) (DELLA) — Descrizione della eruzione del Maggio e Giugno 1806. — *Giorn Encicl. N. 7 Napoli, 1806.*

TORRE DUCA (JUNIORE) (DELLA). — Lettera a Domenico Catalano sulla eruzione del 1806. — *Giornal. Enciclopedico. 1806, in 8°, pp, 155-171.* (C. A.).

TORRE DUCA (JUNIORE) (DELLA)—Catalogue abrégé de la collection vésuvienne de Mr. lé Duc de la Torre de Naples 1820. — *fol. 2 in fol.* (C. A.).

TORRE DUCA DELLA. — Atlante di Vedute de' principali incendij del monte Vesuvio ricavate dagli storici contemporanei ed esistente nel gabinetto del Duca della Torre. — *22 plates in 4°,* (C. A.).

TORRE P. G. M. (DELLA)—Narrazione del torrente di fuoco uscito dal Monte Vesuvio nel 1751. — *Napoli (?), 1751, in 8°, pp. 23.* (O. V.).

TORRE P. G. M. (DELLA)—Storia e fenomeni del Vesuvio.—*Napoli, 1755, in 4°, fol. 4, pp. 120, pl. 8. figs.—In German (?) 1755. — In French, Paris, 1760, in 12°, fol. 4, pp. 120, pl. 8, figs. Also with supplement, Paris, 1760, in 8°, pp. XXIV+399, pl. 6, figs.* (C. A.).

TORRE P. G. M. (DELLA)—Supplemento alla Storia del Vesuvio.— *Napoli, 1761, in 4°, fig, pp. 15, pl. 1.*

TORRE P. G. M. (DELLA). — Histoire du Mont Vésuve et exposition de ses phènomenes. Extracted from: Mélanges d'histoire naturelle. — *Lyon, 1765, in 8°, fig. fol. 14, pl. 1.* (O. V.).

TORRE P. G. M. (DELLA)—Incendio del Vesuvio accaduto l'anno 1766. — *Napoli (?).*

TORRE P. G. M. (DELLA) — Incendio del Vesuvio accaduto il 19 Ottobre 1767. — *Napoli, 1767, in 4°, pp. 30, pl. 1, figs.*

TORRE P. G. M. (DELLA). — Storia e fenomeni del Vesuvio con supplemento. — *Napoli, 1768 in 4°, fol. 3, p. 120+39, pl. 10, figs.* (O. V.).

TORRE P. G. M. (DELLA) — Histoire et phénomènes du Vésuve.— *Naples, 1771, in 8°, pp. 12+298, fol. 3, pl. 11, figs.* (O. V.).

TORTALETTI B. — Sonetto (1631). — *See G. Urbano* (C. A.).

TORRE P. G. M. (DELLA)—Geschichte und Naturbegebenheiten des Vesuvs von den ältesten Zeiten bis zum Jahr 1779. — *Altenburg, 1783, in 8°, pp. 48+222+60, pl. 2. figs.* (O. V.).

TOSCAN B. — Précis du Journal de l'éruption du Vésuve depuis le 11 août jusqu'an 18 sept. 1804. — *Ann. du Museum. Vol. V, pp. 448-461 Napoli (?).*

TOSI C. — De Incendio Vesevi. Ode ed un Sonetto (1631). — *See G. Urbano.* (C. A.).

TRANSARELLI O. — 1631. — *See G. Urbano* (C. A.).

TREGLIOTTA D. — Descrittione dell'incendione del monte Vesuvio e suoi meravigliosi effetti. Principiato la notte delli 15 decembre 1631. — *Napoli, 1632, 8°, pp. 40.*

TROMBELLI G. — Sonetto (1631). — *See G. Urbano* (C. A.).

T. S. A. (Tata according to A. Scacchi). — Relazione dell'ultima eruzione del Vesuvio accaduta in agosto di quest' anno. — *Giornale Enciclopedico d' Italia o sia Memorie Scient. e Lett, Napoli, 1788, in 8°, pp. 16* (C. A.).

TURBOLI D — Supplica et memoria al sig. Duca di Caivano, etc. Con un brevissimo racconto d'alcune sentenze di Seneca, cavata dai libri de Beneficiis, e Provedentia, e d'atre materie gradibili (Erupt. 1631.) — *Napoli, 1632, in 4*, (B. N.).

TURLERI II. — De peregrinatione et Agro Neapolitano. Lib. II. Omnibus peregrinantibus utiles ac necessarii : ac in corum gratium nunc primum editi. — *Argentorati, 1574, in 8,.* (B. N.). *See pp. 104-107.*

TUTINI C. — Memorie della vita, miracoli e culto di S. Gianuario martire. — *Napoli, 1633, in 4°, fol. 4, pp. 141, fol. 3, pl. 1, figs.* (C. A.). *Another edit. 1703, in 4. (B. N.).*

TUTINI C. — Prodigiosi portenti del monte Vesuvio. Invettiva contro gli Spagnuoli in occasione dell' incendio dell' anno 1649. E note riguardanti quella eruzione per Luigi Riccio. — *Archiv. Stor. per le Prov. Nap. Ann. II, Fas. 1, Napoli, 1877, in 4°, pp. 28.* (C. A.).

ULLOA e SEVERINO N.—Lettere erudite. — *Napoli, 1700, in 12°, pp. 24+451+25. See pp. 166-194.*

URSO J. DE. — Vesevi montis epitaphium. — *Napoli, 1632, in 8°, fol. 1* (C. A.).

URSO J. DE — Inscriptiones. — *Napoli, 1642, in fol., fol. 11, pp. 350, Ingrav. frontisp. See pp. 14, 24, 26, 39, 99, 100, 101, 111, 331, 332, 333, 334, 336.*

VACHMESTER M. — Analyse de la Sodalite du Vésuve. — *Annales des Mines, etc. Paris, 1817, Vol. X, pp. 262-263.*

VALDIVIELSO J. DE SILVA. — *See Quinones.*

VALENTINI L. — Voyage médical en Italie fait l'année 1820, précédé d'une excursion au volcan du Mont Vésuve, et aux ruines d'Herculanum et Pompéi. — *Nancy, 1822, in 8°, fol. 2. pp. 166, 2nd Edit. Paris, 1826, in 8°, pp. 399.* (C. A.).

VALENZIANI M. — Dissertazione della vera raccolta o sia Museo di tutte le produzioni del Monte Vesuvio. — *Napoli, ?, fol 7* (C. A.).

VALENZANI M. — Indice spiegativo di tutte le produzioni del Vesuvio, della Solfatara e d' Ischia. — *Napoli, 1783, in 4°, pp. LII, + 135* (C. A.).

VALENZIANI M.—Note de la collection compléte des diverses espéces de productions du Mont Vésuve. — *Without D. or L. 1, fol. in fol.* (O. V.).

VALETTA G. — Epistola de incendio et eruptione montis Vesuvii
anno 1707. — *Philosoph. Transactions of the R. Soc. of
London, Vol. XXVIII, 1713. Venezia, 1793, pp. 59-61.*

VARONIS S. — Vesuviani incendii historiae. Libri tres. — *Neapoli
1634, in 4°, fol. 8, pp. 400, fol. 6.*

VAUQUELIN L. N. —Analyse des cendres du Vésuve, etc. — *Mém.
du Museum d'Hist. Nat. Vol. IX. Paris, 1822, pp. 381-384
Ann. d. Chimie. Vol. XXV, Paris, 1824.*

VAUQUELIN L. N. — Chemische Untersuchung der Asche des Vesuv's. — *Arch. für Ges. Naturw. Vol I, (?),*

VELEZ DE GUEVARA LUIS. — A la Montaña de Soma. Sonetto
(1631). — *Sèe Quinones.*

VENTIGNANO DUCA DI. — Il Vesuvio, Poema. — *Napoli, 1810, in
8°, pp. 126, fol. 1.* (C. A.).

VENUTI D. M. DE. — Descrizione delle prime scoperte dell'antica
città di Ercolano. — *Roma, 1748, in 4°, pp. XXIII + 140.*
(C. A.).

VERNEUIL (DE) E. — Sur l'état du Vésuve au commencement de
Janvier 1858. — *Compt. Rend. d. l'Acad. d. Sc. Vol. XLVI,
Paris, 1858, Bull. d. l. Soc. Géol. de France; Vol. XV,
Paris, 1858, pp. 369-370.*

VERNEUIL DE. — Sur deux ascensions qu'il a faites au sommet
du Vésuve le 30 avril et le 7 mai 1868. — *Bull. Soc. Géol.
2me Sér. t. XXV, pp. 802-810, Paris, 1868.*

VERNEUIL (DE) E.—Sur l'éruption du Vésuve de 1867-68.—*Bull.
d. l. Soc. Géol. de France, Vol. XXV, Paris, 1868.*

VERNEUIL (DE) E. — Sur les phénomènes récents du Vésuve. —
Compt. Rend. d. l'Acad. d. Sc. Paris, 1868.

VERNEUIL (DE) E. — Note sur l'altitude du Vésuve le 26 Avril,
1869. — *Compt. Rend. Acad. Sc. Vol. LXVIII, Paris, 1869.*

VERNEUIL (DE) E. — Sur la dernière éruption du Vésuve.—*Bull.
Soc. Géol. France, 2me Sér. Vol. XXIX, Paris, 1872.*

VESUVINO CRISIPPO. — Dichiarazione genealogica fisico-chimica
naturale, apologetica ed epidemica del Signor Vesuvio Frot-
tola. — ? *in 8°, pp. 16.* (C. A.).

VETRANI A. — Sebethi vindicaie, sive dissertatio de Sebethi antiquitate, nomine, fama, cultu, origine, prisca magnitudine,
decremento, atque alveis, adversus Jacobum Martorellium.—
Neapoli, 1767, in 8°, pp. 8 + 213, pl. II.

VETRANI A. — Il prodromo Vesuviano in cui oltre al nome, origine, etc. del Vesuvio s'esaminano tutti i sistemi dei filosofi,
etc. — *Napoli, 1780, in 8°, fol. 4, pp. 238.*

VILLAYÇAN GARCES G. DE. — El Bolcan que aborto la Montaña de
Soma. Soneto (1631). — *See Quinones.*

VILLEFOSSE (DE) I. — Vue du mont Vésuve et de son éruption
arrivée le 25 Oct. 1751.—*Plate in fol. with engr. descript.*

VIOLA S. (NAP). — Historia del Monte Vesuvio nella quale diffu-
samente si tratta di tutto lo che è occorso in esso dal principio del Mondo sino all'anno 1631 et 1649. Con occasione
del ultima eruttatione di fuoco fatta dal detto Monte a 16 Decembre 1631, et a 28 Novembre 1649. — *Original M. S. in
fol. fol. 3 + 132.* (C. A.).

VIRGILI P. DE. — Al Vesuvio. Poesia. — *Il Vesuvio, strenna pel
Capo d'anno del 1844. Napoli, in 8°, pp. 72.* (C. A.).

VISCARDI F. — Risposta alla lettera dell'Abate Tata de' 21 agosto
per l'eruzione del 1794.—*Napoli, 1794, in !8°, pp. 16, fol. 1,*

VOCOLA A. — Istoria dell'eruzione del Vesuvio accaduta nel mese
di Maggio 1737 scritta per l'Accademia delle Scienze. — *Napoli, 1738.*

VOLPE C. — Breve discorso dell'incendio del monte Vesuvio e degli suoi effetti.— *Napoli, 1632, in 8°, pp. 60.* (O. V.).

VOLPICELLI F. — Il Vesuvio bocca dell'inferno. Leggenda. — *Napoli, 1871, in 8°, pp, 20* (C. A.).

WAGNER P. — 1873. — *See Oesterland.*

WEDDING G. T. A. — Beitrag zu den Untersuchungen der Vesuv-
laven.—*Zeitschr. der Deut. Geol. Gesell. Bd. X, Berlin, 1858,
t. IV, in 8°, pp. 375-411.* (C. A.).

WEDDING G. T. A.—De Vesuvii montis lavis. — *Berolini, 1859,
in 4°, pp. 30.*

WELSCH H. — Warhafftige Reiss-Beschreibung aus eigener Er-
fahrung von Teutschland, Croatien, Italien denen Insuln
Sicilia, Malta, ecc. Nicht venigen bey dennen wunderbahren
brennenden Bergens als dem Vesuvio bey Naples, ecc. Erupt.
1631 etc. — *Stuttgart, 1658, in 4°, fol. 12, pp. 427, 1 portrait.* (C. A.).

WENDTRUP F. —Der Vesuv und die vulcanische Umgebung Neapels. — *Wittemberg, 1860, in 8°, pp. 35.* (C. A.).

WINCKELMANN. — Critical account of the situation and destruction
of Herculanum, Pompei and Stabia by the first Eruption of
mount Vesuvius. — *London, 1771, in 8°, pp. VIII+125.*

ZACCARIA DA NAPOLI. — Discorso filosofico sopra l' incendio del
Monte Vesuvio cominciato a' 16 Decembre 1631, nell' apparir
dell'alba. — *Printed together with Perrotti A.* (C. A.).

ZANNICHELLI G. J. — Considerazioni intorno ad una pioggia di
terra caduta nel Golfo di Venezia, e sopra l'incendio del Ve-

suvio. — *Raccolta di Opuscoli Scientifici e Filosofici. Venezia, 1727, 38, 57, Vol. LI, in 12°, figs. See pp. 87-124
T. XVI.*

ZEZZA BARONE M. — Na chiamata alli peccature. Canzona ncopp'a
l'eruzione de lo Vesuvio a l'anno 1855. — *Napoli, 1855, fol.
2, (C. A.).*

ZITO V. — Sonetti; due (1631) Per lo incendio del Vesuvio negli
scherzi lirici. — *Napoli, 1631, in 12.° See pp. 401-402.*

ZORDA G. — Discorso contro l'opinione dell'assorbimento vulcanico dell'acqua de' pozzi e del mare. — *Napoli, 1805, in 8°,
pp. 15.*

ZORDA G. — Relazione dell'eruzione del Vesuvio del 31 Maggio
1806. — *Napoli, 1806, in 4°, pp. 22.*

ZORDA G. — Continuazione dei fenomeni del Vesuvio dopo l'eruzione del 1806 fino al principio della primavera del 1810. —
Napoli, 1810, in 8°, pp. 16.

ZUPO G. B. — Giornale dell'incendio del Vesuvio dell'anno 1660
con le osservationi matematiche A. C. — *Roma, 1660, in 4°,
pp. 15. (C. A.).*

ZUPO G. B. — Continuatione de successi del prossimo incendio
del Vesuvio con gli effetti della cenere e pietre vomitate da
quello, e con la dichiarazione et espressione delle croce maravigliose apparse in varii luoghi dopo l'incendio. — *Napoli, 1661, in 4°, fol. 11, pl. 1 (C. A.).*

CAMPI PHLEGREAE

AND

CAMPANIAN PLAIN

ABICH H. — Sur la composition du feldspath vitreux et de la ria-
colite (des Campi Flegrei). — *Ann. d. Mines. 3ᵉ, Sér. Vol.
V. Paris, 1834 — Ann. der Phis. und Chem. Band. XXVIII.
Leipzig.*

ABICH H. — Geologische Beobachtungen über die vulkanischen
Erscheinungen und Bildungen in Unter-und Mittel Italien.—
Braunschweig, 1841, in 4°, pp. XI+134 (C. A.).

ACERBI P. FRANCISCI.—Polypodium Apollineum. — *Neapoli, 1674,
fol. VIII, pp. 352* (C. A.).

AGOSTINO (D') L. — Sulle acque termo-minerali balneolane dette
dei Bagnoli di proprietà di Gennaro Masullo.—*Napoli, 1874.*

AMENDUNI G. — Dell'incendio dell'agro puteolano. Epistola di Simone Porzio al Viceré D. Pietro di Toledo. Traduzione italiana preceduta da una illustrazione critica. — *Napoli, 1878,
pp. 24.* (C. A.).

ANCORA (D') GAETANO. — Guida ragionata per le antichità e per
le curiosità naturali di Pozzuoli e dei luoghi circonvicini.—
Napoli, 1792, in 8°, pp. VI+152, pl. 52, engrav. frontispiece
(C. A.). *French trans. by B. de Manville, Naples, 1792, in
8°, pp. VI+142 fol. LI.*

ANONYMOUS. — Wunderbarliche und erschreckliche neue. — *Zei-*

lung so sich neulich auf den XXIII sept. in 1533 in Welschland , nicht fern von Neapolis zugetraegen haben.??, in 8.°, fol. 3.

ANONYMOUS. — Relazione del terremoto accaduto in Napoli il giorno 8 settembre 1694. — *Napoli, 1694.*

ANONYMOUS. — Analisi dell'acqua raccolta dal vapore di una fumarola della Solfatara di Pozzuoli. — *Napoli , 1790, in 8°* , *pp. 27* (C. A.).

ANONYMOUS. — Breve notizia di un viaggiatore sulle incrostazioni silicee termali d'Italia e specialmente di quelle de' Campi Flegrei nel regno di Napoli (Thompson G). — *Giorn. Lett. di Napoli, 1793-1798. Vol. XLI, pp. 32-51.*

ANONYMOUS. — Osservazioni su di un fenomeno avvenuto nel lago di Patria. Lettera 1.ª e 2.ª. — *Napoli, 1796.*

ANONYMOUS. — Relazione del tremuoto sentito in Napoli e altre provincie nel 29 novembre 1732. — *Napoli, 1805.*

ANONYMOUS. — Relazione di un fenomeno avvenuto nel porto di Napoli a 14 dicembre 1798. — *Atti d. Acc. d. Sc. Vol. IX. Siena, 1808.*

ANONYMOUS.—Tableau topographique et historique des îles d'Ischia, Vendatene, Procida, Nisida, du Cap Misène, etc. — *Napoli, 1822, in 8°, pp. VIII+ 216.*

ANONYMOUS. — Monte S. Simone und die Eruption von 1811. — *Morgenblatt, N. 138, pp. 551, (Cit. par Hoff Veränd. II, s. 241), 1823.*

ANONYMOUS. — Extrait d'une lettre sur le [tremblement de terre qui a eu lieu dans l'île d'Ischia le 2 fév. 1828. — *Bibl. Univ. d. Sc. etc. Vol. XXXVII. Genève, 1828.*

ANONYMOUS. — Extrait d'une lettre sur le tremblement de terre qui a eu lieu dans l' île d' Ischia , le 2 févr. 1828. — *Bibl· Univ., March. 1828, pp. 236-240* (C.A.).

ANONYMOUS. — Catalogo della Collezione Orittologica ed Oreognosica del fu chiarissimo Professore Cav. Matteo Tondi Direttore del Museo di Mineralogia di Napoli, ecc.—*Napoli, 1837, in 8°, p. VIII+243.*

ANONYMOUS. — Istituto d'Incoraggiamento di Napoli. Brevi notizie sulle acque minerali della provincia di Napoli. — *Napoli, 1861.*

ANONYMOUS, — Sulle acque balneolane dette di Bagnoli. — *Napoli, 1863.*

ANONYMOUS. — Istituto d'Incoraggiamento di Napoli. Notizie intorno alle acque minerali delle provincie napolitane. — *Napoli, 1865.*

ANONYMOUS. — Don Chisciotte. Catania Casamicciola. — *Catania, 1881.*

ANONYMOUS. — The Earthquake in Ischia.— *Illust. London News N.° 2183, March 19*th *, 1881, p. 271, figs. 4* (C. A).

ANONYMOUS. — Casamicciola nella notte del 28 Luglio 1883. — *3 loose sheets* (C. A.).

ANONYMOUS. — Das Erdbeben auf Ischia am 28 Juli 1883. — *München, 1883, in 8°, pp. 40 with figures* (C. A.),

ANONYMOUS. — Die Erdbeben-katastrophe von Ischia am 28 juli 1883. — *Vienna, 1883, pp. 95. pl. 16, 1 map.* (C. A.).

ANONYMOUS. — Disastri, Ischia-Giava. — *Napoli, 1883, in 8°, pp. 178* (C. A.).

ANONYMOUS. — Relazione della commissione per le prescrizione edilizie dell'Isola d'Ischia istituita dal Ministero dei Lavori Pubblici (*Genala*) dopo il terremoto del luglio 1883. — *Roma, 1883, in 4°, pp. 86, col. pl. 2.*

ANONYMOUS. — Das Mare Morto bei Neapel. — *A plate?* (C. A.).

ANONYMOUS. — Guide du Voyageur pour les Antiquités et curiosités naturelles de Pouzol, et des environs. — *Naples, in 8°, pp. 134 (pp. 17-26: De' Campi Flegrei e della Solfatara).*

ANONYMOUS. — Vue de la Soufrière qui est près de Pozzuole au royame de Naples, appelée Solfatara. — *A plate?* (C. A.).

ANONYMOUS. — Wunderbarliche und erschrockliche newe Zeitung so sich neulich auff den 28 tag Septembris im 1538 jar in Welschland nit fern von Neapolis zugetragen huben. — *?, in 4°,* (B. N.).

ASCOLI (D'). — Earthquake at Naples. — *The Philos. Mag. Vol. XXIII. — London, 1806.*

ARAGO F. — Rapport verbal sur les nouvelles recherches de M. Capocci, sur le phénomène connu de l'érosion du temple de Sérapis à Pouzzoles. — *Compt. rend. d. l'Acad. d. Sc. Vol. IV. Paris, 1854, in 4°, pp. 750-753.*

AUDOT. — Royaume de Naples. — *Paris, 1835, in 8°, pp. 370, pl. 117* (C. A.).

BABBAGE C. — On the geognostical phenomena at the Temple of Scrapis. — *The Edinb. Philos. Journ. Vol. XI. Edinburgh, 1824.*

BABBAGE C. — Observations on the temple of Scrapis at Pozzuoli, near Naples, with remarks on certain cause which may produce Geological Cycles of great extent. — *The American Journ. of Sc. and Arts. Vol. XXVII, New-Haven, 1838. Proceed. of the geol. Soc. of London Vol. II. London 1838.* —

Quarter. Journ. of the geol. Soc. of London. Vol. III. London, 1847.

BABBAGE. CII, — Observations on the Temple of Serapis at Pozzuoli near Naples, with an Attempt to Explain the Causes of the Frequent Elevation and Depression of Large Portions of the.Earth's Surface in Remote Periods, and to Prove that those Causes Continue at the Present Time. « With a supplement: » Conjectures on the Physical Condition of the Surface of the Moon. — *London, 1847, in 8°. pp. 42. 2 Plates.* (C. A.).

BACCHI A. — Elpidiani, Civis Romani. De Thermis, etc. —¬ *Venetiis, 1588, in fol. fig. pp. 48+492+1.*

BALAGUER A. M. — Los estragos del Tremblor, y subterranea conspiracion. — *Napoles, 1697, pp. 360, in 4°.* (C. A.).

BALDACCI L. — Alcune osservazione sul terremoto avvenuto all'isola d'Ischia il 28 luglio 1883. — *Boll. Com. Geol. Il. 1883. 2ª Ser. IV, pp. 157-166. Also « Science » 1883, II, 396-399.*

BARONE G. — ΗΚΑΤΑΣΤΡΟΦΠ ΤΠΣ CASAMICCIOLA, ΤΕΤΡΛΣ-ΤIΧΟΝ. — *Naples, 1883, in 8°, pp. 8.* (C. A.).

BARRAL. — Mémoire sur des roches coquillères trouvées à la cime des Alpes dauphinoises et sur des colonnes d'un temple de Sérapis à Pouzol près Naples. — *Grenoble, 1813.*

BARTOLI S. — Breve ragguaglio dei bagni di Pozzuoli. — *Napoli, 1667, in 4°, pp. 76.* (C. A.).

BARTOLI. — Thermologia Puteolana, — 1679, *2 Vol. pp. 304.*

BENKOWITZ C. J. — Reisen von Neapel in die umliegenden Gegenden, nebst einige Notizen über das letzte Erdbeden in Neapel. — *Berlin, 1806.*

BERTAZZI G. — 1862. — *See Scacchi.*

BERTHIER P. — Analyse de la Pouzzolane de Naples et du trass des bords du Rhin. — *Ann. d. Mines, Vol. I. Paris, 1827.*

BERTOLONI A. — Su di un viaggio a Napoli nella estate del 1834.

BERTONI. — 1874. — *See Macagno.*

BLACHET ET LICANN. — Note sur une substance cristalline, recueilie sur les murs des bains de S. Germano près de Naples. — *Journ. d. Pharm. et d. Sc. accessoires. Vol. XIII. Paris, 1827.*

BLAKE J. F. — A Visit to the Volcanoes of Italy. — *Proceed. Geol. Assoc. London, 1889, Vol. IX, pp. 145-176.*

BOCCONE P. — Osservazioni naturali ove si contengono materia medico-fisiche e di botanica, produzioni naturali, fosfori diversi, fuochi sotteranii d'Italia ed altra curiosità. — *Bologna in 12°, pp. 400 fig.* (C. A.).

BONGHI, MADIA, CASSOLA E DE-RENZI. — Memoria sulle acque termali balneolane. — *Napoli, 1863.*

BORGIA HIERONYMUS. — Incendium ad Averuum lacum horribile pridie calendas octobris MDXXXVIII nocte intempesta exortum. — *Naples, 1538, in 4°, fol. 16.* (C. A.).

BORGIA G. — M. Nuovo. 1538, — *See Giustiniani.*

BORGIAE H.—Massae Lubrensis pontifec, Carmina lirica et heroica quae extant. — *Venetiis, 1666, in 12°, pp. 18+319.*

BOUÉ A. — Ueber Solfataren und Kratercrloschener Vulcane. — *Sitz. d. Kais. Akad. d. Wissens. Vol. XLVIII, 1863, in 8°, pp. 20.* (C. A.).

BRAUNS D. — Das Problem des Terapeums von Pozzuoli. — (*Leopoldina, amtliches Organ der K. leopoldino-carolinischen deutschen Akademie der Naturforscher, 24, Halle, 1888, pp. 15.*

BREISLAK S. — Essais minéralogiques sur la solfatara de Pozzuole. — *Naples, 1792, in 8°, p. 240, mp. I.*

BREISLAK S. — Topografia fisica della Campania. — *Firenze, 1798, in 8°, pp. XII+368, pl. 3. (pp. 225-308).*

BREISLAK S. — Carte physique de la Campanie. — *Dans l'ouvrage: Voyage dans la Campanie. Vol. I. Paris, 1801.*

BREISLAK S. — Voyages physiques et lithologiques dans la Campanie suivis d'une mèmoire sur la constitution physique de Rome. — *Paris, 1801, Leipzig, 1802 (In German.) Vol. I, pp. 300—XVI, pl. 3. Vol. II, pp. 324, pl. 3.*

BREISLAK S. — Notice sur la fontaine de la fumarole à la solfatare de Pouzzoles. — *Journ. des Mines, Vol. XV, Paris, 1803-1804.*

BREISLAK S. — Carta Topografica del cratere di Napoli e dei Campi Flegrei, colla pianta speciale del Vesuvio, secondo le ultime osservazioni del Abte Breislak. — *?, (C. A.).*

BRIVE (DE) A. — Extrait d'un voyage en Italic. Environs de Naples. — *Ann. d. l. Soc. d'Agr. du Puy, 1834.*

BROCCHI G. B. — Notizie di alcune osservazioni fisiche fatte nel tempio di Serapide a Pozzuoli. — *Bibl. Ital. ossia Giorn. d. letter. Sc. etc. Vol. XIV, Milano, 1819.*

BROECK (VAN DEN) E. — On some foraminifera from pleistocene beds in Ischia: preceded by some geological remarks by Ar. W. Waters. — *Quarter. Journ. of the geol. Soc. Vol. XXXIV, N. 134. London, 1878.*

BRUNO FR. SAV. — L'osservatore di Napoli, ossia rassegna delle istituzioni civili, de' pubblici stabilimenti, dei monumenti storici, ed artistici, e delle cose notabili di Napoli, con una breve descrizione de' suoi contorni, ecc. — *Napoli, 1854, in 12°, pp. 8+592. (pp. 379-386).*

BUCH (VON) L. — Scipio Breislak's physiflalische Topographie von Campanien. — *Ann. der Phys. Halle und Leipzig. V Band. 1800.*

BUCH (VON) L.—Ischia.— *Neu Jahrb. der Berg. und Hüttenkunde. Vol. 1. Nürnberg, 1809.*

BUCH (VON) L.—Besuch und Entstehungs weise des Monte Nuovo.— *Zeitschr. d. Deuts. geol. Gesell. 1 Band. Berlin, 1849.*

BUCH (VON) L. — Lettre a Naumann sur sa visite au M.te Nuovo avec M. Pareto. — ?

BULIFON A. — Lettere storiche politiche ed crudite. —. *Pozzuoli 1683, in 8°, pp. 482+VII, raccolta 4ª, pp. 177-188.* (C. A.).

BULIFON A. — Lettera al sig. D. G. F. Paccceo sul terremoto del 5 giugno 1688 in Napoli. — *Napoli, 1697.*

BULIFON A. — Le guide des ètrangers curieux de voir et de connoitre les choses les plus mémorables de Pouzzol, Bayes, Cumes, Misène et autres lieux des environs, de l'abbé Sarnelli, traduite en francais, avec le texte en regard, et la description des vertus et propriétés des bains d'Ischia par J. C. Capaccio. — *Naples, 1699, in 12°, pl. 33, portrait of Bulifon.* (C. A.).

BULIFONE N. — Distinta relazione del danno cagionato dal terremoto del 3 novembre 1706. — *Napoli, 1706.*

BURNET G.—Monte Nuovo. Voyage de Suisse, d'Italie et de quelques endroits d'Allemagne et de France, fait ès années 1685 et 1686. — *Rotterdam, 1690, in 12°, pp. 319.* (C. A.).

CALAMAI L. — Dell'acqua Medici di Castellamare. — *Pisa, 1849.*

CAMPILANZI E. — Sulla corrispondenza dei cangiamenti di livello del mare osservati negli avanzi del tempio di Serapide con quelli avvenuti a Venezia. — *Ann. d. Sc. R. Istit. Lomb. Veneto, Vol. X, Padova e Venezia, 1840.*

CANGIANO LUIGI. — Sul pozzo che si sta forando nel giardino della Reggia di Napoli e di taluni induzioni geologiche di cui è stato occasione. — *Naples, in 4°, pp. 23,* (C. A.).

CANGIANO L. — Sul pozzo forato nel giardino della Reggia di Napoli. — *1847, ?.*

CANGIANO L. — Breve ragguaglio del perforamento dei due pozzi artesiani recentemente compiuti nella città di Napoli. — *Napoli, 1859.*

CANGIANO L. — Sull'attuale condizione delle acque pubbliche in Napoli, e dei modi di migliorarla. — *Napoli, 1859.*

CAPACCI J. C. — Puteolana Historia, accessit ejusdem de Balneis

Libellus. — *Neapoli, 1604, in 4°, fol. 8., pp. 208+88 many figures* (C. A.).

CAPACCIO G. C. — Il Forastiero , dialoghi ne' quali oltre a quel che si ragiona del origine di Napoli, ecc. siti e corpo della Città con tutto il Contorno da Cuma al Promontorio di Minerva, varietà e costumi di habitatori, famiglie nobili, e popolari, con molti Elogii d'huomini illustri, aggiuntavi la cognitione di molte cose. appartenenti all' istoria d' Italia, con particolari relationi per la materia politica con brevità spiegate. — *Dialogo, Napoli, 1634, in 4°, pp, 56+1024.*

CAPANO G. COUNT. — Rapporto dell' Ispettore della Provincia di Napoli (Capano) sul progetto di miglioramento della boscaglia del Monte Nuovo di Pozzuoli. — *Naples, 1823, in 8°, pp. 40* (C. A.).

CAPMARTIN DE CHAUPY ABB. — Découverte de la maison de Campagne d'Horace.—*Rome, vols. III, in 8°, with. topogr. map. Vol. I. p. 101 and follow. de' Vulcani, del Vesuvio de' Campi Flegrei, etc. etc.*

CAPOCCI E. — Nuove ricerche sul noto fenomeno delle colonne perforate dalle folladi nel tempio di Serrpide. — *Il Progr. d. Sc. Lett. ed Arti. Vol. XI. Napoli, 1835.* — *The Edinb. New Philos. Journ. Vol. XIII. Edinburgh, 1837.*

CAPORALI G. — Della acque minerali Campane alla esposizione italiana del 1861. — *Napoli, 1861.*

CAPPA R. — Dell'analisi chimica e delle virtù medicinali dell'acqua termo-minerali di Gurgitello e di Castiglione. — *Naples. 1863, in 8°, pp. 15,* (C. A.).

CAPUA L. DI. — Lezióne intorno alla natura delle Mofete. — *Napoli, 1863, in 4°, fol. VIII, pp. 176, fol. VIII. Also an edition at Cologn.* (C. A.).

CARDONE A. — Saggio di poetici componimenti (Sul funestissimo tremuoto avvenuto in Casamicciola. Ode).—*Naples, 1828, in 8°, fol. 1, pp. 25* (C. A.).

CARTELLI. — Storia della Regione abbruciata in Campagna Felice in cui si tratta il suo sopravvenimento generale, e la descrizione de' luoghi, de' Vulcani, de' Laghi, de' Monti, delle Città litorali e di popoli, etc. — *Napoli, 1787, in 4°, pp. XLIII+382. Pl. I.*

CASTALDO A. — Istoria — Libri IV (in which are recounted the principal events that happened in the kingdom of Naples under the government of the Viceroy D. Pietro di Toledo and the Viceroys his successors till Cardinal Granvela. — *Naples. 1749, in 4°, pp. 21-155,* (C. A.).

14

CASSÓLA F. — 1834. — *See Sementini.*

CASSOLA F. — Analyse des eaux minérales de Castellamare. — *Journ. d. Ch. Méd. etc. Vol. I. Paris, 1835.*

CASSOLA F. — 1803. — *See Bonghi.*

CAVE (LA). — 1840. — *See Costa.*

CESTARI AB. G.—Anecdotti istorici sulle alumiere delli monti Leucogei. — *Napoli, 1790, in 12°,* (B. N.).

CHEVALLEY DE RIVAZ E. — Précis sur les eaux minero-thermales, et les étuves de l'ile d'Ischia. — *Naples, 1831, in 4°, fol 5, pp. 70.* (C. A.). *2nd edit. and 3rd edit. Naples, 1837, in 8°, pp. VIII + 182.*

CHEVALLEY DE RIVAZ E. — Description des eaux minero-thermales d'Ischia. — *1834 and 1837.*

CHEVALLY DE RIVAZ L. — Descrizione delle acque termo-minerale e delle stufe dell'isola d'Ischia. — *Napoli, 1838, in 8°, pp. XII + 276. pl. III* (C. A.).

CHEVALLEY DE RIVAZ E. — Voyage scientifique à Naples. — *Paris, 1843.*

CHEVALLEY DE RIVAZ E. — Lettera al Presidente della R. Accad. d. Sc. in Napoli sopra un terremoto sentito in Casamicciola d'Ischia il 7 Giugno 1852. — *Rend. d. R. Accad. d. Sc. in Napoli., N. S. N.° 3. May and June 1852, pp. 88* (C. A.).

CHEVALLEY DE RIVAZ E. — Terremoto del 7 di giugno 1852 in Casamicciola. — *Rend. d. Acc. Napoli, Maggio 1853.*

CHEVALLEY DE RIVAZ E. — Su di un terremoto ad Ischia. — *Boll. Meteor. Vol. II. Roma, 1863.*

CHUN K. — Das Erdbeben auf Ischia. — *" Illustrirte Zeitung " April 2nd 1881, pp. 265-268, mit 3 vignetten of B. Köhler.* (C. A.).

CIANCIO A. — Ragionamento sulla privativa del Marchese Nunziante nella fabbricazione dell' Allume Vulcanico. — *Napoli, ?. in 4°, p. 60.*

CLAUSON C. — Saggio sulla topografia dell' antica Partenope. — *Napoli, 1889, in 8°, pp. 16, map. 1.*

COLACCI O. DE. — Dialoghi intorno a' tremuoti di quest' anno 1783. — *Napoli, ? 1783, in 8°, pp. 79* (C. A.).

COLLEGNO (DI) G. P. — Contrade vulcaniche delle vicinanze di Napoli. — *Atti d. 6ª Riun. d. Scienz. Ital. nel 1844. Milano, 1845.*

COLLOMB E. — Sur un voyage géologique en Corse, en Sardaigne et aux environs de Naples.—*Bull. d. l. Soc. géol. d. France. Vol. XI, Paris, 1853-54.*

COLOMBO A. — Osservazione sulla conformazione sottomarina del golfo di Napoli. — *Rivista Marittima*, *Ottobre-Dicembre, 1887, pl.*

CONCINA G. — Casamicciola, ossia le acque minerali di Ischia.— *Venezia, 1832.*

CONSIGLIERE (DI) C. — Notizie intorno ad una sostanza particolare che trovansi presso le acque termali d'Ischia, ed intorno ai vapori del Vesuvio. — *Giorn. d. Fis. Ch. e St. Nat. Vol. II, Pavia, 1819.*

CONSTANTIN J. — Voyage scientifique à Naples avec M. Mangendie en 1843. — *Paris, 1844, in 8°, pp. 103* (C. A.).

CONTI. — Saggio di sperimenti su le proprietà chimiche e medicamentose delle acque termo-minerali del Tempio di Serapide. — *Napoli, 1826.*

Contratto costitutivo della compagnia vesuviana.—*Napoli, 1836 in 8°, pp. 30* (C. A.).

COPPOLA M. — 1875-76. — *See Palmieri.*

CORDIER L. — Rapport sur le voyage de M. Constant Prevost à l'île Julia, à Malte, en Sicile. aux îles Lipari et dans les environs de Naples. — *Compt. rend. d. l'Acad. d. Sc. Vol. II. Paris, 1836. Nouv. Ann. des Voy. Paris, 1836.*

COSTA O. G. — Mammiferi viventi e fossile della fauna di Napoli. — *Napoli, 1839.*

COSTA O. G. — Osservazioni ulteriori intorno ai fossili organici di Pozzuoli. — *Napoli, 1853.*

COSTA O. G. — Cenni intorno alle scoperte paleontologiche fatte nel R. di Napoli durante gli anni 1854-55. — *Napoli, 1856.*

COSTA O. G. — Intorno alle scoperte paleontologiche fatte nel regno di Napoli durante gli anni 1857-58. — *Napoli, 1858.*

COVELLI N. — Terremoto in Ischia, 2 febbrajo 1828. — *Two autograph letters to Monticelli, fol. 3* (C. A.).

COVELLI N. — Observations sur le tremblement de terre qui a eu lieu dans l'île d'Ischia le 2 Janv. 1828. — *Bibl. Univ. d. Sc. etc. Vol. XXXIX. Genève 1828. Notiz. aus dem Gebiete der Natur. und Heilkunde XXIII, Band. Erfurt und Veimar, 1829.*

COVELLI N. — Memoria per servire di materiale alla costituzione geognostica della Campania (1827). — *Atti d. Acc. d. Sc. fis. e Mat. Vol. IV. Napoli, 1839, pp. 37.*

COVELLI N.— Cenno sul tremuoto d'Ischia.—*"Il Pontano", N° II, Napoli, 1828-29, pp. 82.* (C. A.). *See also an English translation in Chapter IV of Johnston-Lavis' Monograph of the Earthquakes of Ischia, etc. Naples, 1885*

Cox. — Hints for invalids about to visit Naples, also an account of the mineral Waters of the Bay of Naples. — *London, 1841*

D' Ascia G. — Storia dell'isola d' Ischia. — *Napoli, 1868, in 4°, pp. 517* (C. A.).

Daubeny C. — On the volcanic Strata exposed by a section made on the site of the new thermal spring discovered near the town of Torre dell' Aunnunziata, in the Bay of Naples ; with some remarks on the gases evolved by this and other springs connected with the volcanoes of Campania. — *The Edinb. New Philos. Journ. Vol. XIX. Edinb. 1835. Proceed. of the Geol. Soc. of London. Vol. II. London, 1838.*

Daubrée A. — Rapport sur le tremblement de terre ressenti à Ischia le 28 juillet 1883; causes probables des tremblements de terre. — *Compt. Rend. Ac. Sc. Paris, 1883, Vol. XCVII, pp. 768, Also Revue Scientif. 1883. Vol. XXXII, pp. 165.*

De Angelis G. — Casamicciola e le sue rovine. — *Napoli, 1883, in 8°, pp. 85* (C. A.).

De Ciutiis M. — Casamicciola. — *Naples, 1883, in 8°, pp. 104* (C. A.).

Deecke W. — Il cratere di Fossa Lupara nei Campi Flegrei presso Napoli. — *Boll. Com. Geol. 7-8. Roma, 1888; estratto da (Zeitschrift der deuts. geol. Gesellschaft, XL. Band. I, Heft. pp. 166-172, pl. 1. Berlin, 1888.*

Del Balzo C. — Cronaca del tremuoto di Casamicciola. — *Napoli, 1883, in 8°, pp. 228, fol. IX.* (C. A.). *2nd Edit. pp. 240 + IX.*

Delta (Δ) (pseudon for Forbes J. D. (?). — Remarks on the climate of Naples and its vicinity; with an account of a visit to the Hot springs of la Pisciarella, Nero's Baths. — *The Edinb. Journ. of Sc. Vol. VII. Edinburgh, 1827.*

De Rossi M. S. — Intorno all'odierna fase dei terremoti in Italia e seguitamente sul terremoto in Casamicciola del 4 maggio 1881.—*Boll. Soc. Geograf. Il. 1881, N° 3, in 8°, pp. 25. maps 2* (C. A.).

De Rossi M. S. — Bullettino del Vulcanismo italiano. — *Roma, 1873 to 1888, in 8°,* (C. A.).

Desmoulins. — Vue prise de dessus le Cratère de Monte Nuovo.— *?* (C. A.).

Deville C. Ste. Claire. — Recherches sur les produits des volcans de l'Italie méridionale. — *Compt. Rend. Acad. Sc., T. XLII, pp. 1167-1171, Paris, June 16, 1856* (C. A.).

Deville C. Ste. Claire. — Sur les émanations volcaniques (premier mémoir). — *Compl. Rend. Acad. Sc., T. XLIII, pp. 955-958, Paris, Nov. 17, 1856* (C. A.).

DEVILLE C. Ste. CLAIRE. — Mémoires sur les émanations volcaniques. — *Bull. d. Soc. Géol. 2e , Sér., T. XIV, pp. 254 et suiv. 16 Décem. 1856* (C. A.).

DEVILLE C. Ste. CLAIRE. — Sur les émanations volcaniques des Champs Phlegréens. 3 Lettres à son frère M. H. Sainte Claire Deville.—*Compl. Rend. Acad. Sc. (1re), T. LIV, Paris, Mars. 10, 1862.* — (2me) *Oct. 13, 1862, 3me Nov. 13. 1865* (C. A.).

DIENER. — Das Erdbeben auf der Insel Ischia am 28 Juli 1883.— *Milth. d. geogr. Gesellsch. Wien, 1884.*

DOELTER C. — Die Vulcangruppe der Pontinischen Inseln. —*Denk. d. K. Ak. d. Wissensch. XXXVI Band. Wien; 1875.*

DOELTER C. — Vorläufige Mittheilungen über den geologischen Bau der Pontinischen Inseln. — *Sitzungsb. d. K. Ak. Wiss. LXXI Band. Wien, 1875. Boll. d. R. Com. Geol. d' Italia. Vol. VI, N.° 5-6. Roma, 1875.*

DOELTER C. — Il gruppo vulcanico delle Isole Ponza. — *Mem. d. R. Com. Geol. d'Italia. Vol. III, Part. I. Roma, 1876.*

DOLOMIEU (DE) D. — Mémoire sur les îles Ponces et catalogue raisonné des produits de l' Aetna. Paris 1788. — *Leipzig, 1789.*

DONATO A. DI. — Dell'analisi chimica e delle proprietà medicinali dell'acqua termo-minerale detta *Subveni homini.* — *Naples, 1854, in 8°, pp. 16* (C. A.).

DU BOIS F. — The Earthquakes of Ischia. — *Trans. Seism. Soc. Japan. Vol. VII, pt. I, 1883-84, pp. 16,42 pl. I* (C. A.).

DU BOIS F. — Further Notes on the Earthquakes of Ischia. — *Trans. Seism. Soc. Japan. Vol. VIII, 1885. Yokohama, pp. 95-99* (C. A.).

DUFLOS. — Vue de la Solfatara près de Pouzzole Ancien volcan nommé par Strabon Forum Vulcani. — *Naples, ?* (C. A.).

DUFRÉNOY P. A. — Sur les terrains volcaniques des environs de Naples. — *Compl. Rend. d. l'Acad. d. Sc. Vol. I. Paris, 1835 pp. 4.* — *Notiz. aus. d. Geb. der Natur. und Heilkünde. XLVI Band. Erfurt und Weimar, 1835.* —*The Edinb. New Philos. Jour. Vol. XX, Edinburgh, 1836.* — *Annal. d. Mines. Vol. XI. Paris, 1837. Proc. Verb. d. l'Acad. Philom. Paris, 1837. Paris, 1838, in 8°, fol. 4, pp. 420, pl. 9, figs.*

DUFRÉNOY P. A. — De la manière dont peut se former le terrain des environs de Naples. — *Bull. d. l. Soc. géol. de France, Vol. VIII. Paris, 1836.*

DUFRÉNOY P. A. — Parallèle entre le différents produits volcani-

ques des environs de Naples et rapport entre leur composi-
tion et les phénomènes qui les ont produits. — *Bull. d. l.
Soc. géol. de France. Vol. IX, Paris, 1837-38. — Annal.
d. Chim. Vol. LXIX, Paris, 1838. Annal. d. Mines. Vol.
XIII. Paris, 1838. Compt. Rend. d. l'Acad. d. Sc. Vol. VI.
Paris, 1838.*

EUCHERII DE QUINTIIS. — *See Quintiis.*

FALCO B. DE — Antiquitates Neapolis, Atqnae etc. etc. — *Lug.
Bat. in fol. pp. 48.* — This is the oldest guide of Naples in
which Vesuvius and The Solfatara are spoken of.

FALCONI (DELLI) M. — Dell'incendio di Pozzuoli nel 1538. — *Napoli, 1538, in 4°, fol. 22.* See also Giustiniani L.

FAUJAS DE SAINT FOND B. — Notice sur une espèce de charbon
fossile nouvellement découverte dans le territoire de Naples.—
Annal. du Muséum d'Hist. Natur. Vol. XI, Paris, 1808.

FAZZINI G. — Cenno sulla pozzolana della Baja di Napoli. — *Napoli, 1857, in 4°, pp. 21.*

FERBER. — Lettres à Mr. le Chev. de Born sur la Minéralogie et
sur divers objets d' histoire naturelle de l' Italie, traduit de
l'Allemand, enrichi de notes et d'observations faites sur les
lieux par M. de Dietrich.—*Strasbourg, 1776, in 8°, pp. 16
+ 508.*

FERRERO O. — Relazioni sopra un minerale trovato a Lusciano.—
Annal. d. Staz. Agr. Ann. V. Caserta, 1877.

FERRERO O. E MUSAIO. — Studii ed analisi sopra le roccie vulcaniche costitutive in alcuni punti del territorio della provincia di Caserta. — *Annal. d. Staz. Agr. Ann. VI. 1877,
Caserta, 1878.*

FIESCHI-RAVASCHIERI D^{sa} . — *See Meuricoffre O.*

FLAMMARION C. — Le tremblement de terre d'Ischia. — *Rev. Mens.
d'Astron. Populaire. Paris, 1883, in 4,° N.° 9, pp. 317-
329, 3 fig. (C. A.).*

FONSECA F. — Descrizione e carta geologica dell'Isola d'Ischia.—
Napoli, 1847.

FONSECA F. — Geologia dell'Isola d'Ischia (con carta geologica).—
Firenze, 1870, in 4°, pp. 31 and map.

FORBES. — 1827. — *See Delta (Δ).*

FORBES J. D. — Physical notices of the Bay of Naples. N. 3. On
the district of Posilippo and the Lago d' Agnano. — *The
Edinb. Journ. of Sc. Vol. X, Edinburgh, 1829. Zeitsch fur
Miner. Frankfurt am Main, 1829.*

FORBES J. D. — Physical notices of the Bay of Naples: N. 4. On
the solfatara of Pozzuoli.—*The Edinb. Journ. of Sc. Vol. I,*

New ser. Edinburgh, 1829. Notiz aus dem Geb. der Natur und Heilkunde, XXV. Band. Erfurt und Weimar, 1829.

FORBES J. D. — Physical notices of the Bay of Naples: N. 5. On the temple of Jupiter Serapis at Pozzuoli and the phenomena which it exhibits. — *The Edinb. Journ. of Sc. Vol. I, New ser. Edinburgh, 1829. Jour. d. Géol. Vol. I. Paris, 1830.*

FORBES J. D. — Physical notices of the Bay of Naples: N. 6. District of the Bay of Baia. — *The Edinb. Journ. of Sc. Vol. II. New ser. Edinburgh, 1830.*

FORBES J. D. — Physical notices of the Bay of Naples: N. 7 On the Islands of Procida and Ischia. — *The Edinb. Journ. of Sc. Vol. II, New ser. Edinburgh, 1830. Notiz. a. d. Geb. d. Natur. und Heilkunde. XXVII, Band. Erfurt und Weimar, 1830.*

FORBES J. D. — Physical notices of the Bay of Naples : N. 8. Concluding view of the volcanic formations of the district.— *The Edinb. Journ. of Sc. Vol. III. New ser. Edinburgh , 1831.*

FORTIS A. — Lettera economica su l'attuale stato dell' Allumiera della Solfatara di Pozzuoli. — *?, 1790,* (B. N.).

FORTIS G. B. — Osservazioni litografiche sulle isole di Ventotene e Ponza. — *Mem. d. Acc. d. Sc. Lett. ed Arti di Padova, 1794.*

FOUGEROUX DE BOUDAREY. — Observation sur le lieu appelé Solfatare, situé près de la Ville de Naples. — *Compt. Rend. d. l' Acad. d. Sc. Paris, 1765. T. II, pp. 418-447 , in 12°, pl. III.*

FOUQUE F. — Sur les phénomènes éruptifs de l'Italie méridionale. — *Compt. Rend. Acad. Sc. Paris, 1865. pp. 41-44 (Vesuvius and Solfatara)* (C. A.).

FREDA G. — Sulle masse trachitiche rinvenute nei recenti trafori delle collini di Napoli. — *Rend. R. Acc. Sc. Fis. Mat. Napoli, 1889. Ser. 2, Vol. II, pp. 9.*

FREDA G. — Sulla composizione del piperno trovato nella collina del Vomero e sull'origine probabile di questa roccia.—*Rend. Acc. Sc. Fis. e Mat., S. II, Vol. II, fasc. 6°, Napoli, 1888.*

FUCHS C. W. C. — Ueber die Entstehung der Westkuste von Neapel. — *Verhandl. d. Nat. hist. Medic. Vereins zu Heidelberg, III. Band. Heidelberg, 1865.*

FUCHS C. W. C. — Vulkanische Gebiete Neapels. — *Neu. Jahrb. für Miner. Geogn. etc, Heidelberg, 1865.*

FUCHS C. W. C. — Die Insel Ischia. — *Jahrb. d. K. K. Geol. Reichsanst. XXII, Band. Wien, 1872.*

FUCHS C. W. C. — Monografia geologica dell'Isola d'Ischia. — *Mem. d. R. Com. Geol. d'Italia. Vol. II, Part. 1ª, Firenza, 1873.*

FUCHS C. W. C. — Chemisch-geologische Untersuchung der Insel Ischia. — *Verhand. d. allgm. Schweiz. Naturf. Ges. Chur. 1875.*

GAMBA B. — Lettere descrittive di celebri Italiani. — *2ª. edit. Venezia, 1819, in 8°, pp. 8+262.*

GARRUCCIO G. — Un simposio sul cratere di Baia, disquisizioni archeologiche di Guida da Miseno a Porto Giulio. — *Napoli, 1859, in 8°, pp. 32.* (C. A.).

GATTA L. — L'Italia, sue formazioni, suoi vulcani e terremoti,— *Milano, 1882, in 4°, pp. XV+539, 32 figs,, 3 maps.* (C. A.).

GATTA L. — Considerazioni fisiche sull'isola d'Ischia. — *Boll. Soc. Geol. It. Anno II, fasc. 2°, 1883, in 4°, pp. 10.* (C. A.).

GAVAUDAN G. — Memoria sopra l'uso dei bagni minerali di Gorgitello. — *Napoli, 1845.*

GENOINO G. — Viaggio poetico pe' Campi Flegrei.—*Napoli, 1813. in 16°, pp. 122-3.*

G. F. (FORTUNATO GIUSTINO). — I Campi Flegrei e Pompei. Ricordi dei dintorni di Napoli. — *Napoli, 1870, in 12°, pp. 54.* (C. A.).

GIMBERNAT. (DE) C. — Notice sur les colonnes du temple de Serapis près de Naples qui sont percées jusqu' à une certaine hauteur par les vers marins ou les Pholades. — *Bibl. Univ. d. Sc. etc. Vol. X. Genève, 1819.*

GIMBERNAT (DE) C. — Phénomène observé à Massa Lubrense près Capo Campanella. — *Nouv. Ann. d. Voy. 1820.*

GIMMA G. — Storia naturale delle Gemme, delle Pietre e di tutti i Minerali, ovvero fisica sotterranea. — *Napoli, 1730, 2 vols. in 4°, with antiporto. (Vol. I, pp. 46 + 551; Vol. II, pp. 4+603).*

GIROND A. — Observations sur une mine de fer en sable qui se trouve aux environs de Naples —*Journ. d. Mines, Vol. III, N.° 17, Paris, 1796.*

GIUDICE (DEL) N. — Viaggio medico ad Ischia ed altrove all'oggetto di riconoscere ed analizzare le acque minerali e le stufe. — *Napoli, 1822-25.*

GIUSTINIANI L. — I tre rarissimi opuscoli di Simone Porzio, di Girolamo Borgia, e di Marcantonio delli Falconi, scritti in occasione della celebre eruzione avvenuta in Pozzuoli nell'anno 1538, colle memorie storiche dei sudetti autori. — *Napoli, 1817, in 8°, fol. III, pp. 219.* (C. A.) (B. N.).

GORCEIX II. — Etat du Vésuve et des dégagements gazeux des Champs Phlegréens au mois de Juin 1869. — *Compt. Rend. d. l'Acad. d. Sc. Vol. LXXIV. Paris, 1872.*

GORCEIX. — On the composition of the vapors or gas escaping in the Phlegrean Fields and other places near Vesuvius. — *Am. Journ. of Sc. Ser. III, Vol. IV, 1872, pp. 147.* (C. A.).

GORCEIX II.—Sur les gaz des solfatares des Champs Phlégréens.— *Ann. d. Chim. Vol. XXV. Paris, 1872.*

GOSSELET. — Observations géologiques faites an Italie. — *Lille, 1869, in 8°. pp. 59, pl. VII.* (C. A.). *Compt. Rend. Acad. Sc. T. pp. 417-475, pl. VII.* (C. A.).

GRABLOVITZ G. — Studii marcometrici al Porto d' Ischia. — *An. dell'Ufficio centr. di Meteor. e di Geodinamica, Vol. VIII, pl. IV, Anno 1886, ? 1889, pp. 10, pl. 1.*

GRABLOVITZ G. — Risultati delle osservazioni idrotermiche eseguite al Porto d'Ischia nel 1887. — *Ibid. pp. 12.*

GRABLOVITZ G. — Influenza dello stato orario delle maree sulle sorgive termali del Porto d'Ischia. — *Rend. R. Acc. Lincei, Vol. IV, fasc. 7, 2° semestre, Roma, 1888, pp. 5.*

GRABLOVITZ G. — Risultati delle osservazioni idrotermiche eseguite al Porto d'Ischia nel 1887. — *Ibid. pp. 12.*

GRABLOVITZ G. — Sulle sorgive termali del Porto d'Ischia. — *Ibid. pp, 15,*

GRABLOVITZ G. — Studii preliminari sulle sorgive termali al Porto d'Ischia. — *Ibid. pp. 11.*

GRASSI M. — Relazione storica ed osservazioni sulla eruzione dell'Etna del 1865 e sui terremoti flegrei che la seguirono. — *Catania, 1865.— In French: Bull. Société Géol. d. France. Paris, 1866.*

GUARINI G. — Analisi chimica della sabbia caduta in Napoli la sera del 26 Agosto 1834. — *Atti d. Acc. di Sc. Fis. e Mat. Vol. V. Part. II. Napoli, 1843.*

GUERRA G. — Carta del Cratere esistente tra il Vesuvio e la spiaggia di Cuma. — *Napoli, 1797.* (C. A.).

GUICCIARDINI C. — Mercurius campanus praecipua Campaniae felicis loca indicans et perlustrans.—*Napoli, 1667, in 12°, pp. 274, f. 6.* (C. A.).

GUISCARDI G. — Extrait d'une lettre sur les Etuves de Néron.— *Compt. Rend. Acad. Sc., T. 43, pp. 751-752, 1856.* (C. A.).

GUISCARDI G. — Note sur les émanations gazeuses des Champs Phlégréens. — *Bull. d. l. Soc. Géol. d. France. 2° Série. t. XIV, 1856-57, in 8°, pp. 3.* (C. A.).

GUISCARDI G. — Contribuzioni alla geologia dei Campi Flegrei.—

Rendic. d. R. Acc. d. Sc. Fis. e Mat. Vol. I. Napoli, 1862. — Atti d. R. Acc. d. Sc. Vol. I, Napoli, 1863, pp. 10, pl. 2.

GUISCARDI G. — Sul livello del Mare nel golfo di Pozzuoli. — Rendiconto R. Acc. Sc. Napoli, Fasc. 6° 1865, in 4°, pp. 2. (C. A.).

GUISCARDI G. — Sopra una nuova sorgente d'acqua minerale nella solfatara di Pozzuoli. — Rend. d. R. Acc. d. Sc. Fis. e Mat. Napoli, 1875.

GUISCARDI G. — Il Terremoto d'Ischia del 28 Luglio 1883. — Atti d. R. Accad. d. Sc. Fis. e Mat. di Napoli. Vol. II, 2ª Ser. 1885, in 4°, pp. 8, map. 1. (C. A.).

HAAGEN VON MATHIESEN. — Die Wiederherstellung der Stadt Pozzuoli. — Neues Jahrb. für Miner. etc. Stuttgart, 1816.

HAAGEN VON MATHIESEN. — Ueber die Entstehung des monte Nuovo und die neueste Hecla-Eruption. — Neues Jahrb. für Miner. etc. Vol. VII, Stuttgart, 1847. — Quart. Journ. of the Geolog. Soc. of London. Vol. IV, 1847. — N. Ann. d. Sc. Nat. Vol. VII, Bologna, 1847.

HACKERT F. — La rada di Napoli, 1784. — pl. in fol. mas. (O. V.).

HACKERT F. — Avanzi del tempio di Giove Serapide a Pozzuoli. — 1789, pl. in fol. (O. V.).

HALL B. — On the want of perpendicularity of the standing pillars of the temple of Jupiter Serapis, near Naples. — The Philosoph. Magazine, Vol. VI. London, 1835.

HALLER. — Tableau topographique et historique des isles d'Ischia, Ponza, Ventotene, de Procida et de Nisida, du Cap de Miséne et du Mont Pausilipe par un ultra-montain. — Naples, 1822, in 8°, pp. VIII+216. (C. A.).

HAMILTON W. — Remarque sur la nature du sol de Naples et de ses environs. — Philos. Trans. R. Soc. of London, 1771.

HAMILTON W. — Campi Phlegraei. Observations on the volcanoes of the two Sicilies. — Vol. I, pp. 90, pl. 1, map. 1. Vol. II, pp. 53, pl. 53. Naples, 1776, in folio, with English and French text. (C. A.).

HAMILTON W. — Campi Phlegraci, ou observations sur les volcans des deux Sicilies. — Paris, Lamy, an VII, 2 Vols. in folio with numerous hand coloured plates. (C. A.).

HAMILTON W. — Supplement to the campi Phlegrei being an account of the great eruption of mount Vesuvius in August 1779. — In English and French, Naples, 1779.

HAMILTON W. — Oeuvres complètes, commentées par Giraud-Soulavie. — Paris, 1781.

HAMILTON W. — Neuere Beobachtungen über die Vulkane Italiens und am Rhein, nebst merkwürdigen Bemerkungen des Absts Giraud Soulavii v. G. A. R. — *Frankfurt und Leipzig, 1784 in 8°, pp. XVI+214. map 1.* (C. A.).

HAMILTON W. — Waarneemingen over de Vuurbergen in Italic, Sicilie, en omstreiks den Rhyn als mede over de Aardbeevingen voorgevallen in Italie 1783. — *Amsterdam, 1784, in 8° pp. 552.* (C. A.).

HAMILTON W. — Bericht von gegenwaertigen ins lande des Vesuvs und Beschreibung einer Reise in die Provinz. Abruzzo und nach der Insel Ponza. — *Dresden, 1787.*

HAMILTON W. — Voyages physiques et litologiques dans la Campanie. — *Paris, 1801.*

HOFFMANN F. — Ueber das Albaner Gebirge, den Aetna den Serapis-Tempel von Pozzuoli, und die geognostischen Verhältnisse der Umgegend von Catania. — *Archiv. für Miner. Geogn. Bergbau, und Huttenkunde. III Band. Berlin 1831. The Edinb. New Phil. Journ. Vol. XII, Edinburgh, 1832.*

HOFFMANN F. — Mémoire sur les terrains volcaniques de Naples, de la Sicile, etc. — *Bull. Soc. Géol, T. III, pp. 170-180, 1833* (C. A.).

HÖRNES R. — Aus den phlegraïschen Feldern. — *Wien, 1875.*

HULLMANDEL C. — On the subsidence of the coast near Pozzuoli (1839). — *Proceedings of the Geol. Soc. of London, Vol. III. London, 1842.*

IVANOFF. — Chemische Untersuchung des in Neapel gebraüchlichen Formsandes. — *Oesterreichische Zeitsch. für Berg und Hütten Wesen I, Band. Wien, 1853.*

JAMES C. — Untersuchungen über die Ammonium grotte bei Neapel. (Traduit de la Gazz. Méd. de Paris 1843.) — *Notiz. aus dem Gebiete der Natur. und Heilkunde. XXXVIII Band. Erfurt und Weimar, 1843.*

JANUARIO F. DE. — Felicis Campaniae hilaritas tumulata. — *Napoli, 1632, fol. vol.* (C. A.).

JANUARIO R. — La solfatara di Pozzuoli. — *Annuario Meteor. It., Anno IV, Torino, 1889.*

JATTA G. — Discorso sulla ripartizione Civile, e Chiesastica dell'antico agro Cumano, Misenese, Bajano, e Pozzuolano, sui famosi Campi Flegrei, sul Promontorio di Miseno, sul Monte di Procida, e sul luogo, ove secondo Virgilio fu sepolto Miseno trombettiere di Enea, sulle acque della Bolla, e sull'antico acquedotto che da Serino conduceva l'acqua in Napoli.— *Napoli, 1843, in 8°.*

JASOLINO G. DE. — Rimedii naturali dell'Ischia. — *Napoli*, *1588*, *in 4°*, *pp. XIX + 138, pl. 1. 2nd edit. 1589.*

JERVIS G. — Tesori sotterranei dell' Italia. — *4 Vols. in 8°, Torino, 1874-1888, numerous plates.*

JOHNSTON-LAVIS H. J. — The Earthquake in Ischia. — *"Nature" Vol. XXIII, 1881, p. 497.*

JOHNSTON-LAVIS H. J. — Etude sur l'emplacement des nouvelles villes à l'ile d'Ischia. — *"L'Italie" (Rome) Sept. 15th, 1883.*

JOHNSTON-LAVIS H. J. — Il parere d'uno scienziato. — *"Il Piccolo" (Naples) Sept. 2nd, 1883.*

JOHNSTON-LAVIS H. J. — Le costruzioni a Casamicciola. — *" Il Piccolo" (Naples) Sept. 20th, 1883.*

JOHNSTON-LAVIS H. J. — Notice of the Earthquake of Ischia of March 4th, 1881. — *Reports Brit. Assoc. Advancement of Science, 1883.*

JOHNSTON-LAVIS H. J. — Notice of the Earthquake of Ischia of July 28th, 1883. — *Brit. Assoc. Reports, 1883.*

JOHNSTON-LAVIS H. J. — Notices on the Earthquake of Ischia of 1881 and 1883 with a Map of the Isoseismal — *in 8°, pp. 56, with 1 map. Naples, 1883.*

JOHNSTON-LAVIS H. J. — Observations scientifiques sur le tremblement de terre. — *L'Italie (Rome) Dec. 12th, 1883, The Times (London) Dec. ? 1883.*

JOHNSTON-LAVIS H. J. — Prévision de futures catastrophes dans l'ile d'Ischia. — *"L'Italie" (Rome) Sept. 2nd, 1883.*

JOHNSTON-LAVIS H. J. — Rapport préliminaire sur le tremblement de terre du 28 Juillet 1883 à l'ile d'Ischia. — *" L' Italie , " Sept. 22nd, 1883.*

JOHNSTON-LAVIS H. J. — The Disaster in Ischia. — *"Indianapolis Journal." Sept. 6th, 1883.*

JOHNSTON-LAVIS H. J. — The Disaster in Ischia. — *"Nature" Vol. XXVIII, pp. 346-347.*

JOHNSTON-LAVIS H. J. — The Ischian Earthquake of July 28th 1883. — *" Nature," Vol. XXVIII, 1883, pp. 437-439, with a map.*

JOHNSTON-LAVIS H. J. — Una risposta al Prof. Palmieri. — *" Il Piccolo" (Naples) Sept. 8th, 1883.*

JOHNSTON-LAVIS H. J. — Brevi considerazioni intorno alla relazione del professore L. Palmieri sul terremoto dell'Isola d'Ischia. — *"Il Piccolo" (Naples) March 31st and April 1st 1884.*

JOHNSTON-LAVIS H. J. — Monograph of the Earthquakes of Ischia, a Memoir Dealing with the Seismic Disturbances in that Island from Remotest Times, with Special Observations on those of 1881 and 1883. — *Dulau, London; and Furchhelm, Naples, 1885, in royal 4°, pp. X and 112, with 20 photo-engravings, 2 large maps in color, 3 lithographic plates and 1 chromo-lithographic plate.*

JOHNSTON-LAVIS H. J. — The physical Conditions involved in the injection, extrusion and cooling of Igneous Matter. — *Quart. Journ. Geol. Soc. Lond. Vol. XLI. 1885, pp. 103-106.*

JOHNSTON-LAVIS H. J. — On the Fragmentary ejectamenta of Volcanoes. — *Proceed. Geol. Assoc. Lond. Vol. IX, 1886, pp. 121-132, fig. 3.*

JOHNSTON-LAVIS H. J. — Second Report of the Committee for the Investigation of the Volcanic phenomena of Vesuvius and its neighbourhood. — *Brit. Assoc. Reports, 1886, pp. 3, also "Nature", Vol. XXXIV, 1886, p. 481.*

JOHNSTON-LAVIS H. J.— The relationship of the structure of Igneous Rocks to the conditions of their formation. — *Scientif. Proceed. R. Dublin Soc. Vol. V, N. S., 1886, pp. 112-156.*

JOHNSTON-LAVIS H. J.—Third report of the Committee appointed for the Investigation of the Volcanic phenomena of Vesuvius and its neighbourhood. — *Brit. Assoc. Reports, 1887, pp. 3.*

JOHNSTON-LAVIS H. J. — Fourth report Committee for the Investigation of the Volcanic phenomena of Vesuvius and its neighbourhood. — *Brit. Assoc. Reports, 1889, pp. 7.*

JOHNSTON-LAVIS H. J. — Nuove osservazioni fatte in Napoli e dintorni. — *(Boll. Com. Geol. 11-12). Roma, 1888.*

JOHNSTON-LAVIS H. J. — Fifth Report of the Committee Appointed for the investigation of the Volcanic phenomena of Vesuvius and its neighbourhood. — *Brit. Ass. Reports, 1889, pp. 12 with 5 woodcuts.*

JOHNSTON-LAVIS H. J. — On a remarkable Soladite trachyte lately discovered in Naples, Italy. — *Geol. Mag., Dec. III., Vol. VI., 1889, N. 2. pp. 74-77.*

JOHNSTON-LAVIS H. J. — Viaggio scientifico alle regioni vulcaniche italiane nella ricorrenza del centenario del "Viaggio alle due Sicilie" di Lazzaro Spallanzani. — *(This is the programme of the excursion of the English Geologists that visited the south Italian volcanoes under the direction of the author. It is here included as it contains various new and unpublished observations) Naples, 1889, in 8°, pp. 1-10.*

JOHNSTON-LAVIS H. J.—Nuove osservazioni geologiche in Napoli e suoi dintorni. — *Boll. R. Com. Geol. It. Vol. XXI, N.° 1 and 2, pp. 18-27, fig. 1, pp. 65-68. (A curious error has been made by the translator who seems to have been affec-*

led by a mental Daltonism. (On page 67, line 3, 6 and 9 for "verdi" "read rossi" i. e. for "green" read "red.")

JOHNSTON-LAVIS H. J. — Osservazioni geologiche sulle isole Ventotene e Santo Stefano (Gruppo delle Isole Ponza). — *Boll. R. Com. Geol., Vol. XXI, N.° 1 e 2, pp. 60-64.*

JOHNSTON-LAVIS H. J. — The Ponza Islands. — *Geol. Mag., Dec. III, Vol: VI, 1889, pp. 529-535 with 3 woodcuts.*

JOHNSTON-LAVIS H. J. — Osservazioni geologiche lungo il tracciato del Grande Emissario Fognone di Napoli dalla Pietra sino a Pozzuoli. Relazione alla Società Napoletana degli Ingegneri Costruttori di Napoli. — *Boll. R. Com. Geol., Vol. XXI, 1890, N.° 1 e 2., pp. 18-27 with 1 woodcut.*

JOHNSTON-LAVIS H. J. — Sixth Report of the Committee Appointed for the Investigation of Vesuvius and its neighbourhood. — *Brit. Assoc. Reports, Leeds Meeting, London, 1890.*

JONES E. W. S. — The earthquake at Casamicciola, July 28th 1883. — *Naples, 1883, in 8°, pp. 143. (C. A.).*

JORIO (DE) A. — Ricerche sul tempio di Serapide in Pozzuoli. — *Napoli, 1820, in 4.°, pp. 68, pl. 3.*

JORIO (DE) A. — Guida di Pozzuoli e Contorni. — *Napoli, 1817, in 8°, pp. VII-151. Topogr. map.*

JORIO A. DE. — Pozzuoli und desse Umgenbungen aus dem Italienischen. — *Zurich, 1830, in 8', pp. 100.*

JUDD W. J. — Contribution to the Study of Volcanoes. — *Geol. Mag. Vol. II, N. 133. London, 1876.*

KADEN W. — Die Inseln Ischia in Natur Sitten. und Geschichts Bildern aus Vergangenheit und Gegenwart. — *Lutzen, Prell. no date. In 8°, pp. 115, pl. 4, map. 1 (C. A.).*

KALKOWSKY E. — Der Leucitporhyr vom Averner See. — *Neues Jahrb, für Miner. etc. Stuttgart, 1878.*

KALKOWSKY E. — Ueber den Piperno. — *Zeitsch d. Deutsche Geol. Gesell. XXX Band, Berlin, 1878.*

KALKOWSKY E. — Notice of Mercalli (sulla natura del terremoto Ischiano, 1883). — *New. Jahrb. f. Mineral, etc. 1886, I, pp. 258-259.*

KOSMAN B. A. — De nonullis lavis Averniacis dissertatio inauguratis mineralogica-chimica. — *Statis Saxoniae, 1864, in 8°, (B. N.).*

KRESSNER. — Geographisch-oragraphische Uebersicht über das Vulcanische terrain im Neapolitanischen.—*Berg-und-hüttenmänn. Zeitung, XXII Band, Nordhausen und Leipzig, 1863.*

LALANDE M. DE. — Voyage en Italie, contenant l'histoire, et les anecdotes les plus singulières de l'Italie, etc., etc. — *Genève, 1790, Vols. 7, in 8°, with maps in 4.° pl. XXXV, Vol. VI, pp. 18-39. Solfatara di Pozzuoli.*

LANCELLOTTI F. — Memoria sull'analisi e sintesi dell'acqua solfurea di Napoli. — *Atti d. Soc. Pontan. Vol. II, Napoli, 1812.*

LANZETTA A. — Risposta alla 1ª. di un anonimo sulle osservazioni di un fenomeno avvenuto nel lago di Patria. — *Napoli, 1796.*

LA PIRA. — Memoria sull'origine, analisi ed uso medico delle acque minerali di Terra di Lavoro. — *Caserta, 1820.*

LASAULX VON. — Das Erdbeben von Casamicciola auf Ischia (4 marzo 1881). — *"Humboldt," Stuttgart, Jun 1882, in 4°, pp. 5 (C. A.).*

LAURENTIIS M. DE. — Universae Campaniae Felicis Antiquitates.— *Neapolis, 1826, vols. 2 , in 4°, Parte II, pp. 42 and foll. Monte Nuovo etc.*

LEBERT H. — Le Golfe de Naples et ses volcans et les volcans en général. — *Vevey, Lausanne, etc. 1876, in 8°, pp. 120, pl. 1. (C. A.).*

LECANN L. R. — 1827. — *See Blachet.*

LE RICHE I. — Antiquités des environs de Naples et dissertation qui y sont rélatives par M. J. L. R. — *Naples, 1820, in 8°. fol. 2, pp. 392, maps 3, fol. 3 (C. A.).*

LIBELLUS. — De mirabilibus civitatis Puteolorum et locorum vicinorum : ac de nominibus virtutibusq. balneorum ibidem existentium. — *1507, in 4°, fol. 32. Another edition, Naples, 1475, in 4°, fol. 37 (C. A.).*

LICOPOLI G. — Su d'un pezzo di Legno rinvenuto nel tufo vulcanico appresso Napoli. — *Rend. R. Accad. Sc. Fis. Mat. An. XIII, Napoli, 1874, pp. 141-143.*

LIPPI C. — Fu il fuoco o l' acqua che sotterrò Pompei ed Ercolano?. — *Napoli, 1816.*

LOBLEY L. J. — Mount Vesuvius. A description, Historical and Geological account of the Volcano and its Surroundings. — *London, Roper and Drowley, 1889, in 8°, pp. 385, pl. 20*

LOFFREDI F. — Antiquitas Puteolorum, cum balneorum Agnani, Puteolorum, et Tripercolarum descriptionibus, etc., etc.—*Lug. Bat., in fol. pp. 28.*

LOFFREDO F. — Le antichità di Pozzuoli e luoghi convicini nuovamente raccolte. — *Napoli. 1580, in 12°, fol. 24 (C. A.). Another edit. Napoli, 1675, in 4°, pp. 4+38.*

LOMBARDI A. — Cenno sul tremuoto avvenuto in Tito, ed in altri luoghi della Basilicata il dì 1 Febbraio 1824.—*Potenza 1829.*

LOMBARDUS J. F. — De balneis aliisque miraculis Puteolanis. — *Veneliis, 1566.*

LOMBARDUS J. F. — Synopsis Autorum omnium qui hactanus de balneis, aliisque miraculis Puteolanis scripserunt. — *Napolis, 1557, fol. 124. (C. A.).*

LUCA (DE) S. — Studii fisico-geografici sulla regione da Baia a Castellamare. — *Napoli, 1865.*

LUCA (DE) S. — Osservazioni sulla composizione dell' acqua termale della Solfatara di Pozzuoli. — *Rend. d. R. Acc. d. Sc. Fis. e Mat. Napoli, 1868.*

LUCA (DE) S. — Osservazioni sulla temperatura interna della grande fumarola della Solfatara di Pozzuoli, — *Napoli, 1869.*

LUCA (DE) S. — Ricerche chimiche e terapeutiche sull'acqua termo-minerale della Solfatara di Pozzuoli, — *Rend. d. Acc. d. Sc. Fis. e Mat. Vol. VIII, Napoli, 1869. Journ. de Chim. méd. de Pharm. etc. Vol. VI, Paris, 1870.*

LUCA (DE) S. — Analyse de l'eau thermo-minérale de Pouzzoles.— *Journ. de Pharm. et de Chimie, Vol. XII, Paris, 1870. — Compt. rend. d. l'Acad. d. Sc. Paris, 1870.*

LUCA (DE) S. — Ricerche chimiche sopra una produzione stalatitica della Solfatara di Pozzuoli. — *Rend. d. R. Acc. d. Sc. Fis. e Mat. Vol. X, Napoli, 1871.—Gazz. Chim. ital. Vol. II. Palermo, 1872. — Riv. Scient. Industr. Firenze 1873. Compt. Rend. d. l'Acad. Sc. Vol. LXXVI. Paris, 1873.*

LUCA (DE) S. — Ricerche chimiche sull'allume ricavato dall'acqua termo-minerale della Solfatara di Pozzuoli. — *Napoli, 1871.*

LUCA (DE) S. — Recherches chimiques sur un alun complexe , obtenu de l' eau thermo-minérale de la Solphatare de Pouzzoles. — *Compt. Rend. d. l'Acad. d. Sc. Vol. LXXIV, Paris, 1872.*

LUCA (DE) S. — Sulla composizione dei gaz che svolgonsi dalle fumarole della Solfatara di Pozzuoli. — *Rendic. d. R. Acc. d. Sc. Fis. e Mat. Vol. X, Napoli 1871. Compt. Rend. d. l'Acad. d. Sc. Vol. LXXIV, Paris, 1872. Ann. d. Chim. Vol. XXVI, Paris, 1872. Gazz. Chim. ital. Vol. II. Palermo, 1872.*

LUCA (DE) S. — Ricerche analitiche sopra quattro diverse terre della Solfatara di Pozzuoli. — *Napoli, 1873.*

LUCA (DE) S. — Ricerche analitiche sopra talune produzioni stalammitiche della Solfatara di Pozzuoli. — *Napoli, 1874.*

LUCA (DE) S. — Ricerche chimiche sopra una sostanza legnosa trovata nel tufo vulcanico. — *Napoli, 1874.*

LUCA (DE) S. — Ricerche sperimentali sulla Solfatara di Pozzuoli.—*Rend. d. R. Acc. d. Sc. Fis. e Mal. Napoli, 1874, pp. 104, pl. V.*

LUCA (DE) S. — Sopra una nuova sorgente di acqua termo-minerale, scoperta nella Solfatara di Pozzuoli. — *Rend. d. R. Acc. d. Sc. Fis. e Mal. Napoli, 1875.*

LUCA (DE) S. — Sulla presenza del litio nelle terre e nelle acque della Solfatara di Pozzuoli.—*Rend. d. R. Acc. d. Sc. Fis. e Mal. Napoli, 1875. Riv. Scient. Indust. Firenze, 1875.*

LUCA (DE) S. — Recherches sur la présence du Lithium dans les terres et dans les eaux thermales de la Solfatare de Pouzzole. — *Compt. Rend. d. l'Acad. d. Sc. Vol. LXXXVII, N.° 4. Paris, 1878.*

LUCA (DE) S. — Sulle variazioni di livello dell'acqua termo-minerale nel pozzo della Solfatara di Pozzuoli.—*Rend. d. R. Acc. d. Sc. Fis. e Mal. Ann. XVII. Napoli, 1878.*

LUCA (DE) S. — Sul livello dell'acqua termo-minerale della solfatara di Pozzuoli. — *Napoli, 1878.*

LYELL C. —On the successive changes of the temple of Serapis.— *Notice of the Proceed. at. t. meet. of. t. memb. of the R. Institution Vol. II, London, 1854. Also Amer. Journ. Science, 2nd Ser. Vol. XII, pp. 126-129.*

MACAGNO E BERTONI. — Analisi della terra della Solfatara di Pozzuoli. —*R. Staz. Enol. Sperim. Ann. II, Asti, 1874.*

MACINTOSH C.—On the tides at Naples. — *Quart. Journ. of the Geolog. Soc. of London, Vol. IV, London, 1848.*

MACINTOSH C. — The temple of Serapis at Pozzuoli. — *Atheneum, Page 801. 1848.*

MADIA. — 1853. — *See Bonghi.*

MAFFEI G. C. — Scala naturale, ovvero fantasia dolcissima intorno alle cose occulte e desiderate nella Filosofia. — *Vinegia, 1573, in 8°, fol. 140. Nel secondo grado Cap. I: Cagion perchè in Pozzuolo sono bagni, etc. etc.*

MAJO. — Trattato delle acque acidule che sono nella città di Castellammare di Stabia. — *Napoli, 1754.*

MARANTA B.—De aquae Neapoli, in Luculliano scaturientis (quam ferream vocant) metallica materia, ac viribus epistola — *Neapoli, 1559, in 4. Parch. Very rare.*

MARCELLO M. — Ischia, canti 3. — *Milano 1863.*

MARCHESINO F. — Copia di una lettera di Napoli che contiene li stupendi e gran prodigii apparsi sopra a Pozzolo. — *Napoli, 1538, fol. 4 and engraved frontispiece (C. A.).*

MARCHINAE M. — Virginis Neapolitanae Musa posthuma. — *Nea-*

16

poli, 1701, in 12°, pp. 12-[-111. De incendio Montis Ve-suvii Ode p. 118.

MARIENI L. — 1802. — *See Scacchi.*

MARONE V. — Memoria contenente un breve ragguaglio dell'Isola d'Ischia, — *Napoli, 1847.*

MASELLA E. — Poesie latine istoriche colle note in italiano. -- *Napoli, 1795, in 4°, pp. 74. (Monte Nuovo, 29 Sept. 1538, and Solfatara).*

MASINO DI CALVELLO M. A. — Distinta relatione dell'incendio del sevo Vesuvio alli 16 di Decembre 1631, successo, con la re-latione della città di Pozzuoli, e cause delli terremoti, al tempo di D. Pedro de Toledo Vicerè in questo Regno nel-l'anno 1534 (1538). — *Napoli, 1632, in 4°, pp. 36.* (C. A.).

MAZZELLAE Sc. — Situs et antiquitas Puteolorum, locorumque vicinorum, etc. etc. — *Lugduni Batavorum, in fol. pp. 4+ 92+6. (Earthquakes in Pozzuoli of 1198, 1456 and 1538.*

MAZZELLAE Sc. — Urbium Puteolorum, et Cumarum descriptio, etc. etc.—*Lugduni Batavorum, in fol. pp. 4-[-20+2 Topogr. map.*

MAZZELLA S. — Descrizione di Napoli. — *Napoli, 1586.*

MAZZELLA S. — Sito et antichità della città di Pozzuolo e del suo amenissimo distretto. — *Napoli, 1595, in 8°, pp. 291, figs, map. 1.*

MAZZELLA S. — Opusculum de Balneis Puteolorum Bajarum et Pethecusarum. — *Neapolis, 1593, pp. 43.*

MEDNYANSZY (VON) D.—Beobachtungen in geologischer Beziehung auf einer Reise durch Italien bis Neapel. — *Verhandl. d. Ver. f. Naturk. z. Pressburg, B. IV, 1859.*

MELLONI M. — 1840-42. — *See Piria.*

MÈNE C. ET ROCCATAGLIATA. — Analyses de quelques eaux des sources thermales d'Ischia près Naples. — *Compt. Rend. d. l'Acad. d. Sc. Vol. LXVI, Paris, 1868.*

MERCALLI G. — Sulla natura del terremoto ischiano del 28 Luglio 1883. — *Rendic. d. R. Ist. Lomb. Ser. II, Vol. XVII, fasc. XIX, Milano, 1884, pp. 15,* (C. A.).

MERCALLI G. — L' isola d' Ischia ed il terremoto del 28 luglio 1883. — *Atti R. Ist. Lombardo?, Milano, 1884, in 4°, pp. 55, pl. 2, map. col. 1.*

MEURICOFFRE O. AND FIESCHI-RAVASCHIERI Dssa.—La Carità nell'isola d'Ischia. — *Napoli, 1883, in 12°, pp. 175. (Also a French edition without tables of account).*

MIGLIETTA. — Rapporti sull'uso medicinale delle acque minerali

del Tempio di Serapide in Pozzuoli. — *Napoli, 1818, in 4°,*
pp. 88. (C. A.).

MIGLIETTA. — Acqua minerale del Tempio di Serapide in Pozzuoli. — *Giorn. Arc. d. Sc. etc. Vol. VII. Roma, 1820.*

MISSON M. — Nouveau voyage d'Italie. — *La Haye, 1694 , in 12°, fig.*

M. J. L. R. (LE RICHE). — Antiquités des environs de Naples , et dissertations qui y sont relatives. — *Naples, 1820, in 8°,* (B. N.).

MONTICELLI T. — Recherches sur le territoire de Pozzuoles et des Champs Phlégréens. — *Annales des Mines , etc. Paris, 1816-1826, 30 vols. 1817. (Vol. V. 2ᵉ. série, pp. 293-294).*

MONTICELLI T. — Commentarium. In Agrum Puteolanum camposque phlegraeous Commentarium. — *Napoli, 1826, in 4°, fol. XIV, pp. 25,* (C. A.).

MONTICELLI T. — Opere. - *Napoli 1841-43, Vol. 3 , in 4°, fig. 1, in two vol. (vol. 1°, fol. 7, pp. 295, pl. 2.; vol. 2°. pp. 335 pl. 2* (O. V.).

MORGHEN F. — Pianta del cratere tra Napoli e Cuma , incisa da Filippo Morghen. — ?, (C. A.).

MORO G. — La grotta del Circeo e il tempio di Serapide in Pozzuoli. — *Ateneo Veneto. S. XIII, Vol. II, 1-2-3. Venezia, 1889.*

MALLET R. — The great Neapolitan Earthquake of 1857. — *Il Vols. in 8.°, London , 1862, numerous plates and figures.*

MALLET R. — On some of the conditions influencing the projection of discrete solid materials from Volcanoes, and on the mode in which Pompeii was overwhelmed.—*Jour. Geol. Soc. Ireland, Vol. IV, pt. III, Dublin, 1876.*

MAJONE. — Breve descrizione della Real Città di Somma. — *Napoli, 1702.*

MORGHEN F. — Veduta del Foro di Vulcano denominato la Solfatara. — *? 1 pl. in fol.* (O. V.).

MORMILE G. — Descrizione della Città di Napoli e del suo amenissimo distretto e dell'antichità della città di Pozzuoli. — *Napoli, 1617, and 1625, in 8°, pp. 248, pl. 3, figs. 3ʳᵈ edit. 1670, pp. 251, pl. figs.* (C. A.).

MORMILE G. — Nuovo discorso intorno all'antichità di Napoli, e Pozzuoli. — *Napoli, 1629, in 8°, pp. 69.*

MUELLER A. — Ueber das Vorkommen von reinem Chlorkalium am Vesuv. — *Verhand. d. Nat. Ges. in Basel, 1ᵉˢ, helf 1854, pp. 113-119* (C. A.).

MUNICIPIO DI NAPOLI. — Relazione della Giunta al Consiglio sui provvedimenti adottati per la eruzione del Vesuvio 1872 ed. atti relativi. — *Napoli, 1872, in 4°, pp. 31.* (C. A.).

MÜNTER T. L. — Parerga historico-philologica. De Herculaneo. XV. Vesuvii montis descriptio. — *Gottingae, 1749, in 8°, fol. 4, pp. 128, pl. 1,* (C. A.).

MUSAIO, — 1878. — *See Ferrero.*

NAPOLI-ISCHIA. — Numero unico, pubblicato a beneficio dei danneggiati di Casamicciola e Lacco Ameno dagli studenti della Facoltà di Lettere e Filosofia di Napoli. — *Napoli, 1881, pp. 21.* (C. A.).

NAPOLI-CASAMICCIOLA (NEWSPAPER, 1 NUMBER). — Napoli ai danneggiati di Casamicciola. — *Napoli, 1881, pp. 6,* (C. A.).

NERO E. DEL. — Lettera a Niccolò del Benino sul terremoto di Pozzuoli, dal quale ebbe origine la montagna nuova nel 1538.— *From pp. 93 to 96 of Vol. IX, Archivio Storico Italiano. Firenze, 1846, in 8°,* (C. A.).

NICCOLINI A. — Rapporto sulle acque che invadono il pavimento dell'antico edifizio detto il tempio di Giove Serapide. — *Napoli, 1829, in 4°, pp. 7 + 46, pl. I.*

NICCOLINI A. — Tavola metrica-cronologica delle varie altezze tracciate dalla superficie del mare fra la costa di Amalfi ed il promontorio di Gaeta nel corso di diciannove secoli. — *Napoli, 1839, in 4°, pp. 52,* (C. A.).

NICCOLINI A. — Descrizione della Gran-Terma Puteolana volgarmente detta Tempio di Serapide. — *Napoli, 1846, in 4°, pp. 95, numerous col. and uncol. pl., maps, etc.*

NICOLUCCI G. — Analisi microscopica della pretesa muccilagine che si forma sulle acque termo-minerali del Tambura di Senogalla e della Rete nell'Isola d'Ischia. — *Rend. d. R. Acc. d. Sc. Fis. e Mat. Vol. I. Napoli, 1842.*

NISCO N. — Lettere sull'Isola d'Ischia ed una su Portici cont. in 6 numeri di « Il Diritto ». — *1869.*

NIXON J. — An account of the Temple of Serapis at Pozzuoli in the Kingdom of Naples. — *Phil. Transact. of the R. Soc. of London, Vol. X, 1757, pp. 166-174, pl. 1.*

NOVI G. — Calcarea con *Cardium* contenuta nel tufo di Posilipo?

O' REILLY. — The earthquake of Ischia, July, 28, 1883. — *Nature, 1883, Vol. XVIII, pp. 461.*

ORLICH (VON) L. — Die Insel Ischia. — *Zeitschr. für allgem. Erdkund. Bd. II, Berlin, 1854, pp. 388-416.*

PACICHELLI G. B. — Memorie de' Viaggi per l' Europa Christiana scritte a diversi in occasione de' suoi Ministeri. — *Napoli, 1685, Vols. 5, in 12°. Parte I, pp. 40+743+53. Parte II,*

8+827+40. Parte III, pp. 8+761+27. Parte IV, vol. I, pp. 4+541+20. Parte V, vol. II, pp. 4+438 † 18. (See Parte IV, vol. I, pp. 196 and follow. Della Solfatara.)

PALATINO L. — Storia di Pozzuoli e contorni con breve trattato istorico di Ercolano, Pompei, Stabia e Pesto. — *Napoli, 1826, in 8°, pp. 336, pl. I, maps 2.* (C. A.).

PALLOTTA (DOTT.) G. — Brevi cenni sulla uniformità delle Terme di Casamicciola animate dall'unica acqua di Gurgitello. — *Napoli, 1873.*

PALMERI P. — Le terme del pio monte della Misericordia in Casamicciola (Ischia) dopo il terremoto del 4 maggio 1881. — *Napoli, 1881, in 4°, pp. 11.* (C. A.).

PALMERI P. — Ricerche storiche sul nome e sul luogo e confronti delle analisi delle acque di Gurgitello. — *Napoli, 1879.*

PALMERI P. — Il pozzo artesiano dell'Arenaccia del 1880 confrontato con quello del Palazzo Reale di Napoli del 1847. — *Lo Spettatore del Vesuvio e dei Campi Flegrei. Nuova Serie Vol. 1°. Napoli, 1887, pp. 53-58, pl. I.* (C. A.).

PALMERI P. E COPPOLA M. — Acque minerali del Pio monte della Misericordia in Casamicciola (Ischia). Analisi chimiche delle acque, delle concrezioni e dell'atmosfera delle stufe. — *Napoli, 1875-76.*

PALMIERI L. — Sulle scosse di terremoto avvertite in Napoli il di 24 Giugno 1870. — *Rend. d. R. Acc. d. Sc. fis. e mat. Vol. IX. Napoli, 1870.*

PALMIERI L. — Il Litio scoperto dal Prof. S. De Luca nelle terre della Solfatara, riveduto collo spettroscopio. — *Rend. d. R. Acc. d. Sc. Fis. e Mat. Napoli, 1875.*

PALMIERI L. — Intorno ad un piccolo terremoto accaduto in Napoli il 18 Febbraio. — *Rend. d. R. Acc. d. Sc. Fis. e Mat. Napoli, 1876.*

PALMIERI L. — Sul terremoto di Casamicciola del 4 Marzo 1881.— *Rend. Acc. Sc. Fis. Mat. Napoli, Fasc. 4°, 1881, pp. 8.*

PALMIERI L. — Osservazioni simultanee sul dinamismo del cratere vesuviano e della grande fumarola della Solfatara di Pozzuoli fatte negli anni 1888-89-90. — *Rend. R. Acc. Sc. Fis. Mat. Napoli, 1890, pp. 3.*

PALMIERI L. AND OGLIALORO A. — Sul terremoto dell'isola d'Ischia della sera del 28 Luglio, 1883. — *Napoli, 1884, in 4°, pp. 28, map. 1.*

PANVINI P. — Il forestiere alle antichità e curiosità naturali di Pozzuoli, Cuma, Baja e Miseno in tre giornate, — *Napoli, 1881, in 8°, pp. VIII+156, pl. 53.* (C. A.).

PAOLI (PADRE). — Le antichità di Pozzuoli. — *Napoli, 1768.*

PARENTE M. — Parthenope terraemotu vexata Magnam Matrem publicae securitatis sospitem diligit, et ejusdem dolorum cultui se addicit. Carmen. — *Neapoli, 1830, in 4°, pp. 11.*

PARRINO D. A. — Moderna distintissima descrizione di Napoli, Città nobilissima, antica e fedelissima, e del suo Seno Cratere. — *Aggiunte, osservazioni e correzioni a questo primo tomo della nuova descrizione di Napoli: Napoli, 1703-1704, vol. 2°, in 12°, (Vol. I, pp. 20+438+54+46+2. Vol. II, pp. 16+292+23, XXVIII. pl.*

PARRINO D. A. — Nuova guida dei Forestieri per osservare e godere le curiosità più vaghe della fedelissima gran Napoli. — *Napoli, 1725, in 12°, fol. 18, pp. 382, maps 31, pl. 40,* (C.A.). *Another edition 1751, in 12°, fol. 2., pp. 269, pl. 30, maps 9.* (C. A.).

Parthenope terraemotu vixata, etc. — *See Miscellanea Poetica.* (C. A.).

PASCA W. — Intorno ad una pozzolana rinvenuta presso Itri e delle pozzolane in generale. — *Napoli, Tip. Bonis.*

PASINI L. — Sul pozzo artesiano di Napoli. — *Atti d. I. R. Istit. Ven. di Sc. Lett. e Arti. Venezia, 1825-26.*

PELLEGRINO C. — Discorsi sulla Campania Felice. — *Napoli, 1631, in 4°, fol. 56, pp. 780, pl. 1.* (C. A.).

PER ISCHIA. — « Corriere del Mattino » *of Naples, Loose sheets. pp. 10.* (C. A.).

PETRUCCELLI F. AND PACI G. M. — Memoria chimico-medica su l'acqua termo-minerale del Bagnolo nelle vicinanze di Napoli. — *Napoli, 1832, in 8°, pp. 18.* (C. A.).

PETTERUTI G. — La solfatara di Pozzuoli. — *L' Idrologia medica Ann. II, N. 9, 10 e 11. — Il Golfo di Napoli, Idrol. Med. N. 40. Bassano, 1880.*

PFLAUMERN J. H. A. — Mercurius Italicus hospiti fidus per Italiae et urbes, etc. etc. — *Augustae Vindelicorum, 1625, in 8°, pp. 32+484+2. From pp. 371 and foll. Campi Flegrei.*

PHILIPPI R. A. — Ueber die sub fossilen Lebthier-Reste von Pozzuoli bei Neapel und auf der Insel Ischia. — *Neues Jahrbuch für Miner. Geogn. Geol. etc. Bd. V, Stuttgart, 1837.*

PIGNATELLO A. — Rime date nuovamente alle stampe, e dedicate al sig. Principe di S. Severo. — *Napoli e Gallipoli, 1593, in 4°, pp. 8+94. Sonetto C. alla Solfatara di Pozzuoli, pp. 69.*

PILLA. N. — 1.° e 2.° viaggio geologico per la Campania. — *Napoli, 1814, in 8°.*

PILLA N. — Geologia vulcanica della Campania. — *Napoli, 1823, Vol. II, in 8°. Vol. I, pp. XIX+125, Vol. II, pp. 160.*

PILLA L. — Osservazioni gegnostiche sulla parte settentrionale ed orientale della Campania. — *Ann. Cir. d. R. Due Sicilie. Vol. III, pp. 117-147, Napoli, 1833.*

PILLA L. — Nota sulla questione del Serapeo toccata dal Cav. Tenore. — *Il Progr. d. Sc. etc. Vol XIX, Napoli, 1838.*

PIRIA R. — Sull'azione che alcuni corpi riscaldati esercitano sui vapori che si sviluppano da' fumaioli della Solfatara. — *R. Acc. d. Sc. Napoli, 18 Agosto 1840? pp. 9. (C. A.).*

PIRIA R. ET MELLONI M.—Recherches sur les fumeroles, les solfatares, etc. Compt. Rend. d. l'Acad. d. Sc. Vol. XI. Paris, 1840. — *Ann. d. Chim. Vol. LXXIV, Paris, 1840. — Journ. für Prakt. Chem. Bd. XXII, Leipzig, 1841. — Notiz. aus dem Gebiete der Natur-und Heilkunde, Bd. XVI, Erfurt und Weimar, 1840. — Ann. der Physik. und Chem. Bd. LI, Leipzig, 1842.*

POLI G. S. — Memoria sul tremuoto de' 26 Luglio del corrente anno 1805. — *With 3 fol. Napoli 1806.*

PONZI G. — Osservazioni geologiche fatte lungo la Valle Latina da Roma a Monte Cassino. — *Atti d. Acc. Pont. d. N. Lincei, Vol. I, Roma, 1847-48. — Corrisp. Scient. in Roma per l'avanz. d. Sc. Vol. II, 1853. — N. ann. d. Sc. Nat. Vol. X, 1848.*

PONZI G. — Carta geologica della Valle latina da Roma a Monte Cassino. — *Roma, 1850.*

PORTII S. — *See following.*

PORZIO S. — De conflagratione Agri Puteolani. — *Napoli, 1538, in 8°, pp. 8, frontisp (The only known copy of this edit. is in C. A.). Florentiae, 1551, in 4°, pp. 8. (B. N.) See also Giustiniani L.*

PREVOST C. — Voyage à l'ile Julia, à Malte, en Sicile, aux îles Lipari et dans les environs de Naples. — *Compt. Rend. d. l'Acad. d. Sc. Vol II, Paris, 1836.*

PROCTOR R. A. — Le Vésuve et Ischia. — *Revue Mens. d'Astron. Pop. — Paris, Sept. 1883, pp. 349-343. (C. A.).*

QUINTIIS C. E. — Inarium seu de balneis Pithecusarum; libri sex. *Neapoli, 1726, in 8', fol. 32, pp. 320, pl. 7, frontisp. (C. A.).*

RANIERI A. — Documenti storico-geologici sulle antichità delle acque termali e sulle arene scottanti del littorale dei Maronti nell'isola d'Ischia. — *Napoli, 1871, in 4°, pp. 59.*

RATH (VOM) G. — Mineralogische geognosticae fragmente aus Italien. V. Monte di Cuma, Ischia und Ebene. — *Zeilschr. der Deuts. geol. Gesell. Bd. XVIII, Berlin, 1866.*

RATH (VOM) G. — Tridymit im Neapolitanischen Vulcan Gebiete.— *Ann. d. Phys. und ch. Bd. CXLVII, Leipzig, 1873.* -

RENZI (DE). — 1863. -- *See Bonghi.*

REZRADORE P. — I disastri d'Ischia e di Giava. — *Roma, 1883, in 8°, pp. 35, maps. 2.* (C. A.).

RICCIARDI L. — I tufi vulcanici del Napoletano. — *Atti d. Acc. Gioenia d. Sc. Nat. in Catania, Ser. 3.ª, Vol. XVIII, pp. 10.* (C. A.).

RICCIARDI L. — Sull'allincamento dei vulcani italiani, etc.—*Reggio-Emilia, 1887, in 8°, pp. 10, col. map 1.*

RICCI G. — Analisi chimica dell'acqua ferrata e solfurea di Napoli con un appendice sopra un nuovo liquido vesuviano.— *Napoli, in 8°, fol. 1, pp. 27.* (C. A.). *Giorn. Arcad. d. Sc. etc. Vol. XII, Roma, 1821.*

RIVAZ (CH. D). — Description des eaux minéro-thermales et des étuves de l'île d'Ischia. — *Naples, 1837.*

ROCCATAGLIATA P. — 1868. — *See Mene.*

ROCCATAGLIATA P. E FERRERO O. — Studii analitici sulle acque minerali-termali di Sujo in prov. di Terra di Lavoro. — *Aversa, 1877.*

ROCKWOOD C. G. FR. — The Ischian earthquake of July 28th 1883.— *Am. Journ. Sc. 1883, Vol. XXVI, pp. 473-476.*

ROHAN LE DUC DE. — Formation du Monte Nuovo. — *Voyage du duc de Rohan fait en l'an 1600. Amsterdam, 1646, pp. 102-103.*

ROLLER I. — Un tremblement de terre à Naples et la charité du gouvernement Napolitain. — *Genève, 1860.*

ROMANELLI D. — Viaggio a Pompei, a Pesto e di ritorno ad Ercolano e Pozzuoli. — *Napoli, 1817, vols, 2, in 12°. Parte 1. pp. 228+11, pl. Parte II, pp. 276, II, pl. (Parte II. pp. 116-123. Della Solfatara.).*

ROSSI (DE) M. S. — Fenomeni aurorali e sismici della regione laziale confrontati coi terremoti di Casamicciola, Norcia, e Livorno. — *Boll. d. Vulc. Ital. N. 6, 7 e 8. Roma, 1875.*

ROSSI M. S. (DE). — Comunicazione sulla questione dei segni precursori del terremoto di Casamicciola. — *Boll. Soc. Geol. It. 1883, Vol. II, pp. 217.*

ROSSI M. S. (DE). — Comunicazione sul terremoto di Casamicciola. — *Boll. d. Soc. Geol. It. 1883, Vol. II, pp. 92.*

ROSSINI (DE) P. — Lettera sui terremoti di Monte Olivetto Maggiore. — (?)

ROTH J. — Der Vesuv und die Umgebung von Neapel. — *Berlin, 1857, in 8°, pp. XLIV+540, pl. IX.*

ROTH J. — Zur Geologie der Umgebung von Neapel. — *Gesammt-silz. Akad. Berlin, 10 Nov. 1861, pp. 990-1006.* (C. A.).

ROZET. — Sur les volcans des environs de Naples. — *Bull. Soc. Géol. France, 1844, 2.ᵉ série, t. I, pp. 255-266.* (C. A.).

RUSCONI C. — L' origine atmosferica dei tufi vulcanici della campagna Romana. — *1865.*

RUSSEGER J. — Geognostische Beobachtungen in Rom. — *Neapel, am Aetna, auf den Cyclopen, dem Vesuv, Ischia etc. Neues Jahrb. für Miner. Geogn. u. Geol. Bd. VIII, Stuttgart, 1840.*

SACCHI G. BERTAZZI G. MARIENI L. — Sulla statistica del agro Acerrano e sulle memorie intorno alle acque minerali della Campania del dotto G. Caporali. — *Atti de Ateneo, Vol. XVII, Milano, 1862.*

SAINTE-CLAIRE DEVILLE C. J. — Sur les émanations volcaniques des Champs Phlégréens. — *Compt. rend. d. l' Acad. d. Sc. Vol. LIV, Paris, 1862. 3.º lettre, 1865.*

SAINTE-CLAIRE DEVILLE C. J., LEBLANC F. ET FOUQUÉ F.—Sur les émanatious à gaz combustibles qui se sont échappées des fissures de la lave de 1794 à Torre del Greco, lors de la dernière éruption du Vésuve.—*Compt. rend. d. l'Acad. d. Sc. Vol. LV, Paris. 1862. Id. Vol. LVI, 1863.*

SANCHEZ G. — La Campania sotterranea, e brevi notizie degli edificii scavati entro Roccia delle Due Sicilie etc.—*Napoli 1833, vols 2, in 8°, pp. 2+656.*

SANFELICE A. — Campania. — *Amsteladami, 1656, in 12°, fol. 3, pp. 64, map 1, frontisp.* (C. A.).

SANFELICE A. — Campania notis illustrata. — *Neapoli, 1726, in 4°, pp. 26+256, pl. 1.* (C. A.).

SANFELICE A. — La Campania recata in volgare italiano da Girolamo Aquino Capuano.—*Napoli, 1796, in 8°, pp. LXXI+120, pl. 1, portrait.* (C. A.).

SANFELICII A. — De situ, ac origine Companiae, comprehendens, etc. etc. — *Lug. Bat. pp. 6-12.*

SANTOLI V. M. — De Mephiti et vallibus Anxanti, libri 3. Cum observationibus super nonnullis urbibus Hirpinorum, quorum lapides ed antiquitatum relliquiae illustr. — *with 6 pl. fol. Nap. 1783. Parch.*

SARIIS A. DE — Termologio Puteolana a vantaggio dell'uomo infermo. — *Napoli, 1800, in 8°, pp. XIV+192.* (C. A.).

SARNELLI P. — La vera guida de' forestieri curiosi di vedere e d'intendere le cose più notabili della Real Città di Napoli e

17

del suo amenissimo distretto, etc.—*Napoli, 1685, in 12.ᵃ pp. 22+3+8, numerous plates. 2.ᵈ edit. Nap. 1752, in 12°, pp. 302. 3ʳᵈ edit. 1788, in 12°, pp. VIII+396, pl. 13. Several other editions.*

SCACCHI A. — Della Voltaite, nuova specie di minerale trovato nella Solfatara di Pozzuoli. — *Antologia di Sc. Nat. Vol. I, Napoli, 1841, in 8°, pp. 67-71.*

SCACCHI A. — Notizie geologiche sulle conchiglie che si trovan fossili nell'Isola d'Ischia e lungo la spiaggia tra Pozzuoli e Monte Nuovo. — *Antol. di Scienze Naturali, Vol. I (only) Naples, 1841, in 8°, pp. 33-48.*

SCACCHI A. — Osservazioni critiche sul modo come fu seppellita l'antica Pompei. — *Lettera al Cav. Avellino, Boll. Archeol. Napolit. N.° 6. Napoli, marzo 1843.*

SCACCHI A. — Sulla origine del tufo della Campania (in nota).— *Rend. della R. Acc. delle Scienze di Napoli, fasc. III, 1842.*

SCACCHI A. — Lezioni di Geologia.—*Napoli, 1843, in 8°, pp. 178, From pp. 155 to 174: Vulcani di Roccamonfina, Campi ed Isole Flegree, M.ᵗᵉ Somma e Vesuvio.*

SCACCHI A. — Voltaïte und Periklase, zwei neue Mineralien; mit Bemerkungen von Kobell. — *München, Gelehrte, Anz XVI, 1843, pp. 345-348, Erdm. Journal für. prakt. Chemie, XVII, 1843, pp. 486-489.*

SCACCHI A. — Notizie geologiche dei vulcani della Campania e-stratte dalle Lezioni di geologia. — *Napoli, 1844, in 8°.*

SCACCHI A. — Campi ed Isole Flegree, etc. — *See: Napoli e i luoghi celebri delle sue vicinanze II vols. Napoli, 1845, in 4°, pp. 361-413.*

SCACCHI A. — Descrizione delle carte geologiche dei Campi Flegrei. — *Atti della Settima Adunanza degli Scienziati Italiani tenuta in Napoli, 1845. Seconda Parte, Napoli, 1846, pp. 1176-1181, pl. 2.*

SCACCHI A. — Notice sur le gisement et sur la cristallisation de la Sodalite des environs de Naples. — *Ann. d. Mines. 4ᵐᵉ Sér. Vol. XII, Paris, 1847, pp. 385-389, figs. 11-14 of pl. 3.*

SCACCHI A. — Memorie geologiche sulla Campania. — *Napoli, 1849, in fol. pp. 131, pl. 4. Rend. d. R. Acc. d. Sc. Fis. e Mat. Napoli, Vol. VIII, 1849, pp. 41-65, 115-140, 234-261, pl. 3, 317-335. Vol. IX, 1850, pp. 84-114, pl. 1. Zeitschr. d. Deuts. geol. Gesell. Bd. IV, Berlin, 1852; Journ. für prakt. Ch. Bd. LV, Leipzig, 1852.*

Scacchi A. — Osservazioni di fenditure aperte nelle pianure di Aversa il 21 settembre 1852. — *Lettera al Sig. A. Perrey. 1852 ?. (Original M. S. in C. A.).*

Scacchi A. — Ueber die Substanzen die sich in den Fumarolen der Phlegreischen Feldern bilden. — *Abdruck a d. Zeitschr. d. deutschen geologischen Gesellschaft , Bd. IV , 1852, pp. 162-189, figs. 1-6 of pl. 3.*

Scacchi A. — La regione vulcanica fluorifera della Campania. — *Atti Acc. Sc. Napoli , S. II, Vol. II, Napoli, 1885, pl. 3. Annali Soc. It. Sc. (1880-81-82). Mem. Mat. Fis. Soc. It. Sc. Ser. 3ᵃ, T. VI, Napoli, 1887, pp. XV-XXV.*

Scacchi A. — Notizie delle fenditure apertesi nella pianura di Aversa nell'autunno del 1852 e del piperno per le medesime messo allo scoperto. — *Rend. d. R. Acc: d. Sc. Fis. Mat. Anno XX, Napoli, 1881, pp. 159-161.*

Scacchi A. — Notizie preliminari intorno ai proietti vulcanici del tufo di Nocera e di Sarno. — *Trans. d. R. Acc. dei Lincei 1881, Vol. V, Ser. 3ᵃ, pp. 270-273.*

Scacchi A.—Breve notizia dei vulcani fluoriferi della Campania.— *Rend. R. Accad. Sc. Fis. Mat. Ann. XXI, 1822, pp. 201-204. Annali Soc. It. Sc. (1878-79). Mem. Mat. Fis. It. Sc. Ser. 3ᵃ, T. IV, Napoli, 1882, pp. XIII-XX.*

Scacchi A. — Sul legno carbonizzato del tufo di Lanzara.—*Rend. R. Accad. Sc. Fis. Mat. An. XXI, 1882, pp. 176-182.*

Scacchi A. — Il vulcanetto di Puccianello.— *Rend. Acc. Sc. Fis. e Mat. S. 2ᵃ, Vol. II, n. 12°, Napoli, 1888, Ser. 2ᵃ, N.° 7, pp. 471-480, pl. 4.*

Scacchi A. — Seconda appendice alla memoria intitolata: La regione vulcanica fluorifera della Campania. — *Rend. Acc. Sc. Fis. e Mat. S. 2ᵃ, Vol. II, An. XXVII. Napoli , 1888, pp. 130-133.*

Scacchi A. — Sulle ossa fossili trovate nel tufo dei vulcani fluoriferi.della Campania. — *Atti R. Acc. Sc. Fis. e Mat. S. 2ᵃ, Vol. III, n. 3°. Napoli, 1888. pp. 2.*

Scacchi A. — La regione vulcanica fluorifera della Campania. Seconda edizione etc. — *Mem. R. Com. Geol. d'It., Vol. IV, Pl. 1°, Florence, 1890, Fol. I, pp. 48, pl 4. (C. A.).*

Schafhäult C. E. — Ueber den gegenwärtigen Zustand des Vesuv und sein Verhältniss zu den Phlegräischen gefielden.— *Gel. Anz.; heransgeg v. Mitgl. d. Königl. Bayer. Akad. d. Viss. Bd. XX, München, 1845.*

Schiavoni F. — Relazione all'Accademia Pontaniana intorno allo studio delle Maree compiuto sul littorale di Napoli per dedurre il livello medio del mare. — *Napoli, 1867, in fol. pl. 8.*

SCHMIDT J. — Die Eruption des Vesuvs in Mai 1855. « Nebst beiträgen zur Topographie des Vesuv, der Phlegräischen Crater Roccamonfina und der alten Vulkane in Kirchenstaate. — *Wien, 1856, in 8°, pp. 212, with figs.* (C. A.).

SCHMIDT J. — Nove Höhen « Bestimmungen am Vesuv, in den phlegräischen Feldern, zu Roccamonfina und in Albaner gebirge. — *Wien, 1856, in 4°, pp. 41.* (C. A.).

SCHULTZ A. W. F. — Die Heilquellen bei Neapel. — *Berlin, 1837.*

SCIVOLETTO P. — 1870. — *See Luca (de) S.*

SCROPE (POULETT) G. — Notice on the geology of the Ponza Isles (1824). — *Trans. of the Geol. Soc. of London, Vol. II, London, 1829. Zeitschr. für Mineral. Bd. V. Frankfurt am Mein, 1829.*

SCROPE (POULETT) G. — On the volcanic district of Naples. (1827). — *Proceedings of the Geol. Soc. of London. Vol. I, 1826-33. — Trans. of the geol. Soc. of London, Vol. II, London 1829. — Zeitschr. für Mineralogie, Bd. V, Frankfurt am Mein, 1829. Bull. d. Sc. Nat. et d. Géol. T. XIV, pp. 312-414.*

SEMENTINI L. ET VULPES. — Analyse des eaux minérales de Castellamare. — *Journ. d. Ch. Médic., d. Pharm. etc. Vol. I Paris, 1835. — Ann. d. Ch. und Pharm. Bd. XV, Lemgo und Heidelberg, 1835.*

SEMENTINI, VOLPE E CASSOLA. — Analisi e proprietà medicinali delle acque minerali di Castellamare. — *Napoli, 1834.*

SEMMOLA G. — Delle Mofete del lago di Agnano. — *Rend. d. R. Acc. d. Sc. Fis. e Mat. Vol. VI, Napoli, 1846. Ann. d. Ch. Vol. VI, Milano, 1848.*

SEMMOLA M. — Analisi chimica delle acque potabili dei dintorni del Vesuvio e del Somma. — *Napoli, 1857.*

SERPIERI A. — Il terremoto dell'isola d'Ischia del 28 Luglio 1883. — *Rimini, 1883, in 8°, pp. 14.* (C. A.).

SIANO (DE) F. — Brevi e succinte notizie di storia naturale e civile dell'isola d'Ischia. — *?. in 8°, pp. V+106.* (C. A.).

SPADA-LAVINI A. — Passage du Mémoire de M. C. Puggaard sur la presqu'île de Sorrento. — *Bull. d. l. Soc. Géol. de France. Vol. XV, Paris, 1857-58.*

SPADA-LAVINI A. — Sur l'âge des tufs de l'île d'Ischia. — *Bull. d. l. Soc. Géol. d. France. Vol. XV, 2e Sér. Paris, 1858.*

SPALLANZANI L. — Travels in the Two Sicilies and some parts

of the Apennines. — *Translated from the Original Italian. 4 vols. with 11 plates. London, 1798.*

SWAINE. — 1832. — *See Taylor.*

TARCAGNOTA G. — Del sito et lodi della città di Napoli. — *Napoli, 1566, in 8°, fol. 12+174.* (C. A.).

TAYLOR A. AND SWAINE. — An account of the grotta del Cane; with remarks on suffocation by carbonic acid.—*The Med. and Phys. Journ. Vol. LXVIII, London, 1832 — Notizen aus dem Gebiete der Natur. und Heilkunde. Bd. XXXVI, Erfurt und Weimar, 1833.*

TCHIHATCHEFF (DE) P. — Coup d'œil sur la constitution géologique des provinces méridionales du Roy. de Naples et observations sur les environs de Nice. Avec carte Géologique de S. Germano (Cassino) jusqu' à l' extremité méridionale de la Calabre. — *Berlin, 1842.*

TELLSII BERNARDINI CONSENTINI. — De hisquae in Acre fiunt, et de terraemotibus. Liber unicus. — *Neapoli, 1570, in 4°, fol. 13.* (C. A.).

TENORE M. — Cenno sulla Geografia fisica, e botanica del Regno di Napoli.— *Napoli, 1827, in 8°, pp. 124+II Geograph. maps.*

TENORE M. — Relazione del Viaggio fatto in alcuni luoghi di Abruzzo Citeriore nella state del 1831. — *Napoli, 1832, in 8°. pp. 132, Geographical map.*

TENORE M. — Ragguagli di alcune peregrinazioni effettuate in diversi luoghi della provincia di Napoli e di Terra di Lavoro nella primavera e nell'estate del 1832. — *Il Progr. d. Sc. etc. Vol. IV, V, VI, Napoli, 1833, in 8°, pp. 84.*

TENORE M. — Intorno ad un passo degli « Elementi di Geologia » del Sig. Lyell relativo al Serapeo di Pozzuoli. — *Rend. d. R. Acc. d. Sc. Fis. e Mat. Vol. I, Napoli, 1842.*

TENORE M. — Polvere caduta in Napoli colla pioggia nella notte del 9 al 10 novembre 1842. — *Ann. d. Fis. e Ch. etc. Vol. XI, Torino, 1843.*

THOMPSON G. — Breve notizia di un viaggiatore sulle incrostazioni silicee termali d'Italia e specialmente di quelle dei Campi flegrei. — *Without d. or l. Napoli, 1795, in 8°, pp. 35.* (O. V.).

THOMPSON G. — Topografia fisica della Campania. — (?).

TISSANDIER. — Le tremblement de terre d' Ischia du 28 juillet 1883. Rapport de la Commission. — *La Nature, 1885, I, pp. 91-94, map 1.*

TOLEDO (DA) P. — Ragionamento del terremoto del nuovo Monte etc. nell'anno 1538. — *Napoli, 1539, in 4°, fol. 16, pl. 1.*

TOLEDO (A) P. JACOBI. — De Puteolani acris natura, epistola. — *Napoli, 1544, fol. 4.* (C. A.).

TONDI M. — Catalogo delle collezioni orittologica ed orcognostica del fu prof. M. Tondi. — *Napoli, 1837.*

TORCIA M. — Tremblement de terre du 5 Janvier 1783 — ?

TRIBOLET (DE) M. — Ischia et Java en 1883. — *Conférence académique, Neuchâtel, 1884.*

TURLERI H. — De peregrinatione et Agro Neapolitano. Lib. II. Omnibus peregrinantibus utiles ac necessarii: ac in corum gratium nunc primum editi. — *Argentorati, 1574, in 8°. See pp. 85-86.* (B. N.).

Ursi J. B. — Inscriptiones. — *Napoli, 1642, in fol., fol. 11, pp. 350. See pp. 14, 24, 26, 39, 99, 100, 101, 111, 331, 332, 333, 334, 386.* — (C. A.).

VALENZIANI M. — Indice spiegato di tutte le produzioni del Vesuvio, della Solfatara e d'Ischia. — *Napoli, 1783, in 4°, pp. LII + 135.* (C. A.).

VALPES. — 1834. — *See Sementini.*

VELAIN C. — Le tremblement de terre d'Ischia du 28 juillet 1883. — *La Nature, 1883, II, pp. 183-187, pl. 2, map 2.*

VELAIN C. — Les cataclysmes volcaniques de 1883; Ischia, Krakatao, Alaska. — *Bull. Hebd. N.° 288 et 289, Oct. 1885 de l'Assoc. Sc. d. France. Paris, 1885, in 8°, pp. 27 et figs,* (C. A.).

VERDE M. E REALE N.—Dell'analisi chimica di una nuova acqua termo-minerale nel comune di Forio d'Ischia, prec. da una descriz. natur dell'isola. — *Napoli, 1866.*

VETRANI A. — Sebethi vindiciae, sive dissertatio de Sebethi antiquitate, nomine, fama, cultu, origine, prisca magnitudine, decremento, atque alveis, adversus Jacobum Martorellium.— *Neapoli, 1766, in 8°, pp. 8 + 213, pl. II.*

VILLAMENA F. — Ager Puteolanes, sivè prospectus ciusdem insigniares. — *Roma, 1652, in 4°, pl. 24.* (C. A.).

VILLANO G. (NAP). — Le croniche dell'inclita città di Napoli con li bagni di Pozzuoli et Ischia. — *In Raccolta di varii libri overo opuscoli d'historie del Regno di Napoli. Napoli, 1090, in 4°, pp. 120.* (C. A).

VIRGILIO G. — I Campi Flegrei. Ricordanze. — *Napoli, 1877, in 4°, pp. 24.* (C. A.).

VIZIOLI F. — Intorno le acque minerali del golfo di Napoli. — *Notizie. Dal Morgagni. Napoli, giugno 1869.*

VOLPE. — 1835. — *See Sementini.*

WALTHER J. — I vulcani sottomarini del Golfo di Napoli.—(*Boll. Com. Geol.*, *9-19*). *Roma, 1886, pp. 360-370, pl. 1.*

WALTHER J. AND SCHIRLITZ P. — Studi Geologici sul golfo di Napoli. — *Boll. Com. Geol. It., 1886, pp. 383-396.*

WALTHER J. AND SCHIRLITZ P. — Studien zur Geologie des Golfes von Neapel. — *Zeits. d. Geol. Gesells., Bd. 38, pp. 295-342.*

WATERS A. W. — Remarks on the recent geology of Italy suggested by a short visit to Sicily, Calabria and Ischia.—*Trans. of the Geol. Soc. Vol. XIV, Manchester, 1877.*

WEBER C. — De Agro et vino Falerno. — *Marburgi, 1855.*

WENTRUP F. — Der Vesuv und die Vulkanische Umgebung Neapels. — *Wittemberg, 1860.*

WERTHER. — Ein Ausflug zur Solfatara bei Pozzuoli. — *Schrift. d. K. ph.oecko n. Gesell. Koenigsberg, X Jahr. 1869.*

WOLF H. — Suite von mineralien aus dem vulcanischen gebiete Neapels und Siciliens. — *Verb. d. K. K. geol. Reichsanst. Wien, 1870.*

YOUNG — The gas of the grotta del Cane near Naples. — *Journ. of the Ch. Soc. London, 1878.*

ZINNO S. — Sulle industrie delle roccie e minerali dei campi Flegrei. — *Il Piria, Ann. 1, N.° 2, Napoli, 1872.*

ZINNO S. — Nuova analisi delle acque minerali delle terme Manzi in Casamicciola d'Ischia, con brevi riflessioni del dott. B. Paoni. — *Napoli, 1880.*

ROCCAMONFINA

AND

SUJO

ABICH H. — Ueber Ehrebungskatere und der Volcan von Rocca-monfina. — *Berlin, 1841.*

ANDREA D. — Anno Domini 1688. Immani Terremotu Furente Munimen Hoc Concussum; etc. — *A photograph in my own collection of a fresco incription at Sessa Aurunca of 1693.*

BREISLAK S. — Topografia fisica della Campania. — *Firenze, 1798, in 8°, pp. XII+368+III pl. (See pp. 69 to 104.)*

BUCCA L. — Il monte di Roccamonfina, studio petrografico.—*Boll. Com. Geol., 7-8. Roma, 1886, pp. 245-266.*

CASORIA. — Analisi delle acque di Sujo. — *See Roccatagliata.*

FERRERO L. O. — Sopra i Metamorfismi chimici che le roccie esistenti nei pressi delle acque di Sujo presentano, in dipendenza delle mofete e sorgive locali. — *See : Roccatagliata, Analisi, etc. pp. 21-45.*

FERRERO L. O. — Cenni stratigrafici e geologici sul luogo di Sujo, — *Ibid. pp. 47-51.*

FERRERO L. O. — Analisi delle più rinomate acque mincro-termali di Sujo. — *Ibid. pp. 65-99.*

FUSCO M. DE. — Le acque di Sujo sulla sponda destra del Garigliano. — *Il Movimento Medico-Chirurgico, An. XVI, fasc. 5-6 Napoli, 1884, pp. 7.*

18

FUSCO M. DE. — Le acque de Sujo. — *Napoli*, *1889*, *in 8°, pp.*
8. (*Chiefly medical*).

GATTULA. — Hist. Abbatiae Cassinensis. — *See T. II, p. 759.*

JERVIS G. — Tesori sotterranei dell'Italia. — *IV Vols. in 8°, To-rino, 1874-1888, numerous plates.*

JOHNSTON-LAVIS. II. J. — The Relationship of the Structure of
Igneous Rocks to the Conditions of their Formation. — *Scien-tif. Proceed R. Dublin Soc., Vol. V. 1886, N. S., pp. 112-156.*

JOHNSTON-LAVIS. II. J. — Viaggio scientifico alle regioni vulcaniche italiane nella ricorrenza del centenario del " Viaggio alle
due Sicilie" di Lazzaro Spallanzani. (This is the programme of
the excursion of the English geologists that visited the south
Italian volcanoes under the direction of the author. It is here
included as it contains various new and unpublished observations). — *Naples, 1889, in 8°, pp. 1-10.*

JOHNSTON-LAVIS H. J. — Excursion to the South Italian Vol.
canoes. — *Proceed. Geol. Assoc. Lond. 1890, Vol. XI, N.° 8,
pp. 389-423. See pp. 412-415.*

MODERNI P. — Note geologiche sul gruppo vulcanico di Roccamonfina. — *Boll. Com. Geol. It. Roma, 1887, Vol. VIII, pp.
74-100, geol. map 1.*

MONACO V. — Saggio Analitico ed uso Medico delle acque medi-
cinali fredde e termali di Sujo in Terra di Lavoro. — *Piedi-monte di Cassino?, 1798.*

PEROTTA G. — Storia del Regno di Napoli. — *Napoli, 1837, in 4°.*

PERROTTA G. — La Sede degli Aurunci antichissimi d'Italia, etc. etc. —
*Napoli, 1737, in 4°, with portrait, pp. 33+366+15. From
pp. 147 to 155. Tremuoto spaventevole da cui fu gravemente
crollata la Rocca Monfina nell'anno 1728, etc.*

PILLA L. — Geologia vulcanica della Campania. — *Napoli, 1823.*

PILLA L. — Sur le groupe vulcanique de Rocca Monfina. — *Compt.
Rend. d. l' Acad. d. Sc. Vol. X, Paris, 1840. — Ann. d.
mines. Vol. XVIII, Paris, 1840. — Neues Jahrb. für Miner.
Geol. Geogn. u. Petrefk. Bd. IX, Stuttgart, 1841.*

PILLA L. — Analyse du Mémoire intitulé: Application de la théorie des cratères de soulèvement au volcan de Rocca Monfina,
dans la Campania. — *Bull. Soc. Géol. France, 1842, pp. 402-
403.* (C. A.).

PILLA L. — Applicazione della teorica dei crateri di sollevamento
al vulcano di Rocca Monfina. — *Atti d. 3ª Riun. d. Sc.
Ital. Firenze, 1841. — Mém. d. l. Soc. Géol. de France. —
Vol. 1, Paris, 1844.*

PILLA L. — Sur quelques minéraux recueillis au Vésuve et à la

Roccamonfina. — *Compl. Rend. Acad. Sc. Vol. XXI, Paris,*
1845.

PILLA N. — Saggio litologico dei vulcani estinti di Rocca Monfina,
Sessa e Teano.—*Napoli, 1795, in 8°, pp. XIII+77, maps 2.*

PLINIUS. — Lib. 2, Cap. 103.

RATH G. (VOM) — Zwei Gesteine der Rocca Monfina. — *Zeitschr.*
der Deuts. geol. Gesell. Berlin, 1873.

RICCIARDI L. — Sull' allineamento dei vulcani italiani, etc.—*Reg-*
gio-Emilia, 1887, in 8°, pp. 10, col. map. 1.

ROCCATAGLIATA P., FERRERO L. O., CASORIA.— Analisi delle acque Minero-Termali di Sujo in Provincia de Terra di Lavoro. — *Aversa, 1877, pp. 99 + 2, 4 col. pl. maps 2.*

SCHMIDT J. — Die Eruption des Vesuvs in Mai 1855. — Nebst bei-·
trägen zur Topographie des Vesuv, der Phlegräischen Crater,
Roccamonfina und der alten Vulkane im Kirckenstaàte.—*Wien,*
1856, in 8°, pp. 212 with figs. (C. A.).

SCACCHI A. — Lezioni di Geologia. — *Napoli, 1843, in 8°, pp. 178,*
From pp. 155 to 174: Vulcani di Roccamonfina, Campi
ed Isole Flegree, M. Somma e Vesuvio.

SCHMIDT J. — Neue Höhen-Bestimmungen am Vesuv, in den phlegräischen Feldern, zu Roccamonfina und in Albaner Gebirge.—
Wien, 1856, in 4°, pp. 41.

TARTARO E FIORILLO. — Analisi delle acque di Sujo. Società Economica della Provincia di Caserta. — *? 1866.*

TENORE G. — Ragguaglio di un breve viaggio geologico alla contrada vulcanica di Sessa e di Roccambnfina. — *Ann. d. Acc.*
d. Asp. Nat. Vol. II, Napoli, 1844.

WOLFFSOHN L.—Sujo on the Garigliano.—*Gentleman's Magazine,*
Sept. 1890 pp. 265-279. In Italian in L'Araldo. Giornale
di Terra di Lavoro, Feb. e Marzo 1891.

ZARLENGA F.—Le Acque di Sujo? — See: *Filiatre Sebezio, Set-*
temb. 1852.

ALBAN HILLS

ABBATI (DEGLI) F. — Del suolo fisico di Roma e suoi contorni, sua origine e sua trasformazione. — *Cosenza, 1865.*

ABICH H. — Geologische Beobachtungen über die vulcanischen Erscheinungen und Bildungen in Unter und Mittel-Italien.— *Braunschweig, 1841.*

ANCA F. — Sull' elefante africano rinvenuto fra i fossili post-pliocenici presso di Roma. — *Atti della Reale Accademia dei Lincei; vol. XXV, Roma, 1872.*

ANONYMOUS. — La Géologie de Rome. — *Revue Britannique, N. 12, Paris, 1867.*

BAGLIVI G. — De terrae motu romano et urbium adjacentium anno 1703. Opera omia medico-practico et anatomica. — *Lipsiae, 1828, Vol. II, pp. 192-265.* (C. A.).

BARBIERI G. — I vulcani Cimino e Vulsinio. — *Viterbo, 1877.*

BARLOCCI S. — Ricerche fisico-chimiche sul lago Sabatino e sulle sorgenti d'acque minerali che scaturiscono ne' suoi contorni.— *Roma, 1816.*

BARLOCCI S. — Giornale arcadico di scienze, ecc. ; vol. XLVI.— *Roma, 1830.*

BARLOCCI S. — (Terza Edizione). — *Roma, 1843.*

BELLEVUE (FLEURIAU DE). — Mémoire sur les cristaux microscopiques et en particulier sur la séméline, la mélilite, la pseudo-sommité et les selce-romano. — *Journal de physique, de chimie et de l' histoire naturelle, par J. C. de Lamétherie et Ducrotay de Blainville, vol. LI, Paris, 1800.*

BLEICHER (DR.). — Essai d' une monographie géologique du Mont sacré — *Bulletin de la Société d' Histoire naturelle' de Colmar ; 2 e. année, Colmar, 1861.*

BLEICHER (DR.) — Recherches géologiques faites dans les environs de Rome. — *Bulletin de la Société d' Histoire naturelle de Colmar; 6 e. année, Colmar, 1865.*

BLEICHER. — Sur la géologie des environs de Rome. — *Bull. d. l. Soc. Géol. d. France. 2.e Sér. vol. XXIII, Paris, 1866.*

BONWICK J. — The volcanic rocks of Roma and Victoria compared. — *Victoria Roy. Soc. Trans. Vol. VII, 1866, Melbourne, 1866.*

BORKOWSKY (S. DUNIN) — Geognostische Beobachtungen in der Gegend von Rom. — *Taschenbuch für die gesammte Mineralogie , von K. C. Leonhard ; vol. X, Frankfurt-am-Main, 1816.*

BREISLAK Sc. — Voyages physiques et lythologiques dans la Campanie, suivis d' une mémoire sur la constitution physique de Rome. — *Paris, 1801.*

BROCCHI G. B. — Sopra una sostanza fossile contenuta nella lava di Capo di Bove presso Roma.— *Giornale di fisica, chimica e storia naturale, diretto da L. Brugnatelli; 1ª serie, vol. VI. Pavia, 1814.*

BROCCHI G. B. — Catalogo ragionato di una raccolta di roccie disposte con ordine geografico per servire alla geognosia d' Italia. — *Milano, 1817.*

BROCCHI G. B. — Osservazioni sulla corrente di lava di Capo di Bove presso Roma e su quella delle Fratocchie sotto Albano.— *Biblioteca Italiana, ossia Giorn. di lett., scienze, ecc,; vol. VII, Milano, 1817.*

BROCCHI G. B. — Risposta a una lettera del sig. Riccioli intorno all' olivina della lava basaltina di Capo di Bove — *Biblioteca Italiana, ossia Giorn. di Lett. scienze, ecc.; vol. VIII, Milano, 1817.*

BROCCHI G. B. — Dell' antica condizione della superficie del suolo di Roma. — *Roma, 1820.*

BROCCHI G. B.—Dello stato fisico (geologico) del suolo di Roma e carta fisica geologica del medesimo. — *Roma, 1820.*

BROCCHI G. B. — Dello stato fisico del suolo di Roma ad illustrazione della carta geognostica di questa città. — *Roma, 1820.*

BROCCHI G. B. — Memoria sopra la storia fisica del bacino di Roma. — *Ann. d. Sc. Fis. e Mat. Roma, luglio, 1850.*

BROGNIART A. — On the freshwater formation of the environs of Rome (Translation). — *The Philos. Magaz. or Ann. of Chem. Mathem. Astron. Nat. Hist. and general Sc.; by R. Taylor and R. Phillips, Vol. II, London, 1827.*

BUCH (L. VON). — Geognostische Uebersicht der Gegend von Rom. — *Neue Schriften der Gesellschaft naturforschender Freunde zu Berlin; Vol. III, Berlin, 1801.*

CAPPELLO A. — Saggio sulla topografia fisica del suolo di Tivoli. — *Giorn. Accad. d. Sc. ecc. Vol. XXIII, Roma, 1827.*

CAPPELLO A. — Reflexions géologiques sur les événements arrivés récemment dans le cours de l'Aniene. — *Bull. d. Sc. Nat. et d. Geol. par le Baron de Ferussac, Vol. XVI, Paris, 1829.*

CAPPELLO A. — Riflessioni geologiche sugli avvenimenti recentemente accaduti nel corso dell'Aniene. — *Atti d. R. Acc. dei Lincei, Roma, 1828. Id. Giorn. Arcad. d. Letter. Sc. ecc. Vol. XXXV, Roma, 1827. Id. 2ª. ediz. Opusc. Scelti Scient. Roma, 1830.*

CAPPELLO A. — Ulteriori schiarimenti intorno al fiume Aniene presso Tivoli. — *Giorn. Arcad. di letter. Sc. ecc. Vol. LX. Roma, 1832.*

CARPI P. — Lettera al Brocchi contenente nuove notizie sulla corrente di lava di Capo di Bove. — *Biblioteca Italiana, ossia Giorn. di Lett., Scienze, ecc.; Vol. VII. Milano, 1817.*

CARPI P. — Osservazioni chimico-mineralogiche sopra alcune sostanze che si trovano nella lava di Capo di Bove. — *Biblioteca Italiana, ossia Giorn. di Lett., Scienze, ecc.; Vol. XXV, Milano, 1827.*

CARPI P. — Sopra un'antica corrente di lava scoperta nelle vicinanze di Roma. — *Giornale Arcadico di Scienze, ecc.; Vol. XLI). Roma, 1829.*

CARLUCCI C. — Sulle condizioni fisiche e stato civile della provincia romana. — *Relazione esposta al consiglio superiore di sanità di Roma). Roma, 1876.*

CAVAZZI A. — Analisi chimica completa della pozzolana di S. Paolo di Roma e della pozzolana delle maremme toscane. — *Bologna, 1875.*

CERMELLI P. M. — Carte corografiche e memorie riguardanti le pietre, le miniere e i fossili, per servire alla storia naturale delle provincie del Patrimonio, Sabina, Lazio, Marittima, Campagna e dell'Agro Romano. — *Napoli, 1782.*

CESELLI L. — Esposizione descrittiva ed analitica su i minerali dei dintorni di Roma, e della quiritina nuovo minerale. — *Corrispondenza scientifica in Roma*; *Vol. VII*, *n.° 30-31*). *Roma, 1866.*

CESELLI M. — Tavola topografica e climatologica di Roma e sua campagna. — *Roma, 1875.*

CESELLI M. — Sui prodotti minerali utili della Provincia Romana. — *Roma, 1877.*

CESELLI M. — La Giovane Roma, rivista economica amministrativa, ecc.; anno II, n.° 17. — *Roma, 1877.*

CESELLI M. — Giorn. Il Popolo Romano, anno V, n.° 246 e 248. — *Roma, 1877.*

CESELLI M. — Sui prodotti minerali utili delle provincie romane. — *Giorn. Il Popolo Romano , N.° 246 e 248, Roma, 1877.*

CLEMENT MULLER J. J. — Documents historiques et géologiques sur le lac d'Albano. — *Bulletin de la Soc. géol. de France, 2e . Série, Vol. XI, Paris, 1853-54.*

CLERICI E. — Il travertino di Fiano Romano. — *Boll. Com. Geol., 3-4. Roma, 1887.*

CLERICI E. — La vitis vinifera, fossile nei dintorni di Roma. — *Boll. Soc. Geol., VI, 3. Roma, 1887.*

CLERICI E. — Sopra alcuni fossili recentemente trovati nel tufo grigio di Peperino presso Roma. — *Boll. Soc. Geol., VI, 1, Roma, 1887.*

CLERICI E. — Sopra i resti di castoro finora rinvenuti nei dintorni di Roma. — *Boll. Com. Geol., 9-10, Roma, 1887.*

COHN F. — Ueber die Entstehung des Travertin in den Wasserfällen von Tivoli. — *Jahrbuch für Miner. Geogn. Geol. und Petrefakt. von Leonhard und Bronn. Stuttgart, 1864.*

CONDAMINE (DE LA). — Extrait d'un journal de voyage en Italie.— (*Histoire de l'Académie R. des Sciences : Mém. de Mathématiques; année 1757, Paris, 1762.*

CONTARINI G. B. — Bibliografia geologica e paleontologica della provincia di Roma, pubblicata per cura del R. Ufficio Geologico. — *Roma, 1886, in 8°, pp. 116.*

DAVIES W. — Pilgrimage on the Tiber, from mouth to source, with notices of his tributaries. — *London, 1873,*

DEGLI ABBATI FR. — Del suolo fisico di Roma e suoi contorni, · sua origine e trasformazione. — *Cosenza, 1869.*

DELTA (PSEUD. FOR FORBES J. D.). — On the cold Caves of the monte Testaccio at Rome. — *The Edinburgh Journ. of Sc. exhib. a view of the Progr. of Discov. in Nat. Philos. Chem.*

Min. Geol. etc. cond. by David Brewster, Vol. VIII, Edinburgh, 1828,

DELTA (PSEUD. FOR FORBES J. D.). — Observations on the style of buildings employed in Ancient Italy and the materials used in the city of Rome.—*The Edinburgh Journ. of Sc. Vol. IX, Edinburgh, 1828.*

DEMARCHI L. — I prodotti minerali della provincia di Roma. — *Annali di Statistica; serie 3ª, vol. II. Roma, 1882.*

DESOR E. — Compte rendu d'une excursion faite à une ancienne nécropole des Monts Albans, recouverte par un dépôt volcanique. — *Bull. d. l. Soc. d. Sc. Nat. d. Neuchâtel, Vol. XI, Cah. I, Neuchâtel, 1877.*

DI SUCCI P. — Dell'antico e presente stato della campagna di Roma in rapporto alla salubrità dell' aria e alla fertilità del suolo. — *Roma, 1878.*

DI TUCCI P. — Saggio di studi geologici sui peperini del Lazio.— *Atti della R. Acc. dei Lincei; Memorie della Classe di Sc. Fis. Mat. e Nat.; ser. 3ª, vol. VI, Roma, 1879.*

ESCHINARDI F. — Descrizione di Roma e dell' Agro Romano. — *Roma, 1750.*

FERBER J. J. — Briefe aus Welschland über natürliche Merkwurdigkeiten dieses Landes. — *Traduz. del barone Dietrich, Prag, 1773.*

FLOTTES M. L. — Géologie des environs de Rome. — *Bulletin de la Soc. d' hist. nat. d. Toulouse : 13º an., 3e fasc. Toulouse, 1879.*

FORBES J. — Ueber die Vulcane. — *Latiums, 1849.*

FORBES J. D. — On the volcanic formation of Monte Albano. — *The Edinb. New phil. Journ. , vol. XLVIII, Edinburgh, 1850.*

FORBES J. D. — (In extract). — *Neus Jahrb. für Miner. etc., B. XIX, Stuttgart, 1851.*

FORTIS G. B. — Dei vulcani spenti della Maremma Romana. — *Venezia, 1772.*

FOUGEROUX DE BOUDAREY A. D. — Mémoire sur les solfatares des environs de Rome. — *Histoire de l'Acad. des Sciences. Mémoires de mathém.; année 1770, Paris, 1773.*

FUCHS C. W. C. — Ueber die erlöschenen Vulcane in Mittel-Italien. — *Verhandl. des Naturhist. Medic. Vereins zu Heidelberg. III, Band, Heidelberg, 1862-63.*

GATTI A. — Discorso sull' Agro Romano e cenni economico-statistici sullo stato pontificio. — *Roma, 1840.*

GISMONDI C. — Osservazioni sopra alcuni particolari minerali dei

19

contorni di Roma. — *Biblioteca Italiana*, *ossia Giorn. di Lett. sc. ed Arti; Vol. V, Milano, 1817.*

GISMONDI C.—(Leondhard's Taschenbuch; B. XI).—*Stuttgart, 1817.*

GIORDANO F. — Cenni sulla costituzione geologica della Campagna romana. — *Boll. d. R. Com. Geol. d' Italia , anno II. Firenze, 1871.*

GIORDANO F. — Cenni sulle condizioni fisico-economiche di Roma e suo territorio — *Firenze, 1871.*

GIORDANO F. — Condizioni topografiche e fisiche di Roma e Campagna romana (Monografia archeologica e statistica di Roma e della Campagna romana). — *Roma, 1878.*

GMELIN L. — Observationes oryctognosticae et chimicae de Hanyna et de quibusdam fossilibus quae cum hac concreta inveniuntur, praemissis animadversionibus geologicis de montibus Latii veteris. — *Heidelbergae, 1814.*

GMELIN L. — Oryktognostische und chemische Beobachtungen über den Haüyn und einige mit ihm vorkommenden Fossilien, neben geognostischen Bemerkungen über die Berge des alten Latiums. — *Journ. für Ch. und Phys. J. S. C. Schweigger. XV Band. Nürnberg, 1815. — Ann. of Philos. or Magaz. of Chem. Min. etc. by T. Thomson. Vol. IV, London, 1814.*

GMELIN L. — Carte géologique des environs d'Albano. — *Tubingen, 1816.*

GOSSELET G. — Observations géologiques faites en Italie. — *Mémoires de la Soc. Imp. des Sciences, de l'Agriculture et des Arts de Lille; 3° série, Vol. 6. Lille, 1869.*

HACQUET B. — Ueber Versteinerungen des ausgebrannten Vulkans bei Rom offenbar in dortigem Basaltuff. — *Leipzig, 1780.*

HECKE (VAN DEN). — 1854. — See *Rayneval (De) A.*

HESSENBERG F. — Haüyn von Marino am Albanergebirge bei Rom (Mineralogische Notizen von F. Hessenberg. Neue Folge, II. V). — *Frankfurt am Main, 1868.*

HOFFMANN F. — Ueber die Beschaffenheit des römischen Bodens, nebst einigen allgemeinen Betrachtungen über den geognostischen Charakter Italiens. — *Annalen der Physik und Chemie , von J. C. Poggendorf ; B. XVI. Leipzig, 1829. The Edinburgh Philosophical Journal, by Jameson: Vol. VIII, Edinburgh, 1830.*

HOFFMANN F. — Ueber das Albaner Gebirge, den Aetna, den Serapis Tempel von Pozzuoli und die geognostischen Verhältnisse der Umgegend von Catania. — *Arch. für Min. Geogn. Bergbau und Hüttenkunde von C. J. B. Karsten. III Band.*

Berlin *1831*. — *The Edinburgh New Philos. Journ. Vol. XII, Edinburgh, 1832.*

HOFFMANN F.—Geognostische Beobachtungen, gesammelt auf einer Reise durch Italien und Sicilien in den Jahren 1830 bis 1832.— *Karsten's Archiv. für Min., Geogn., etc.; Bd. XIII. Berlin, 1839.*

INDES (LES FRÈRES).—Sur la formation des tufs et sur une caverne à ossements des environs de Rome. — *Bull. de la Soc. Géolog. de France; 2e Sér., vol. XXVI, Paris, 1869. 2e édition — Béthune, 1875.*

INDES (LES FRÈRES). — Sur la formation des tufs des environs de Rome — *Bull. de la Soc. géolog. de France; 2e série, vol. XXVII, Paris, 1870. 2e édition, Béthune, 1875.*

JERVIS G. — The mineral ressources of central Italy. — *London, 1862. — Id. — London, 1868.*

JERVIS G. — I tesori sotterranei dell' Italia. Parte seconda : Regione dell'Appennino e vulcani attivi e spenti dipendentivi.— *Torino, 1874.*

JOHNSTON-LAVIS H. J. — Viaggio scientifico alle regioni vulcaniche italiane nella ricorrenza del centenario del « Viaggio alle due Sicilie » di Lazzaro Spallanzani. (This is the programme of the excursion of the English geologists that visited the south Italian volcanoes under the direction of the author. It is here included as it contains various new and unpublished observations). — *Naples, 1889. in 8°, pp. 1-10.*

JUDD J. W. — The great crateric lakes of Central Italy — *The Geological Magazine, n. 134, London, 1875.*

KARRER F. — Der Boden der Hauptstädte Europas. Rome — *Wien, 1881.*

KELLER F. — Contributo allo studio delle roccie magnetiche dei dintorni di Roma. — *Rend. Acc. Lincei, 1ª sem., Vol. V, 7. Roma, 1889.*

KIRCHER A. — Latium idest nova et parallela Latii tum veteris, tum novi descriptio. — *Amestelaedami, 1671.*

KLAPROTH M. H. — Chemische Untersuchung des krystallisirten schwarzen Augits von Frascati. — *Journal für die Chemie und Phisik. von A. F. Gehlen; B. V, Berlin, 1808. — Annales de Chimie ou Recueil de Mémoires et les Arts qui en dépendent; Vol. LXVI. Paris, 1808 — Journal of Natural Philosophy, Chemistry, and the Arts, by W. Nicholson; Vol. XXXVII. — London 1810.*

KLEIN C. — Beiträge zur Kenntniss des Leucits. — *Ibidem, Jahr. 1885, II B, Stuttgart, 1885.*

KLEIN C. — Optische Studien am Leucit. — *Neus Jahrb. für Min.,* *Geolog. etc; III. Beit. B, Stuttgart, 1885.*

KLITSCHE DE LA GRANGE A. — Sulla formazione dei tufi vulcanici nell'Agro romano e nel Viterbese. — *Roma, 1884.*

KOBELL (VON) I. — Ueber den Spadait, eine neue Mineral Species und über den Wollastonit von Capo di Bove. — *Gelehrte Anzeig. herausg. von Mitgliedern der Königl. Bajerischen Akad. der Wissensch. XVII Band, München, 1843. Journ. für prakt. Ch. von Otto Limè Erdmann, XXX Band. Leipzig 1844. Giorn. Arcad. d. Sc. ecc. Vol. XCIX, Roma, 1844.*

LAPI G. G. — Lezione accademica intorno l'origine dei due laghi Albano e Nemorense. — *Roma, 1781.*

LAPPARENT (DE). — 1869. — *See Dellese.*

LARTEL ED. — Sur les débris fossiles de divers éléphants découverts aux environs de Rome. — *Bull. de la Soc. Géol. de France; 2° série, vol. XV). Paris, 1857-58.*

LUDWIG R. — Geologische Bilder aus Italien. — *Bull. de la Soc. Imp. des Naturalistes de Moscou; an. 1874, n. 1. Moscou, 1875.*

MANTOVANI P. — Descrizione mineralogica dei Vulcani Laziali.— *Roma, 1868.*

MANTOVANI P. — 1872. — *See Verneuil (de) E.*

MANTOVANI P. — Descrizione geologica della Campagna romana. — *Torino, 1874.*

MANTOVANI P. — Escursione fatta dalla sezione romana del Club alpino italiano al Monte Pila nell'aprile 1876. — *Corrispondenza scientifica in Roma; Vol. VIII, n. 29, Roma, 1876.*

MANTOVANI P. — Is Man tertiary? The antiquity of Man in the Roman Country. — *Geol. Magaz. Vol. IV, 1877.*

MANTOVANI P. — Descrizione geologica dei Monti Laziali. — *Annuario del R. Liceo E. Q. Visconti per l'anno scolastico 1876-77). Roma, 1878.*

MANTOVANI P. — Uno sguardo alla costituzione geologica del suolo romano. — *Monografia Archeologica e Statistica della città di Roma e della Campagna romana. Roma, 1878.*

MARTINORI E. — I vulcani laziali. (Rassegna di alpinismo; an. 2°, n.° 9). — *Rocca S. Casciano (Firenze), 1880.*

MAURO FR. — Ricerche chimiche sulle lave di Montecompatri, del Tuscolo, di Villa Lancellotti e di Monte Pila. — *Atti della R. Acc. dei Lincei; Transunti, ser. 3, Vol. IV, fasc. 7, Roma, 1880.*

MEDICI-SPADA (DE) L. — Sopra alcune specie minerali non in prima osservate nello Stato Pontificio. — *Racc. di lett. ecc. in-*

*torno alla Fis. ed alla Mat. d. C. Palumba. Vol. I. Roma,
1845.*

MEDICI SPADA (DE) L. E PONZI G. — Profilo teorico dimostrante
la disposizione dei terreni della Campagna romana. -- *Roma,
1845.*

MELI R. — Rinvenimenti d'ossa fossili nei dintorni di Roma. —
*Boll. del R. Com. geol. d'Italia, Vol. XII, n.° 11-12. Roma,
1881.*

MELI R. — Notizie ed osservazioni sui resti organici rinvenuti
nei tufi leucitici della provincia di Roma. — *Boll. del R. Com.
Geol. d'Italia; Vol. XXII, n.° 9-10. Roma, 1881.*

MELI R. — Ulteriori notizie ed osservazioni sui resti fossili rinvenuti nei tufi vulcanici della provincia di Roma. — *Boll.
del R. Com. geol. d'Italia; Vol. XIII, n.° 9-10 e 11-12.
Roma, 1882.*

MELI R. — Bibliografie riguardanti le acque potabili e minerali
della provincia di Roma. — *Roma, 1885, in 8°, pp. 108.*

MOROZZO (DI) C. L. — Sopra i denti fossili di un elefante trovato
nelle vicinanze di Roma (1802) ed Analisi chimica di un
dente fossile fatta dal dott. Morecchini. — *Memorie di Ma-
tematica e Fisica della Società Italiana delle Scienze; Vol.
X, p. 1, Modena, 1803.*

MORTILLET G. DE. — Géologie des environs de Rome. — *Atti della
Società Italiana di Scienze Naturali, Vol. VI, Milano,
1864.*

MURCHISON R. J. — Ueber die älteren vulkanischen Gebilde im
Kirchenstaate, und über die Spalten velchen in Toscana heisse
Dämpfe entsteigen und deren Beziehung zu alten Eruptions
und Bruche Linien. — *Stuttgart, 1851. In English, Quart
Journ. Geol. Soc. Vol. VI, 1858, pp. 281-310.*

NECKER L. A. (DE SAUSSURE). — Note sur la Gismondine de Carpi
et sur un nouveau minéral (Berzeline) des environs de Rome. — *Bibliot. Universelle des Sciences; 1re s., Vol. XLV,
Genève, 1831.*

PENTLAND G. B. — On the geology of the country about Rome.—
London, 1859.

PERREAU L. — Il sottosuolo dell' Agro romano. — *Roma, 1884.
nel giorn. Il Popolo Romano, A. XII, N. 339. — Roma, 1884.*

PETRINI. — Gabinetto mineralogico del collegio Nazzareno. — *Roma, 1791.*

PICCINI A. — Analisi di un' augite del Lazio delle vicinanze di
Roma. — *Atti della R. Acc. dei Lincei; Trasunti, ser. 3,
Vol. IV, fasc. 7, Roma, 1880.*

PILLA L. — Osservazioni geognostiche che possonsi faro lungo la strada da Napoli a Vienna attraversando lo Stato romano, la Toscana, lo.Stato veneto, la Carintia, la Stiria ed Austria.— *Napoli, 1834.*

PONZI G. — 1843. — *See Medici Spada L.*

PONZI G. — Osservazioni geologiche fatte lungo la Valle Latina da Roma a Monte Cassino.— *Atti dell' Accademia Pontificia dei Nuovi Lincei, anno I, Roma, 1848.*

PONZI G. — Storia fisica del bacino di Roma.—*Atti d. Acc. Pontif. d. N. Lincei. Vol. II, Roma, 1849.*

PONZI G. — Mémoire sur la zone volcanique d'Italie. — *Bulletin de la Soc. Géol. de France ; 2e série , Vol. VII, Paris, 1849-50.*

PONZI G. — Sopra un nuovo cratere vulcanico nelle vicinanze di Roma. — *Atti d. Acc. Pontific. d. N. Lincei, Vol. IV, Roma, 1850-51.*

PONZI G. — Descrizione della carta geologica della provincia di Roma. — *Atti d. Acc. Pontific. d. N. Lincei, Vol. IV, Roma, 1851.*

PONZI G. — Sulla corrente di lava e sopra un nuovo cratere vulcanico nelle vicinanze di Roma. — *Atti dell'Accademia pontificia dei Nuovi Lincei, anno IV, Roma, 1851.*

PONZI G. — Sopra un nuovo cono vulcanico rinvenuto nella Valle di Cona. — *Atti dell'Accademia pontificia dei Nuovi Lincei, anno V, Roma, 1852.*

PONZI G. — Sulla Valle Latina. Appendice alla memoria pubblicata nella Sessione XVII del 31 dicembre 1848. — *Atti dell'Accademia pontificia dei Nuovi Lincei, anno IV , Roma, 1852.*

PONZI G. — Descrizione della Carta geologica della Cormaca di Roma. — *Atti dell' Accademia pontificia dei Nuovi Lincei, anno VI, Roma, 1855.*

PONZI G. — Sui terremoti avvenuti in Frascati nei mesi di maggio e giugno 1855. — *Atti dell' Accademia pont. dei Nuovi Lincei, anno VI, Roma, 1855.*

PONZI G. — Nota sulla carta geologica della provincia di Frosinone e Velletri. — *Atti d. Acc. Pont. d. N. Lincei, Vol. XI, Roma, 1857-58.*

PONZI G. — Sullo stato fisico del suolo di Roma. — *Giornale arcadico di scienze, ecc.; nuova serie, Vol. IX. Roma, 1858.*

PONZI G. — Storia naturale del Lazio. — *Giornale Arcadico di Scienze ecc; nuova serie, Vol XII, Roma, 1858.*

PONZI G. — Sulle correnti di lava scoperte dal taglio della ferrovia di Albano. — *Atti d. Acc. Pontif. d. N. Lincei, Vol. XII. Roma, 1858-59*.

Ponzi G. — Carta geologica dei Monti vulcanici del Lazio — *Atti d. Acc. Pont. d. N. Lincei, Vol. XIII. Roma, 1859-60.*

Ponzi G. — Storia geologica del Tevere. — *Giornale Arcadico di Scienze, ecc.; nuova serie, Vol. XVIII, Roma, 1860.*

Ponzi G. — Carta geologica dei monti vulcanici del Lazio. — *(Atti dell'Acc. Pont. dei Nuovi Lincei, anno XIII, Roma, 1861·*

Ponzi G. — Catalogo ragionato di una collezione di materiali da costruzione e di marmi da decorazioni dello Stato Pontificio. — *Roma, 1862.*

Ponzi G. — Sopra i diversi periodi eruttivi determinati nell'Italia centrale. — *Atti dell'Accad. pont. dei Nuovi Lincei, anno XVII, Roma, 1864.*

Ponzi G. — Quadro geologico dell'Italia centrale. — *Atti dell'Acc. pont. dei Nuovi Lincei, anno XLX, Roma, 1866.*

Ponzi G. — Memoria sulla storia fisica del bacino di Roma, da servire d'appendice all'opera: Il suolo fisico di Roma di G. B. Brocchi (Estratto dagli Annali di Sc. Matem. e Fisiche, pubbl. in Roma, luglio 1800). — *Atti dell'Acc. pont. dei Nuovi Lincei, anno XX, Roma, 1867.*

Ponzi G. — Le volcanisme romain. Remarques sur les observations faites en Italie par M. Gosselet. — *Bulletin de la Société géol. de France; 2e série, Vol. XXVI, Paris, 1869.*

Ponzi G. — Storia fisica dell'Italia Centrale. — *Atti della R. Acc. dei Lincei, Vol. XXIV, Roma, 1871.*

Ponzi G. — Carta geologica del bacino di Roma. — *Bollettino della Società geografica italiana, Vol. VIII, Roma. 1872.*

Ponzi G. — Del bacino di Roma e della sua natura. — *Annali del Ministero d'Agr., Ind. e Comm. Firenze-Genova, 1872.*

Ponzi G. — Les relations de l'homme préhistorique avec les phénomènes géologiques de l'Italie centrale. — *Comptes rendus du Congrès international d'anthropologie, etc., 1871. Session V, Bologne, 1873.*

Ponzi G. — Il bacino di Roma (In the collection entitled: Studi sulla geografia naturale e civile d'Italia.) — *Roma, 1875.*

Ponzi G. — Storia dei vulcani laziali. — *Atti della R. Acc. dei Lincei, serie 2°, vol. I, Roma, 1875.*

Ponzi G. — Sulle epoche del vulcanismo italiano. — *Atti d. R. Acc. d. Lincei, Vol. II, fasc. I, Transunti, Roma, 1878.*

Ponzi G. — I tufi vulcanici della Tuscia romana. — *R. Acc. d. Lincei, Ann. CCLXXVIII (1880-81). Ser. 3ª, Mem. d. Class. d. Sc. Fis. Mat. e Nat. Vol. IX. Roma, 1881.*

PONZI G. — Sui tufi vulcanici della Tuscia Romana a fine di togliere qualunque discordanza di opinione emessa sulla loro origine, diffusione ed età. — *Atti d. R. Acc. d. Lincei, Ser.* 3ª, *Trans. Vol. V, Roma, 1881.*

PONZI G. — Intorno alla sezione geologica scoperta al Tavolato sulla via Appia Nuova, nella costruzione del tramway per Marino, con una nota dell'ing. R. Meli, sulle fenditure delle mura del Pantheon — *Atti della R. Acc. dei Lincei; Mem. della Classe di Sc. Fis. Mat. e Nat.; Serie 3ª, Vol. XII, Roma, 1882.*

PONZI G. — Sulle ossa fossili rinvenute nella cava dei tufi vulcanici della Sedia del Diavolo sulla via Nomentana presso Roma.— *Boll. del R. Comitato geologico d'Italia, Serie 2ª, vol. VI, n. 3-4, Roma, 1883.*

PONZI G. — Conglomerato del Tavolato; pozzo artesiano nella lava di Capo di Bove, storia dei vulcani laziali accresciuta e corretta. — *Atti della R. Acc. dei Lincei; Mem. della classe di Sc. Fis. Mat. e Nat.; serie 4ª, vol. I, Roma, 1885.*

PONZI G. — Contribuzione alla geologia dei Vulcani Laziali sul cratere tuscolano. — *Atti della R. Acc. dei Lincei, Rendidiconti, serie 4ª, vol. I, Roma, 1885.*

PONZI G. E MELI R. — Molluschi fossili del Monte Mario presso Roma. — *Mem. Acc. Lincei, S. IV, vol. III, Roma, 1887.*

PROCACCINI RICCI V. — Descrizione metodica di alquanti prodotti dei vulcani spenti nello Stato Romano. — *Firenze, 1820.*

PROCACCINI RICCI V. — Viaggio ai vulcani spenti dello Stato Romano. — *Firenze, 1821-24.*

RATH (G. VOM). — Mineralogische-geognostische Fragmente aus Italien. I, Rom und die römische Campagna. — *Zeitschrift der deutsch. geolog. Gesellschaft, B. XVIII, Berlin, 1866.*

RATH (G. VOM). — Mineralogische-geognostische Fragmente aus Italien II. Das Albaner Gebirge. — *Zeitschrift der deutsch. geolog. Gesellschaft. B. XVIII, Berlin, 1866.*

RICCIARDI L. — Sull'allineamento dei vulcani italiani, etc. — *Reggio-Emilia, 1887, in 8º, pp. 10, col. map. 1.*

RICCI VITO PROCACCINI. — Viaggi ai Vulcani spenti d'Italia nello stato Romano verso il Mediterraneo. — *Vol. 2, in 8º, Firenze, 1814. (O. V.).*

RIGACCI C. — L'origine atmosferica dei Tufi vulcanici della campagna romana trovata dall'Abate Carlo Rusconi il dì 11 novembre 1864. — *Roma, 1865.*

ROSSI (DE) M. S. — Analisi geologica ed architettonica delle catacombe di Roma (Nota inscrita nell' opera di G. B. De Rossi intitolata " Roma Sotterranea " Vol. I. — *Roma, 1864.*

Rossi (DE) M. S. — Etudes géologico-archéologiques sur le sol Romain. — *Bull. d. l. Soc. géo!. d. France, 2e Sér. Vol. XXIV, Paris, 1867.*

Rossi (DE) M. S. — Saggio degli studii géologico-archeologici fatti nella Campagna Romana. — *Roma, 1867.*

Rossi (DE) M. S. — (In estratto nella Zeitschrift der deuts. geol. Gesellschaft. B. XXII). — *Berlin. 1870.*

Rossi (DE) M. S. — Nuova ed importante scoperta fatta nella necropoli preistorica dei Colli Albani coperta dalle eruzioni del Vulcano Laziale. — *Giorn.* " *L'Opinione* " N.° *12, Roma, 1871.*

Rossi (DE) M. S. — Le fratture vulcaniche laziali ed i terremoti del genn. 1873. — *Atti d. Acc. Pont. d. N. Lincei. Ann. XXVI, Sess. II, Roma, 1873.*

Rossi (DE) M. S. — Intorno al seppellimento vulcanico della necropoli ed abitazioni albane. — *Bull. d. Vulcano Ital. Ann. I, fasc. VIII, Roma, 1874.*

Rusconi C. — Sulla origine atmosferica dei tufi vulcanici della Campagna romana.—*Corrisp. Scient. in Roma per lo avanz. d. Sc. ecc. Vol. VII, N. 19-20. Roma, 1865.* — *Bull. d. l. Sc. géol. de France, 2e Sér. Vol. XXII, Paris, 1865.*

Rusconi C. et Mortillet G. (DE). — Sur l'âge des tufs volcaniques de la Campagne de Rome. — *Bulletin de la Soc. géol. de France; 2ª Série, Vol. XXII, Paris, 1865.*

Rusconi C. — Nuovo deposito di ossa fossili trovato nella Campagna romana. — *Corrispondenza scientifica in Roma; Vol. VII, n. 38, Roma, 1867.*

Russegger J. — Geognostische Beobachtungen in Rom, Neapel, am Aetna, auf den Cyclopen, dem Vesuv, Ischia ecc. — *N. Jahrb. f. Miner., Geogn., Geol. und Petrefk. von Leonhard und Bronn. VIII,' Band. Stuttgart, 18·10.*

Salmon U. P. — Mémoire sur un fragment de basalte volcanique tiré de Borghetto, territoire de Rome. — *Rome, 1799.*

Saussure (Necker de) L. A. — Note sur la gismondine de Carpi et sur un nouveau minéral (Berzeline) des environs de Rome. — *Bibl. Univ. des Sc. Belles lett. et Arts, faisant suite à la Bibl. Britann. rédigée à Genève. Part. d. Sc. 1e Sér. Vol. XLV, Genève, 1831.*

Scacchi A. — I composti fluorici dei vulcani del Lazio.— *Rend. Acc. Sc. Napoli, Ser. II, Vol. I, II, Napoli, 1887.*

Schmidt J. T. Y. — Die Eruption des Vesuvs in ihren Phänomaenen in Maj 1855, nebst. Ansichten und Profilen der Vulkane des phlegräischen Gebietes, Roccamonfina's und des Albaner Gebirges. — *Wien und Olmütz, 1856.*

SESTINI F. — Studio sui tufi della Campagna romana. — *Bollettino del Comizio agrario di Roma; anno IV, n. 4. Roma, 1873.*

SESTINI F. — Analisi diverse. Travertino della Campagna romana; minerale manganesifero di Subiaco. — *Boll. d. Com. Agr. d. Roma. N. 3-4. Roma, 1874.*

SICKLER F. CH. L. — Pantogramme ou vue descriptive générale de la Campagne de Rome. — *Rome, 1821.*

SICKLER F. CH. L. — Plan topographique de la Campagne de Rome, considérée sous les rapports de la géologie et des antiquités. — *Rome, 1821.*

SILLIMAN B. (JUNIOR). — Miscellaneous notes from Europe: 1° Present condition of Vesuvius; 2° Grotta del Cane and Lake Agnano; 3° Sulphur Lake of the Campagna, near Tivoli; 4° Meteorological observatory of Mount Vesuvius; 5° Light for illumination obtained from the burning of hydrogen. — *The Amer. Journ. of Sc. and Arts; by B. Silliman, and Dana. Vol. XII, 2d Ser. New-Haven, 1851.*

SPADONI P. — Osservazioni mineralo-vulcaniche fatte in un viaggio per l'antico Lazio. — *Macerata, 1802.*

STRÜVER G. — Ueber das Albaner Gebirge und über Somma Bomben mit der schönsten zonen-Structur. — *Neues Jahrb. für Min., Geol. und Palaeont., von G. Leonhard und H. B. Geinitz; Jahrg. 1875, Stuttgart, 1875.*

STRÜVER G. — Ueber die erste Abtheilung seiner Studien über die Mineralien des Albaner Gebirge. — *Neues Jahrb. für Min., Geol. und Palaeont., von G. Leonhard, und H. B. Geinitz; Jahrg. 1876, Stuttgart, 1876.*

STRÜVER G. — Studi sui minerali del Lazio. Parte 1.ª e 2.ª — *Atti d. R. Acc. dei Lincei; Memorie della classe di Scienze Fis. Mat. e Nat., serie 2ª, Vol. III e serie 3ª, Vol. I Roma, 1876 e 1877.*

STRÜVER G. — Studi petrografici sul Lazio. — *Atti della R. Acc. dei Lincei; Memorie della classe di Scienze Fis., Mat. e Nat., ser. 3ª, Vol. I, Roma, 1877.*

STRÜVER G. — Forsterite di Baccano. — *Rend. Acc. Lincei, S. IV, Vol. II, fasc. 13°, Roma, 1886.*

TERRIGI G. — Le formazioni vulcaniche del bacino romano considerate nella loro fisica costituzione e giacitura. — *Atti della*

R. Accad. dei Lincei; Memorie della classe di Scienze Fis., Mat. e Nat.; serie 3ª, Vol. X, Roma, 1881.

TERRIGI G. — Ricerche microscopiche fatte sopra frammenti di marna inclusi nei peperini laziali. — *Bollettino del R. Comitato Geologico d'Italia, serie 2ª, Vol. VI, n. 5-6, Roma, 1885.*

TERRIGI G. — Relazione della commissione per lo studio delle acque del sottosuolo della città di Roma. — *Boll. R. Acc. Medica, Anno XIII, 6, Roma, 1887.*

TSCHERMAK G. — Ueber Leucit von Acquacetosa bei Rom. — *Mineralogische Mitheilungen; Jarhgang, 1876, Heft. 1, Wien, 1876.*

TUCCI (DI) P. — Saggio di studii geologici sui peperini del Lazio. — *Atti d. R. Acc. d. Lincei; Mem. d. class. d. Sc. Fis., Mat. e Nat. Ser. 3ª, Vol. VI, Roma, 1879.*

TUCCIMEI G. A. — La geologia del Lazio. — *La Rassegna Italiana, Roma, 1882.*

TUCCIMEI G. A. — Sulla costituzione geologica del Colle Esquilino in Roma. — *Atti dell' Acc. Pont. dei Nuovi Lincei; Anno XXXVII, Sess. 4ª. Roma, 1884.*— *Estratto dalle Memorie dell'Acc. Pont. dei Nuovi Lincei, Vol. I, Roma, 1884.*

TUCCIMEI G. A. — Contribuzione alla geologia dell'interno di Roma. — *Memorie Acc. Pont. dei Nuovi Lincei, Vol. I, Roma, 1886.*

UZIELLI G. — Sopra lo zircone della costa tirrena. — *Atti della R. Acc. dei Lincei; Memorie della Classe di Sc. Fis., Mat. e Nat., Serie 2ª, Vol. III, Roma, 1876.*

VERRI A. — Sulla cronologia dei vulcani tirreni, e sull'idrografia della Val di Chiana anteriormente al periodo pliocenico. — *Rendiconto del R. Istituto Lombardo di Scienze, Lettere ed Arti, Vol. XI, Milano, 1878.*

VERRI A. — Due parole sui tufi leucitici dei vulcani tirreni. — *Bollettino della Società Geologica Italiana, Vol. II, Roma, 1883.*

VERRI A. — Sui tufi dei vulcani tirreni. — *Boll. Soc. Geol., V. Roma, 1886.*

VOLPICELLI P. — Sulla Memoria del Sig. Cav. M. S. De Rossi intitolata: Analisi geologica ed architettonica delle catacombe romane. — *Atti dell' Acc. Pont. dei Nuovi Lincei, Anno XVIII, Roma, 1865.*

ZEZI P. — Escursione ai monti Laziali. — *Programma della R. Scuola d'Applicazione per gli ingegneri in Roma. Roma, 1877.*

Zezi P. — Indice bibliografico dèlle pubblicazioni italiane e straniere riguardanti la mineralogia, la geologia e la paleontologia di Roma, con un appendice per le acque potabili, termali e minerali. — *Estr. d. l. Mon. Arch. e Stat. di Roma e d. Camp. Rom. Roma, 1878.*

APPENDIX
LIPARI ISLANDS

ACUNTE G. — Veduta del fondo del Cratere del volcano nell'Isola di Vulcano il 31 agosto 1840. — *A pencil and sepia sketch-plan in my own collection.*

STURDZA D. DIM. — Insulele Liparice. Conferintă tinută in sedinta Adunării generale de la 25 Februarie 1890. — *Buletin Societalea Geografié Română, Anul. al XI*, Trim. 1, Bucuresci, 1890, pp. 78-91.*

ETNA

CHAIX E. — Une course à l'Etna—*Genève, 1890. See also: Bull. Americ. Geogr. Soc. Vol. XIII, March, 1891, pp. 92-101, pl. 3.*

VESUVIUS

ABICH. H. — Geologische Beobachtungen und Bildungen in Unter und Mittel-Italien. — *Braunschweig, 1841, in 4°, fol. 5, pp. 134 + 11 + 3, table 1, pl. 5, figs.* (O. V.).

ALSARII CRUCII VINCENTII. — Vesuvius (1631).—*See Alzario della Croce.*

ALVINO F. — La penisola di Sorrento descritta. — *Napoli, 1842, in 8°, figs.* (B. N.).

AMARO F. (DE) — Acrumni (Vesuvii) anni 1822. Epistola. — *Neapoli, 1823, in 4°, pp. 12.* (*Frontisp. with view of Vesuvius engr. by Spani*).

AMARO F. DE. — Ode (de Vesuvio). — *Neapoli, 1824, in 4°, pp. 7.* (B. N.).

AMBROGIO LEONE NOLANO. — La storia di Nola—*In folio, Venezia, 1514.* — *(In this book is the oldest figure of Vesuvius, and on the authority of this author alone depend for evidence those who count an eruption of the mountain in 1500.)* (O. V.).

AMBROSIO F. (DE). — La Torre del Greco. (Erupt. 1861). —- *Napoli, 1862, in 8°, pp. 8.* (B. N.).

ANONYMOUS. — La morte di Plinio nell'incendio del monte Vesuvio e l'effetto che fece. — *Napoli, 1632, in 8°, fol. 2.* (O. V.).

ANONYMOUS. — Madrigale sopra l'incendio del Vesuvio. — *Napoli, 1632, in 4°, pp. 51. See Perrotti A.*

ANONYMOUS. — Principio e progressi del fuoco osservati giorno per giorno dalli 3 fino alli 25 di Luglio di questo anno 1660 ed esposti alla curiosità de' forestieri. (O. V.).

ANONYMOUS. — Continuazione de' successi del prossimo incendio del Vesuvio con gli effetti della cenere e pietre da quello vomitate, e con la dichiarazione ed espressione delle croci meravigliose apparse in varii luoghi dopo l'incendio. — *Napoli, 1661. (Palmieri says of these two pamphlets " both are very rare, the Duca della Torre did not possess them: no bibliographer of Vesuvius had seen them. Mecatti must have seen the second, because he reproduces the view of the crater contained in it, but cites it with the title "Giornale dell'incendio del Vesuvio" and declares the author to be Dr G. Carpano, or elsewhere he attributes it to Macrino and dedicated to Carpano. See Supo P.* (O. V.).

ANONYMOUS. — Giornale dell'incendio del Vesuvio dell'anno 1660. Con le osservationi matematiche al molto illustre e molto eccellente signor mio Padrone osservandissimo il Signor ·D. Gius. Carpano Dottore dell'una e dell'altra legge e nella sapienza di Roma primario professore. A. C. — *Roma, 1660. (This is the Roman edition of the other two articles under the head of "Anonymous" reffering to the erupt. of 1660. The author was padre Supo as proved by a M. S. See Supo. P.* (O. V.).

ANONYMOUS. — Continuatione de successi del prossimo incendio del Vesuvio, con gli effetti della cenere; e pietre da quello vomitate, etc. — *Napoli, 1661, in 4°, pl. 1.* (B. N.).

ANONYMOUS. — Vera e distinta relazione dell'incendio ed eruzione del Monte Vesuvio comininciato al primo di Luglio per fino li 13 del presente anno 1701. Per quello che n'ha ocularmente osservato, e diligentemente notato un Curioso de' Deputati della Terra di Ottajano. — *Napoli, ?, in 8° pl. 2.* (B. N.).

ANONYMOUS. — Declectus Scriptorum rerum Neapolitanarum qui

populorum, etc. etc. — *Neapoli, 1735, in fol. 6+986+36,* (*pp. 6 lo 10, Vesuvius*) *with topogr. map, etc.*).

ANONYMOUS. — Una descrizione del Vesuvio dopo l'eruzione del 1737. — *M. S.* (O. V.).

ANONYMOUS. — The Natural History of Mount Vesuvius, with the explanation of the various Phenomena that usually attend the Eruptions of this celebrated Volcano. — *Trans. from the the orig. It., composed by the R. Accad. of Sc. at Naples, by order of the king of the Two Sicilies. London, 1743, in 12°, pl. 2.*

ANONYMOUS. — Notizie del memorabile scoprimento dell' antica città Ercolano , etc. , fino al corrente anno 1718 , etc. (Per Anton. F. Gori, — *Firenze, 1748, in 8°,* (B. N.).

ANONYMOUS. — Vero e distinto Ragguaglio di ciò che operossi dal Procurador Fiscale, etc. in render vuota la Regal Polveriera della Torre nel dì 7 dello scaduto mese di Dicembre su' l' terribil annunzio, che una spaventevol Fiumana di fuoco scoppiata dal Monte Vesuvio incaminavasi al di lei danno, e sterminio, etc. — *Napoli, 1755, in 4.°* (B. N.).

ANONYMOUS. — Veduta del Vesuvio con l'epigrafe: Oblate ad Sebeti pontem simulacro Januari Coelestis patroni contra erumpentes flammas stetit incendium vesuvianum; tantique beneficii ergo anno 1767 statua martyri dicatur quam cultor eius Gregorius Roccus Dominicanus populo in spem salutis demonstrat. — *Pl. in 4°.* (O. V.).

ANONYMOUS. — Voyage d' un Françoi en Italie. — *1769 , in 8°, Vol. 7°, fol. 2, pp. 475.* (O. V.).

ANONYMOUS. — Recherches sur les ruines d' Herculanum. — *Paris, 1770.* (O. V.).

ANONYMOUS. — Dissertatio de Vesuvio 1773.— *The writing seems original but the name of the author cannot be understood,* M. S. in (O. V.).

ANONYMOUS. — Delle Mofete eccitate dall'Incendio del Vesuvio.— *See Anonymous, Dei Vulcani o Monti Ignivomi, 1779.*

ANONYMOUS. — Prospetto del Vesuvio dal Palazzo Reale. — *See Anonymous, Dei Vulcani o Monti Ignivomi, 1779.*

ANONYMOUS. — Vedute del Vesuvio iu grande eruzione.— *Without date (1779 ?) R. in fol.* (O. V.).

ANONYMOUS. — Breve descrizione geografica del Regno di Sicilia. — *Palermo, 1787, in 4.°* (B. N.).

ANONYMOUS. — Istruzione al forastiere, e al dilettante intorno a quanto di antico e di raro si contiene nel Museo del Real Convento di S. Caterina a Formiello de' P.P. Domenicani

Lombardi in questa Città di Napoli. — *Napoli, 1791, in 4°,
pp. 19 (mentions the existence of about three hundred spe-
cimens of different lavas, etc. from Vesuvius, in that museum)* (B. N.).

ANONYMOUS. — Compendio delle Transazioni filosofiche, ecc. —
Giornale Letterario di Napoli per servire di continuazione al-
l'Analisi ragionata de' libri nuovi.—*Napoli, 1793, Vol. CXII,
in 8°, (vol. V, pp. 78-89.*

ANONYMOUS. — Considerazioni su i prodotti del Vesuvio. — Compendio delle transazioni filosofiche della Società Reale di
Londra. — *Compiled and illustrated by Gebelin. Venezia ,
1793, 20 Vols. in 8°, fig. Vol. XVI, pp. 491-495.*

ANONYMOUS. — Le montagne di basalto sono prodotti vulcanici,
o effetti di una cristallizzazione? Compendio delle transazioni filosofiche della Società Reale di Londra. — *Compiled and
illustrated by Gebelin. Venezia, 1793; 20 vols. in 8°, fig.
Vol. XVI, pp. 495-498.*

ANONYMOUS. — Avviso al pubblico sull'analisi della cenere eruttata dal Vesuvio (nel dì 16 di Giugno 1794). — *?, loose sheet.*
(B. N.).

ANONYMOUS. — Erupt. 1794. — *See Gazzetta Civica Napoletana,
N.° 25 and 26, 21 Giugno 1794, in 4°, pp. 163-176.*

ANONYMOUS. — Due lettere concernenti la morte di Plinio il Vecchio, etc. ed a proposito dell' ultima eruzione de' 15 Giugno
1794 di cui da valente Persona anonima si da succinta relazione, con descrizione de' danni da essa cagionati e figure in
rame. — *Napoli, 1794. in 8°, pl. 1.*

ANONYMOUS. — La Storia dell'anno 1751. — *Amsterdam , in 8°,
Estratta per l'eruzione del 1755, pp. 7, Ibid. dell'anno 1760,
pp. 2. Ibid. dell'anno 1779 , pp. 3 , Ibid. dell' anno 1794,
pp. 5. (O. V.).*

ANONYMOUS. — La Torre del Greco. — *Ode , without Date or l.
(1794) fol. 1, in fol. (O. V.).*

ANONYMOUS. — Lettera seconda del danno accaduto nel paese
detto Somma non già del foco; ma di acqua pietre arena e
saette che hanno spianato detto paese con Ottajano sin'oggi
li 27 giugno 1794. (O. V.).

ANONYMOUS. — Nuova descrizione de' danni cagionati dal Monte
Vesuvio dalla sera de' 15 sino al giorno 28 giugno dell'anno
1794. — *? in 8°, pp. 8. (O. V.).*

ANONYMOUS. — Nuova istoria di una grazia particolare ottenuta
da Dio alla città di Napoli, per intercessione di Maria Ss. ed
il glorioso S. Gennaro per il Terremuoto sortito la sera dei

12 Giugno, e la grand eruzione del Monte Vissuvio la sera
de' 15 del sudetto mese; giorno di Domenica alle ore 2 della
notte del 1791. A qual' effetto allagò di foco molti villaggi
intorno stendendosi sino al mare con rovinare la gran Terra
della Torre del Greco. — ?, *in 12*. (B. N.).

ANONYMOUS. — Prodigioso miracolo del nostro gran santone e difensore S. Gennaro di averci liberati dall'incendio del Vesuvio e dal terremoto nell'anno 1794. — ? (O. V.).

ANONYMOUS. — Relazione fisico-storica della eruzione Vesuviana
de 15 Giugno 1794. — *Gazzetta Civica Napoletana*, N. 25
and 26, pp. 171-175. (B. N.).

ANONYMOUS. — Risposta di un regnicolo ad un suo amico in Napoli sull'eruzione del Vesuvio. — *Napoli, 1794, in 8°, pp. 7*
(B. N.), (O. V.).

ANONYMOUS. — Veduta del Vesuvio in eruzione con fuga dei Tor-
resi. Probably of the year 1794, eccellently drawn. — *Pl. R.
in fol.* (O. V.).

ANONYMOUS. — Tragedia Vesuviana. — *In latin verse 1794* M. S.
(O. V.).

ANONYMOUS. — Veduta del monte Vesuvio disegnata dal mare di-
rimpetto alla Torre del Greco dopo che la medesima fu quasi
interamente distrutta dalla formidabile eruzione dei 15 giugno 1794. — *Pl. in R. fol. with description* (O. V.).

ANONYMOUS. — Dissertationis Isagogicae ad Herculanensium voluminum explanationem, Pars prima. — *Neapolis, 1797, in fo-
glio, pp. 5+104*.

ANONYMOUS. — Breve catalogo di alcuni prodotti ritrovati nell'ultima eruzione del Vesuvio. — *Giorn. Lett. di Napoli, 1793-
1798, Vol. XLI, pp. 51-55*.

ANONYMOUS. — Raccolta di tutte le vedute che esistevano nel Gabinetto del Duca della Torre rappresentanti l' eruzioni del
Monte Vesuvio fin oggi accadute, etc. — *Napoli, 1805, in
oblong folio, pp. 20, pl. 50*. (B. N.).

ANONYMOUS. — Recueil de toutes les vues qui existaient dans le
cabinet du Duc de la Tour et qui représentent les incendies
du mont Vésuve arrivés jusqu' à présent. — *Naples, 1805, in
4.° fol. 2, pp. 20, pl. 25*. (O. V.).

ANONYMOUS. — Giornale di Napoli. — *Fogli N.° 19 dal dì 21 ottobre al dì 11 novembre, 1822*. (O. V.).

ANONYMOUS — Descrizione del Viaggio pittorico, storico e geografico da Roma a Napoli, e suoi contorni. — *Napoli, 1824, in
8°, pp. 190+4. From pp. 187 to 190: Pianta di una parte
dell'antico Cratere ed il Vesuvio.*

ANONYMOUS. — Eruzione di cristalli di Leucite avvenuta sul Vesuvio. — *Annali Civ. d. Due Sicilie. Napoli, 1833-47. Vol. XLIV, pp. 62-66.*

ANONYMOUS. — Parere su le Facoltà salutifere dell'acqua termominerali Vesuviana Nunziante. — *Annali Civili del Regno delle Due Sicilie. Napoli, 1833-1847, Vol. VI, pp. 109-111.*

ANONYMOUS.—Il Vesuvio.—*Album di Roma, 1834, pp. 105-107.*

ANONYMOUS. — Catalogo della collezione Orittologica ed Oreognosica del fu chiarissimo Professore Cav. Matteo Tondi Direttore del Museo di Mineralogia di Napoli, etc. — *Napoli, 1837, in 8°, pp. VIII+243.*

ANONYMOUS. — Guide nouveau de Naples en abrégé, (érupt 1834).— *Naples, 1841, in 12°, pp. 84.*

ANONYMOUS. — Memoria per la remissione della strada che dal comune di Resina conduce al Monte Vesuvio — *Napoli, 1841, in 4.° pp. 20.*

ANONYMOUS. — Souvenir du Vésuve. — *Naples, 1841, in 8°, pp. 12.* (O. V.).

ANONYMOUS. — Tavola cronologica delle principali eruzioni del Vesuvio dall'anno 79 al 1850. — *Atti R. Ist. d. Incorag. a Sc. Nat. Napoli, Vol. IX, al pp. 8, fol. 6.*

ANONYMOUS. — Giornale di Napoli. — *Fogli n. 24. 22 decembre, 1854 e 1 a 31 maggio 1855.* (O. V.).

ANONYMOUS. — Preghiera al glorioso martire S. Gennaro. (Erupt. 1855). — *Napoli, 1855. Loose sheet.*

ANONYMOUS. — Collection complète ou liste des différentes productions du Mont Vésuve. Raccolta compita o sia lista delle differenti produzioni del Monte Vesuvio che si trovano presso il signor Nicola Amitrano. — *? in 4.°* (B. N.).

ANONYMOUS. — De Vesuvio Monte nunc scimus fuisse Franciscum Mariano Suaresium Bibliotecharium Cardinalis Barberini. Copiato dal sig. Camillo Minieri dal M. S. che si conserva nella Biblioteca Brancacciana. *M. S.* (O. V.).

ANONYMOUS. — Dissertazione della vera raccolta o sia museo di tutte le produzioni del monte Vesuvio, etc. — *?. in 4°.* (B. N.).

ANONYMOUS. — Epigrammata leges et carmina insculpta in Villula et hortulo Joh. Donatus Rogadeus; eques hicrosolymitanus cinto praedio extruxit Villulam, Leges de Villula et circunerario regundis in XII Tabulas digestae. — *Carmen continens Breviarium legum ad hospites. — Carmen de Vesuvio in fol. pp. 24.*

ANONYMOUS. — Il Vesuvio Anacreontica. — *? in 8°, pl. 1.* (B. N.).

ANONYMOUS. — I Napolitani al cospetto delle nazioni civili, con appendice contenenti, etc. — *See Francesco II.*

ANONYMOUS. — La miracolosa immagine di nostra Signora del Carmine della Torre del Greco. — (O. V.).

ANONYMOUS. — La Torre del Greco. Ode. — *Napoli, loose sheet.*

ANONYMOUS. — Pianta delli confini tra Portici e S. Giorgio a Cremano fatta per ordine del Presidente della S. R. C. di S. Chiara. — *Pl. in R. fol.* (O. V.).

ANONYMOUS. — Relazione del Vesuvio. — *?, in 4°.* (B. N.).

ANONYMOUS. — Sostanze date fuori o sviluppate nelle eruzioni del Vesuvio. — *Il Propagatore delle Scienze Nat. Anno I, P. II, pp. 366.* (B. N.).

ANONYMOUS. — Supplica alla Maestà del Re delle Due Sicilie (Carlo III.) in nome de' possessori de' territori ne' contorni del Vesuvio. — *Without l. or date.* (O. V.).

ANONYMOUS. — Touchant le M. Vésuve. — *Extract from Mélan. ges d'Histoire naturelle, Vol. IV. Without date. in 8°, fol. 13.* (O. V.).

ANONYMOUS.— Veduta del Vesuvio.— *Three small views attached together.* (O. V.).

ANONYMOUS. — Vesuviani Incendii Elogium. — *?. in 4°.* (B. N.).

ANONYMOUS. — Vulcani di Europa. — *Il Propagatore delle Sc. Nat. Anno I, Pt. II. pp. 328.*

ARACRI G. — Altra relazione della pioggia di cenere avvenuta in Calabria ulteriore nel detto giorno (27 marzo 1800). — *Mem. Soc. Pontaniana di Napoli. T. 1, 1810, pp. 167-170.* (B. N.).

ARAGO F. — Liste de Volcans actuellement enflammés. — *Annu. d. Bur. Longit. année 1824, pp. 167-189.* (C. A.).

BACCHI A. — Elpidiani, Civis Romani. De Thermis, etc. — *1588. Venetiis, in fol. fig. pp. 48+49+21.*

BACCI A. — De thermis; libri septem, etc. — *Venetiis, 1571, in fol.* (B. N.).

BERGMANN T. — Dei prodotti Vulcanici considerati chimicamente con note di Dolomieu. — *Napoli, ?, in 8°, pp. 254, tables 2.* (O. V.).

BIASE (DI). — Sonetto od ode sul Vesuvio. — *In the "Poesie" of that author.*

BLONDI F. F. — De Roma instaurata, de Italia Illustrata, de gestis Venetorum, Imperatorum Rom. Vitae, et Conflagratio Vesaevi Montis ex Dione. — *Venetiis, 1510, in fol. pp. 3+146.*

BOCCOSI F. — Centurie poetiche. Centuria II, piacevole. — *Napoli, 1714, in 8°, pp. 8+100+5. Sonetto LXXX.*

BRACCHI D. A. — Una gita al Vesuvio nella notte del 10 al 20 Maggio (1855). — *Poliorama Pittoresco, N.° 16, with figs.* (O. V.).

BRACCI G. — Veduta del Vesuvio interiore nel 1755, pl. in 4.°— *It is the 22ᵈ pl. of the Raccolta delle più interessanti vedute della città di Napoli e luoghi circonvicini disegnate da Giuseppe Bracci ed incise in 30 rami da Antonio Cardoni.* (O. V.).

BREISLAK SCIP. — Topografia fisica della Campania. — *Firenze, 1798, in 8°, pp. XII+368,III pl. (pp. 104-203).*

BRIGNOLE (COMTE DE). — Lettre a S. E. le Comm. Bianchini, avec réponse. — *? 1858, in 8°, pp. 3.*

BROOKE. — On Montecellite. — *Philos. Mag. 1831.*

BRUNO F. S. — Regolamento del Prefetto di Polizia per le Guide volgarmente dette Ciceroni. — *L'Osservatore di Napoli, 1854, in 12°.*

BRYDONE M. — Voyage en Sicile et à Malthe traduit de l'anglais par M. Demeunier. — *Amsterdam et Paris, 1775, vol 2° in 8°; Vol. I. pp. XVI+419; Vol. 2, pp. 400+4; another edition accurately corrected, on the second english edition. By M. B. P. A. N. Neuchatel, 1776, contains pp. 263-272 a: Lettre de M. Brydoné au P. della Torre sur une éruption du Vésuve.*

BYLANDT (LE COMTE DE). — Résumé préliminaire de l'ouvrage sur le Vésuve. — *Naples, 1833, in 8°.* (B. N.).

CHAIX E. — Une course à l'Etna. — *Genève, 1890. See also: Bull. Americ. Geogr. Soc. Vol. XXIII, March, 1891, pp. 92-101, pl. 3.*

CAESII B. — Mutinensis è Soc. Jesu. Mineralogia, sive naturalis philosophiae Thesauri etc. — *Lugduni, in fol. pp. 16+626+ 69. See pp. 118-122, Vesuvio, etc.*

CALÀ C., DUCA DI DIANO E MARCHESE DI RAMONTE. — Memorie historiche dell'apparitione delle Croci prodigiose. — *Napoli, 1661, in 4°, pp. 12+189+25.*

CAMPOLONGO E. — La Vulcanide. — *Napoli, 1766, in 8°, pp. 52.*

CAPACII J. C. — Neapolitanae Historiae. Tomus primus in quo antiquitas, etc.—*Neapoli, 1607, in 4°, pp. 24+900. From pp. 449 to 460, del Vesuvio etc.*

CAPACII J. C. —Historiae, etc.—*Neapoli, 1771, vol. II, in 4°, figs. Vol. I, pp. 7+312. See Vol. II, pp. 500. See Vol. I, pp. 78-93, del Vesuvio, etc.*

CAPACII J. C. — Antiquitates et Historiae Neapolitanae. — *Lug-*

duni Batavorum, in fol. on 2 columns, with portrait, pp. 8+194+6. pl. X.

CAPECE-MINUTOLO F. — Al sempre invitto protettore della Città di Napoli, S. Gennaro. — Sonetto. — ?, loose sheet. (B. N.).

CAPECE-MINUTOLO F. — Per l'eruzione del Vesuvio accaduta ai 15 Giugno 1794. Canzone. — ?, loose sheet. (B. N,).

CAPMARTIN DE CHAUPY ABB. — Découverte de la Maison de Campagne d'Horace. — Rome, vols. III, in 8°, with topogr. map. Vol. I, pp. 102, and follow. de' Vulcani, del Vesuvio, etc.

CAPOCCI E. — Investigazioni delle interne masse vulcaniche dai loro effetti sulla gravità. — Atti del Reale Istituto d'Incoraggiamento alle scienze naturali di Napoli. Napoli, 1811-1863. Vol. IX, pp. 215-229.

CAPOCCI E. — Viaggio alla Meta, al Morrone ed alla Majella. — Annali Civile del Regno delle Due Sicilie. Napoli, 1833-47, Vol. VI, pp. 112-125.

CAPOCCI E. — Su di un poco noto fenomeno vulcanico. — Rend. R. Acc. Sc. Fis. Mat. Napoli, T. V, 1846, pp. 14-18.

CAPOCCI E. — Catalogo de' tremuoti avvenuti nella parte conti. nentale del Regno delle Due Sicilie posti in raffronto con le eruzioni vulcaniche ed altri fenomeni cosmici, tellurici, meteorici, etc. — Atti d. R. Ist. d. Incoraggiamento alle Sc. Nat. Napoli. Vol. IX, pp. 335-421. (B. N.). Vol. X, pp. 293-327.

CARLES, FLAMINIO MARTINO DI.—Ottave sopra l'incendio del Monte Vesuvio. — Napoli, 1632, in 12°. (B. N.).

CARLETTI N. — Storia della regione abbruciata in Campagna Felice in cui si tratta il suo sopravvenimento generale, e la descrizione de' luoghi, de' Vulcani, de' Laghi, de' Monti, delle Città litorali, e de' popoli che vi furono e vi sono, etc.—Napoli, 1787, in 4°, pp. XLIII+382, pl. 1. (B.N.).

CARUSI G. M. — Tre passeggiate al Vesuvio. — Napoli, 1858. (O. V.).

CASTALDI-CERASI J. — Inscriptiones in solenni celebritate Divi Januarii Curiae Montanae vertente secennio. — Napoli, 1798, 'in 4°.

CERASI. — 1798. — See Castaldi-Cerasi.

CHICCHIO F. X. (DE). — Dissertatio de Vesuvio — sub disciplina d. Josephi Vairo Academiae Neapolitanae Lectoris qui finem imposuit Kalendis Iuniis in vesperis corporis Christi anno bisextili 1768, M. S. (O. V.).

CORRADO M. — Descrizione (Nuova) de' danni cagionati dal Monte Vesuvio dalla sera de' 15, sino al giorno 28 di Giugno dell'anno 1794, c della somma religiosità de' cittadini napolitani. — *Napoli, 1794, in 12°.* (B. N.).

CRISCONII P. A. — Vescvi Montis clogica inscriptio. — *Napoli, 1634, loose sheet.* (B. N.).

DAU LUIGI. — Dettaglio dell'antico stato cd cruzione del Vesuvio colla relazione dell'cruzione de' 15 giugno 1794 di F. M. D. C. A. T. —? *in 8° pp. 16.* (O. V.).

DOGLIONI N. — Anfitcatro d'Europa ctc. — *Venetia, 1623, in 4° with portrait. pp. 72-1377. See pp. 993. Dell' Ethna detto Mongibello e sua historia. pp. 694. Del Monte di Somma, e sua historia.*

DOMIZI F. S. (RINALDO). — Prodigioso miracolo del nostro gran difensore S. Gennaro d'avcrci liberati dall'incendio del Vcsuvio, e dal terremoto la sera del dì 15 Giugno 1794. — *Napoli, 1794, in 8°.* (B. N.).

DURINI B. — Conghiettura geologica sulla cagione dei Vulcani. — *Dal Giornale Enciclopedico di Napoli, 1841; in 8°, pp. 23.* (O. V.).

FABIO GIORDANO. — De Vesuvio Monte. — *Copied by Miniere M. S.* (O. V.).

FALCO B. (DE). — Antiquitates Neapolis, atquae etc. — *Lug. Bat. in fol. pp. 48.—This is the oldest guide of Naples in which Vesuvius and La Solfatara are mentioned.*

FAZZINI G. — Cenno sulla pozzolana della Baia di Napoli. — *Napoli, 1857, in 4°.*

FREDA G. — Sulla composizione di alcune recenti lave vesuviane. — *Gazz. Chim. It., Anno XIX, 1. Palermo, 1889.*

FREDA G. — Sulla costituzione chimica delle sublimazioni saline vesuviane. — *Gazz. Chim. It., Anno XIX, 1. Palermo, 1889.*

GENNARO A. DI. — 1.ª lettera. Raccolta di monumenti sopra l'cruzione del Vesuvio seguita nell'agosto, 1779. — *Giornale delle Arti e del Commercio, Vol. I, Macerata, 1780, in 8°, al pp. 141 and following.* (O. V.).

GENOVESI AB. — Raccolta di lettere scientifiche ed erudite dell'Ab. (Genovesi). — *Napoli, 1780, in 8°, pp. 247.* Letter 7th, Account of the last eruption of Vesuvius, 1779. At the end of this are eight verses of P. Ant. de Sanctis, selected from the work of the same entitled : Il Mostruoso parto del Monte Vesuvio ora dal volgo detto il Monte Diavolo la cui mostruosità è qui descritta. — *Napoli, 1632.*

GIOENI G. — Saggio di litologia Vesuviana. — *Napoli, 1791* (O. V.).

GIORDANO G. — Fossili Marini sul Vesuvio. — ? (O. V.).

GIOVENE G.—Discorso meteorologico-campestre per l'anno 1794.—?, *in 4°*, (B. N.).

GIROS S. — Continuazione delle notizie riguardanti il Vesuvio.—?, *in 8ª*, (B. N.).

GIUDICE (F. DEL). — Brevi considerazioni intorno ad alcuni più costanti fenomeni vesuviani. — *Atti del Reale Istituto d'Incoraggiamento alle scienze naturali di Napoli. Napoli, 1811-1863, Vol. IX, pp. 1-67.*

MALLET R. — On some of the conditions influencing the projection of discrete solid materials from volcanoes and on the mode in which Pompei was overwhelmed. — *Journ. R. Soc. Ireland, Vol. XIV, pt. 3, 1876, in 8°, pp. 144-169.* (C. A.).

SCLOPIS F. — Prospetto Generale della città di Napoli dedicata a Sua Ecc.ª Giorgianna, Viccecontessa Spencer. — *Two panoramas in 6 sheets engraved Erupt. 1760. (Collect. of M.ʳ Tell Meuricoffre of Naples.*

www.ingramcontent.com/pod-product-compliance
Lightning Source LLC
Chambersburg PA
CBHW021526210326
41599CB00012B/1394